AF527813

PHYSIOLOGY OF TRANSPORT

ADVANCES IN PLANT PHYSIOLOGY SERIES—9

Physiology of Transport

D. K. Arora
Seema Gupta

Anmol Publications Pvt Ltd
NEW DELHI-110002 (INDIA)

Anmol Publications Pvt. Ltd.
4374/4B, Ansari Road, Daryanganj,
New Delhi-110002

Physiology of Transport

First Edition 1996

. ISBN 81-7488-243-X

PRINTED IN INDIA

Published by J. L. Kumar for Anmol Publications, Pvt. Ltd. New Delhi and printed at Chetna Offset Press, Delhi.

Preface

'Physiology of Transport' offers a comprehensive account of the fundamental principles of plant physiology. It incorporates and organizes the information concerning Physiology of Transport from relevant authentic sources. It is an humble attempt to present a correct and precise account of the subject matter of this important branch of Botany which forms an integral part of the studies undertaken by the undergraduate and post-graduate students of this subject. The principles have been outlined in considerable detail alongwith this fundamental facts and theories that explain the life processes of plants. The main object of the present book is to provide a comprehensive and well illustrated account of the prescribed subject matter in a very simple and lucid language and concise manner yet—still maintaining its desired academic standard.

Most of illustrations have been taken from professional papers, standard books and journals. Subject matter has also been obtained from a large number of standard books, reviews and research papers by internationally renowned workers and established authors. That is why there can be no claim of originality except in the manner of treatment.

Though the authors have taken special care to present a current account, yet they are fully aware of their limitations, and the readers may come across the mistakes of various types. For all types of mistakes they extend their due apology.

The authors express their thanks to their friends and colleagues whose continuous inspirations have invited them to bring out this text.

Contents include : Sieve Tubes, Mechanism of Mineral Salt Absorption, Water in Relation of Plants, Movement of Water in Plants, Translocation of Water, Translocation in Fungi, Loss of Water in Plants, and many other important chapters.

Authors

Contents

1

Sieve Tubes

CHARACTERISTIC FEATURES OF SIEVE ELEMENTS

Phloem tissue consists of several types of cell, but only one, the sieve element, is modified for the massive longitudinal translocation of sucrose and similar solutes, which is the primary function of the phloem. In this chapter the term sieve element will be used to describe the single cell which, after complex development in its cytoplasm, is linked to similar cells to form the sieve tube system which is continuous from leaf to root in all plants with phloem.

The system can transport in both directions, depending on conditions. Radioactive tracer work demonstrates the speed of the translocation and measurements of mass accumulation by storage organs show that the sieve tube is capable of transporting large amounts over several months. The sieve tube fluid contents can be shown to have a very limited range of solutes in similar proportion whatever the plant (sucrose or compound of sucrose 0.3—0.9 M ; protein 1—10 mg ml^{-1} ; K^+ 20—80 mM ; Na^+ 0.06—0.3 mM ; Ca^{++} 0.2—0.5 mM ; Mg^{++} 2—23 mM ; phosphate 3—10 mM).

The range of sieve tube structure as seen by electron microscopy is as limited as the variation in solute content. We intend to present the view of 'typical' mature sieve tube structure using sycamore (*Acer pseudoplatanus*) stem phloem as an example, and then consider evidence for, and objections to that view.

Here we consider a mature sieve element as one which has the following characteristics.

1. A diameter equivalent to that of a transverse end wall sieve plate, with a plasmalemma closely apposed to the wall.

2. A sieve plate which, if free of callose, would have sieve pores pierced through the middle lamella of the average diameter for the species (0.1 to 15 μm each and usually occupying between 50 and 60% of the total plate area). Callose is present on the sieve pores, occluding them to a variable degree. The callose is bounded by a unit membrane continuous with the wall plasmalemma.

3. Plastids and mitochondria each with an intact double membrane boundary. The plastids usually contain starch grains and the mitochondrial inner membrane is folded into cristae in the normal manner.

4. Flattened cisternae arranged singly, or in parallel or patterned aggregates occur sporadically on the walls. The minimal width of the cisternae is within the normal range found for endoplasmic reticulum in adjacent parenchyma or cambial cells.

5. Phloem protein (P protein or 'slime') fibrils 8—10 nM in diameter and often with a 8—10 nm transverse banding occur in a meshwork in the lumen of the sieve tube and through the sieve pores. In the sieve pores the P protein fibrils may be packed much more tightly than in the sieve tube. Such fibrils have not been conclusively shown to exist in gymnosperms.

6. The only pores through the wall of the sieve tube connect to specialised cells developmentally and/or functionally related to the sieve tube which die when the sieve element dies, normally after one year. These pores characteristically have many fine branches on the specialised [albuminous (gymnosperms) : companion (angiosperms)] cell side and a single wide pore, usually bounded by callose, on the sieve tube side.

7. There may be remnants of a nucleus consisting of dense clumps of material, but these are rare in angiosperms, more frequent in gymnosperms.

8. There are no golgi bodies, ribosomes, microtubules or tonoplast.

EVIDENCE FOR THE CURRENT VIEW OF PHLOEM STRUCTURE

The arguments for and against such a structure are as follows :

Uniqueness of the system

Because of the continuity of the sieve tube system and its considerable internal pressure (turgor pressure), any cutting or killing process must introduce disturbances into the system, and delicate structures involved in the translocation may not be preserved. Thus it is necessary to establish criteria for recognition of good fixation in the sieve tube which lacks many of the usual organelles.

One of the main criteria for judging good fixation in an unusual cell is that organelles of predictable structure should be at least as well preserved as in adjacent cells with 'normal' structure, *e.g.* phloem parenchyma. The only two 'predictable' organelles in the sieve tube are the mitochondria and plastids. The only common structure which *all* plastids share is an intact double outer membrane. Thus any electron micrograph claiming to display an accurate image of the sieve tube must have well preserved mitochondria and intact plastids. Starch grains have so far always been found to be bounded by the double membrane of a plastid if the plant cell was respiring normally at the time of fixation. Free starch grains or swollen mitochondria are therefore indications that the sieve tube is badly preserved and the other structures correspondingly dubious.

The tissue surrounding the sieve tube, consisting of parenchyma and companion cells, should also display good fixation as judged by the structure and disposition of the various organelles if the sieve tube image is to be considered equivalent to that *in vivo*.

Freeze etch studies of the phloem using intact plants, which avoid the need for chemical fixation and embedding procedures, are bedevilled by many technical difficulties. Johnson's (1968, 1973) results have been interpreted as showing the absence of any continuous longitudinal structures running through the lumen of the sieve tube, and the presence of P protein fibrils passing longitudinally through the sieve pores.

Active translocation

There is no method at present of determining whether a sieve tube observed on a micrograph was translocating at the moment of fixation. However, it seems logical to examine systems which are under conditions which on physiological grounds should promote an active translocation at the moment of fixation. Conditions which drastically reduce the amount of translocation can produce a structure peculiar to that particular state. This is the main argument against the universality of structures found, for example, in incubated severed segments of plants, wilted or cooled plants, plants with their leaves removed and left in the dark, or phloem nodules in callus tissue. Such structures may represent the true state under those conditions but there is no guarantee that the same structure is responsible for normal high translocation rates.

Developmental evidence

Many people would disagree with the picture of the sieve tube on the grounds that there are no longitudinal structures through which it is claimed translocation occurs. These structures are said to consist of P protein or endoplasmic reticulum or both. When the sieve tube is translocating such structures may be too delicate to survive fixation and subsequent electron microscope processing because of their continuity and peculiar osmotic surroundings or content. However, if they are necessary for translocation to occur, there should be indications of their formation during the development of the sieve tube before the sieve pores are fully pierced, when the sieve element is not part of a continuous osmotically stressed system. No indication of any such longitudinal structures have been found in developmental studies of adequately fixed phloem. The two suggested components for such longitudinal strands are the P protein and the endoplasmic reticulum. In *Acer*, for which a developmental series is presented in Plates, the P protein first appears as individual tubules or fibrils, aggregates into compact bodies and finally these disperse to produce the P protein meshwork. No longitudinally continuous structures are seen.

Rough endoplasmic reticulum is frequent at the very early stages of development. Occasionally short longitudinal cisternate are seen but these fragment and aggregate into discrete masses during development. In *Acer* some of these aggregates display a characteristic patterned structure when the sieve pore is nothing more than a narrow plasmodesma and this structure can still be seen in the mature state. This is also true of the more frequent flat stacks of membrane along the wall, save that they tend to form rather later but still before the sieve pores appear functional. Such continuity of structure does not support the view that these aggregates in the mature sieve tube represent the remnants of a continuous longitudinal strand system destroyed by fixation and/or processing.

There is no way yet of deciding at what stage in development a sieve element starts translocating. However until the sieve pores are at least two to three times the diameter of the plasmodesmata from which they develop, no significant translocation would be possible. Kollman (1964) and Wark and Chambers (1965) have suggested that the developing sieve tube is active in traslocation, but this would seem very unlikely since the widening of the plasmodesma to form the pore does not take place until a late developmental stage. In angiosperms the middle lamella can be seen to form a barrier even though the remainder of the potential pore site is only occupied by callose which may or may not be present *in vivo*. The removal of this middle lamella material seems to be the last stage in development, after the disappearance of the nucleus, dispersal of the P protein, and aggregation of the endoplasmic reticulum into stacks along the wall. Correlations between organelle changes such as these are best observed on entire sieve elements. Micrographs claiming to be of mature or translo cating sieve elements must at least include a sieve plate pore to allow independent estimation of the possibility of transloca-tion.

The greater frequency of plastids and mitochondria along the side walls close to the region of the sieve plates is another feature of the mature sieve element which is also seen during development before the pores are pierced.

In those angiosperms for which a developmental study is available the nuclear material is completely degraded prior to the loss of the nuclear envelop. In gymnosperms the densely stained areas in the nucleus generally thought to represent the chromosomal material darken progressively and become more and more compact. Eventually only a homogeneous or granular dense elongate mass remains. Such masses would be difficult to distinguish from P protein bodies at the light microscope level.

In a developmental study of several conifers (*Pinus, Ginkgo, Taxus, Cedrus, Funiperus* and *Metasequoia*) scientist have found as yet no trace of bodies equivalent to the P protein bodies of the angiosperm. What is common too all is a development of large aggregates of endoplasmic reticulum and again these may well be taken for P protein bodies at a light microscope level. Murmanis and Evert (1966) claimed to have demonstrated P protein (called slime body) at the electron microscope level in developing and mature sieve elements in *Pinus strobus*. However, the structures had none of the characteristics of the angiosperm P protein and the fixation quality was too poor overall to allow a decision as to just what they did represent.

Srivastava & O'Brien (1966) could find no P protein, only traces of fibrils, in the mature sieve tube of *P. strobus*. These fibrils had no visible substructure and seem more likely to have originated from the fibrillar whorls which develop inside the sieve tube plastids of several gymnosperm and angiosperm species. With good fixation such material should remain inside the plastids.

A recent study of *Welwitschia* phloem development revealed no P protein at any stage, and the few mosses and the one fern investigated with the electron microscope so far, do not have P protein either.

Callose

The only agreement about the amount of callose (a 1—3, β-linked glucan) around the sieve plate pores is its variability depending on fixation methods, conditions prior to fixation and the species involved. Esau's comprehensive review of the light microscope work describes a range from none at all in tree bark phloem, freshly sampled, to massive deposits on material wounded first and/or killed slowly.

Cronshaw and Anderson (1969) have shown that uncut plants killed rapidly in liquid nitrogen and subsequently fixed in cold fixative have far less callose on the pores than material sectioned into an aldehyde fixative at room temperature. Unfortunately the preservation quality in the former case is not good, in that the sieve tube plastids are ruptured, so that the result is not a conclusive proof that there is little callose *in vivo*. Most of the electron microscope studies involving cutting the plant into the fixative show a considerable range in the amount of callose on any one sieve plate. The pores show a range from 30 or 40% occluded to complete blockage by callose. Most of the evidence indicates that the more rapid the killing procedure the less the callose. The evidence suggests that in the actively translocating sieve tube between 10 and 20% of the pore volume is occluded by callose, and that in the 5 to 15 minutes taken for the fixative agent to fully penetrate to the otherwise undisturbed sieve tube more callose is laid down as the plant reacts to the trauma induced.

The function would be to prevent leaks from a wounded sieve tube system, but only as part of the plugging mechanism.

P Protein

P protein is found in anglosperms in two basic forms. In the young sieve element the most common form is a 1—10 μm diameter aggregate of closely packed 20-22 nm diameter hollow tubes (called P_1 protein hereafter) which have no recognisable precursor. This compact mass (or masses) disperses over a period which ends with the piercing of the sieve plate pores. The substructure changes from a hollow tube to an 8-12 nm wide fibril usually showing a 10 nm banding.

Occasionally an endoplasmic reticulum cisternum may surround the very early compact mass, but no organelle has been seen to be consistently associated with the dispersing P protein at any later stage. It has been suggested that the banded 10 nm fibril is composed of two helically wound subfibrils each 4 nm wide. There is some support for this from preliminary studies of negatively stained isolated P protein which has been shown to contain only a few major protein constituents.

The banded fibril (called P_2 protein hereafter) is capable of aggregation in response to unfavourable conditions as can be seen by examination of sieve elements at the cut ends of well-fixed stem pieces. Near a cut end the fibrils associate together in the sieve element lumen to form strands of varying width which are not seen in the continuation of the same sieve tube deeper in the block on the same section. The plasmalemma in such strand-containing sieve tubes is usually ballooned away from the wall, indicating poor preservation which is shared by other adjacent cells which are cut into during fixation. It is only when the poor quality of the adjacent fixation is ignored that strands can be accepted as real. This is also true in the instances where strands have been demonstrated in sieve tubes with free starch grains or burst mitochondria.

If the fixation of adjacent cells and sieve tube organelles is good then the P protein is homogeneously distributed across the width of the sieve element, although usually in much higher concentration near the transverse sieve plates than at the middle of the sieve element. In the centre of a cut stem segment the sieve elements have about equal amounts of P protein at either end. At the cut end of the sieve tube there is usually more P protein at that end of the sieve element nearest the cut. In gymnosperms, which lack P protein, aggregates of endoplasmic reticulum are associated with the sieve plate areas in the same way as P protein in angiosperms.

The function of the P_2 protein fibril is still obscure, but the two major suggestions are that it is either involved in producing some or all of the motive force causing translocation or that it is a material designed for rapid transient blocking of the sieve pores in response to injury (as a primary response subsequently replaced by secondary callose deposition).

Microtubules and microfilaments are the two cytoplasmic structures possibly involved in cytoplasmic streaming or directed movement in plant and animal cells. Microtubules are thought to provide a cell skeleton while microfilaments are implicated in generating the movement from chemical energy within or upon that skeleton. A comparison of the properties and morphology of microtubules, micro-filaments and the two forms of P protein can be seen in Table 1.1.

There are few similarities between microfilaments and the P_2 protein fibril found in mature angiosperm sieve tubes, even if one accepts the suggestion of Parthasarathy and Muhlethaler (1969) that the P_2 protein fibril is really two subfibrils (of less than microfilament dimensions) wound helically round one another. The two subfibrils in a P_2 protein fibril could then react with one another to produce a directed movement in the surrounding fluid. The fact that translocation requires the sieve tube to be metabolically active along its entire length can be interpreted as demonstrating that the system of producing flow is also universally present in the sieve tube coincidentally with the P protein. The experimental evidence concerning the effect on translocation of cytochalasin B (which inhibits most of the other demonstrably microfilament-based movement systems) is conflicting. However, the lack of any P protein or its equivalent in gymnosperms, ferns, mosses and some monocots, makes P protein an unlikely basis for the generation of an appreciable amount of force for translocation flow.

Table 1.1. Comparison of cellular filament systems.

	Diameter	Form of grouping	Occurrence	Effects of Colchicine	Effects of Cyto-chalasin B
Micro-tubules	22-24 nm* tubular 9 nm hole 13 sub-units	Usually single or 2D or 3D arrays	Most cell types+ DSE Not in SE	Disrupts structure	No effect
Micro-filaments	4-8 nm* solid, or ? tubular	Parallel bands or bundles	Many cell types+ DSE Rare in SE	No effect	Stops cyto-plasmic streaming ? Disrupts structure
Compact P_1 protein	20-22 nm* tubular 6 nm hole ? 6 sub-units	Tightly packed masses, fine cross links	Phloem only DSE, SE ? rare in companion cell	No effect	No effect
Dispersed P_2 protein	8-10 nm ? solid Longitudinal 8-10 nm repeat	Random meshwork can aggre-gate laterally	Phloem only DSE, SE	No effect	? Effect on trans-location uncertain No struc-tural change

Abbreviations DSE, developing sieve element ; SE, sieve element.
*All of indeterminate length, straight or slightly curved.

With most methods of cutting and fixation, plugs of aggregated P protein are seen within any part of the sieve pore not blocked by callose. If there is far less callose *in vivo*, as argued above, then the P protein would be more loosely distribute l in the pore. In practice, whatever the amount of callose observed the P protein plug is usually of equivalent density. Only under conditions of rapid killing and/or fixation of the plant are the pores loosely packed with P_2 protein fibrils. These conditions are designed to inactive the wound sealing activity, thus it is possible that the P protein may act as a rapid seal for cut or wounded sieve elements. [In gymnosperms, some monocots, mosses and ferns this immediate sealing function might be performed by membrane vesicles. The much narrower pores of gymnosperms (0.1—1 μm) would be more easily blocked by callose than those of angiosperms (0.1—15μm).

This sealing function would explain the preferential distribution of P protein close to the sieve plate. The observation of reversible inhibition of translocation by local applications of metabolic inhibitors can be explained by reversible aggregation of the P protein in the sieve pores. The sieve tube needs to be a living conduit, so that membrane semi-permeability may be maintained and P protein kept in a dispersed state.

Work by Milburn (1970) does suggest that exudation from cut stems is controlled by two sealing systems rather than only by callose deposition. None of the available evidence rules out P protein being a plugging mechanism, but it is difficult to see how the suggestion can be positively verified. Spanner's (1971) proposal that the longitudinally oriented, presumably negatively charged, P protein in the pore would act to magnify the electro-osmotic potential induced by a flow of K^+ ions through the pore might well be true. But electro-osmosis would only work *efficiently* for neutral sugar molecules and cations *or* anions but not both. There is ample evidence that both do move together in the sieve tube, which seems to rule out electro-osmotic mechanisms as a major source of motive force for translocation.

In summary :

1. No good evidence for strands of more than one P_2 protein fibril in the lumen of well preserved sieve tubes.

2. Little evidence for a P protein involvement in movement generation.

3. No evidence to count against the morphological observation that P protein acts as a sieve pore plugging mechanism.

Endoplasmic reticulum in the sieve element

The form and origin of the membranous material in the sieve tube is best understood by reference to its development.

The various membrane formations originate from aggregates of rough endoplasmic reticulum. The ribosomes are lost when the aggregation occurs. The aggregates occur along the walls of the sieve tube, and the width of individual cisternae is not greater than that of endoplasmic reticulum in adjacent cells in well fixed tissue.

Spanner and Jones (1970) have suggested that the stacks of parallel cisternae at right angles to the wall are in fact extensions of, or are continuous with, the plasmalemma, forming what they term a 'brush order'. Their micrographs show grazing sections of sieve tube wall with the plasmalemma invisible as it is cut too obliquely to see. Sections at right angles to the plasmalemma show the unit membrane structure, and such sections show no continuity of plasmalemma and membrane stack. Indeed, the transversely sectioned membranes of the stack and plasmalemma are of different width and appearance. All true 'brush border' plasmalemma extensions can be shown to be continuous with the normal plasmalemma and are morphologically identical with it.

In some mature gymnosperm sieve tubes, isolated, though elongated, endoplasmic reticulum cisternae are more closely associated with the mitochondria than in angiosperms. This relationship can be seen during development, but there is no indication that the membranes are related in any way to P protein as Esau has suggested.

Thaine's strands

Thaine (1964) has observed and photographed, through the light microscope, strand-like formation in sieve tubes in free-hand sections of stem segments. The strands are up to 5 μm wide and run from sieve plate to sieve plate, particles move within these strands and across the sieve plate. The strands are preserved by glutaraldehyde, but Thaine claims that they do no survive dehydration with alcohol and thus are not visible on electron micrographs, however well fixed the rest of the tissue. Negative staining can show elongated structures of equivalent size in sieve tube exudates.

The manipulation required to observe such strands would seem bound to cause considerable trauma to the sieve tube system and consequent aggregation of P protein prior to observation. Any residual flow in the sieve tube might well produce the movement observed around the aggregated P protein. Aggregation of P protein in sieve tube exudate would be expected, the structures found bearing no necessary relation to the state of aggregation *in vivo.*

The most recent form suggested by Thaine for *in vivo* strands is a longitudinal bundle of endoplasmic reticulum tubes bounded by an elastic web of P protein which causes peristaltic flow in the tubes by cyclic contraction. The strands run between sieve pores. Such structures would have to be assembled before a sieve element could start conduction, and yet no trace of them has been found in any developmental study.

Fensom (1972), who suggests smaller strands from observations of phloem still attached at both ends to the plant system, has recently published two micrographs showing a small degree of aggregation of P protein with one or two membranous cisternae included. The form of the cisternae is said to indicate that peristalsis was occurring *in vivo.* To produce such pictures a special 'rapid action fixative' is necessary and must be injected close to the intact phloem sieve tube. It is possible to explain the aggregation of the P protein as a response to bad fixation. There is no evidence that the 'rapid fixation' is capable of fixing the rest of the tissue or plastids or mitochondria in the sieve tube. An occasional cisternum of endoplasmic reticulum may well be washed into the lumen and oriented by the translocation flow in the meshwork of P protein ; its presence does not necessarily indicate a whole system of similar tubes *in vivo.*

Sieve element loading

Transfer cells with large increases in wall surface area have been described around xylem and phloem tissues by several scientists but no one has demonstrated any specialization of the sieve element plasmalemma, even in areas where massive loading or unloading of the sieve element would be expected. The sugar must either cross the plasmalemma of the sieve tube by some energy-requiring process or pass through the pores on the sieve element wall. In the vast majority of critically examined cases, sieve elements are connected only to other sieve elements or to specialised companion (angiosperms) or albuminous (gymnosperm) cells which die with the sieve element. The pores between sieve element and associated cell are branche many times on the associated cell side. On the sieve element side a single wide callose lined pore (angiosperms) or, several narrower callose lined pores (gymnosperms) are present. The endoplasmic reticulum of the associated cell is continuous with tubular traces in the branched pores which widen out and join together to form membranous material at the centre of the wall. This may be continuous with the membrane material on the sieve element side of the pore. Since the associated cells do have simple pore connections to other parenchymal cells it is theoretically possible for sugar to pass via the endoplasmic reticulum and plasmodesmatal traces (developmentally derived from endoplasmic reticulum) from or to anywhere in the plant, to or from the sieve tube. No evidence is as yet available to distinguish pore loading of sieve tubes from plasmalemma loading.

Living sieve tube for translocation

If the callose on the sieve pores is minimal, there is sufficient gradient of osmotic pressure in those sieve tube systems that have been investigated to drive a mass flow system through open pores. Inhibition of flow by local application of metabolic inhibitors can be explained if the P protein/callose system reversibly blocks the sieve pores under such conditions. With a local application of dinitrophenol, Qureshi and Spanner (1973) could not demonstrate any callose deposition to account for the translocation blockage, but the state of the P protein was not examined. Experiments to test the hypothesis of P protein plugging will require sampling methods which could not themselves induce aggregation of P protein. The normal explanation of the living conduit requirement is that energy is needed all along the sieve tube to overcome the resistance of the P protein-plugged and/or callosed sieve pores.

Local breakdown of the semipermeability of the plasmalemma by metabolic inhibition should not stop mass flow, but merely cause a local leakage very small in amount compared to the total translocated.

It has been argued above that neither P protein movement nor electro-osmosis is a satisfactory explanation for the production of flow with the characteristics of phloem translocation. However, the concept of a dynamic equilibrium between dispersed and aggregated P protein maintained by metabolic energy would allow a rapid reversible plugging of the pores under certain conditions (wounding or other stresses), and a fine control of the (mass) flow rate under normal conditions. The P protein which has been isolated from exudates should be amenable to investigation for such reversible reaggregation properties.

Bidirectional translocation in the same sieve element

The only anatomical system which can allow this is Thaine's semipermeable strands, each of which is independent. Two tracers can be demonstrated in the same sieve tube from separate applications above and below the (aphid or autoradiographic) sampling position. However, one cannot rule out the possibility of lateral diffusion of the tracers from adjacent sieve tubes translocating in opposite directions (which only requires a separate source and sink for each sieve tube). Also there is the possibility of transfer of tracer applied in solution to a leaf or stem via the xylem transpiration stream and subsequent diffusion from the xylem to the phloem. As yet there is no experiment which demonstrates bidirectional transport in a single mature sieve tube which is free of the possibility of recirculation of one of the labels prior to its appearance in the sieve tube.

2

Mechanism of Mineral Salt Absorption

Plants require a large number of inorganic elements derived from minerals or mineralized by decay of organic matter. The mineral nutrients are taken up in the form of ions and incorporated into the plant mass or stored in the cell sap. After the combustion of the organic dry matter in the laboratory, the inorganic compounds remain as ash. In the ashes of plants, one finds all of the chemical elements occurring in the lithosphere. Some of these are essential for life ; such substances include the *macronutrients* N, P, S, K, Ca, Mg and Fe, large quantities of which are required, as well as the *trace elements* or *micronutrients* Mn, Zn, Cu, Mo, B and Cl. In addition there are elements that are essential only for certain plant groups : Na for the Chenopodiaceae, Co for the Fabales with symbionts, Al for the ferns, Si for the diatoms, and Se for some planktonic algae.

As has been known since the time of J. Liebig, a nutrient available in inadequate concentration is a yield-limiting factor (the *law of the factor in minimum*). However, a nutrient "in minimum" is not the only the factor that determines yield. If metabolism is to be well regulated, production of new tissue rapid, and development unimpaired, both macronutrients and trace elements must be taken up by the plant not only in sufficient quantities but in suitable proportions (rules describing this relationship have been developed by scientists in agriculture and forestry). The various species of plants differ considerably in their nutrient requirements. The requirements of agricultural plants have

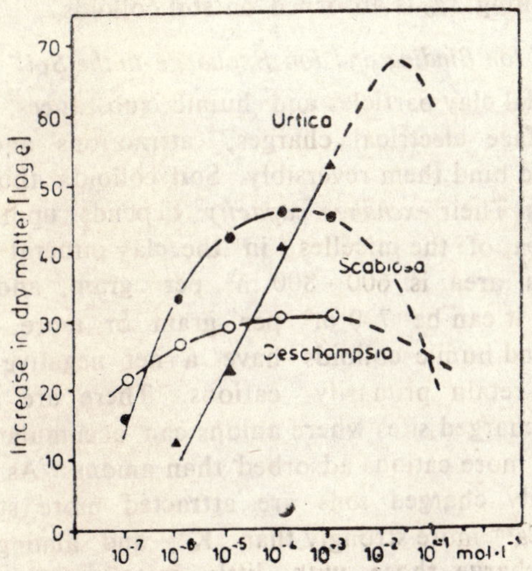

Phosphate concentration in external midium

Fig. 2.1. The effect of phosphate supply on dry-matter production by a ruderal plant (*Urtica dioica*), a meadow plant (*Scabiosa columbaria*) and a plant adapted to nutrient-poor acid soils.

been studied in considerable detail, but little experimental research has been done on the specific needs of wild plants—even though it is precisely this sort of investigation that can provide important insights into the causes of characteristic patterns of distribution within the earth's plant cover. An example of the differing phosphate requirements of species with contrasting characteristics in respect to mineral nutrition.

THE SOIL AS A NUTRIENT SOURCE FOR PLANTS

The Mineral Nutrients in the Soil

Plant nutrients occurs in the soil in both dissolved and bound form. Only a tiny fraction (less than 0.2%) of the nutrient supply

is dissolved in the soil water. **Most of** the remainder, almost 98%, is bound in organic detritus, humus, and relatively insoluble inorganic compounds or incorporated in minerals. These constitute a nutrient reserve which becomes available very slowly as a result of weathering and mineralization of humus. The remaining 2% is absorbed on soil colloids.

Adsorptive Ion Binding and Ion Exchange in the Soil

Colloidal clay particles and humic substances, because of their surface electrical charges, attractions and molecular dipoles and bind them reversibly. Soil colloids thus act as ion exchangers. Their *exchange capacity* depends upon the active surface area of the micelles ; in the clay mineral montmorillonite this area is 600—800 m^2 per gram, and in humic substances it can be 700 m^2 per gram or more. Both clay minerals and humic colloids have a net negative charge, so that they retain primarily cations. There are also certain positively charged sites where anions can accumulate, but there are always more cations adsorbed than anions. As a rule, the more highly charged ions are attracted more strongly—for example, Ca^{2+} more strongly than K^+—and among ions with the same charge those with little water of hydration are retained more firmly than strongly hydrated ions. By this kind of adsorptive binding, masses of ions accumulate on the surface of the much swollen micelles of clay and humus.

This coating of ions amounts to an intermediate stage between the fixed soil phase and solutions in the soil. If ions are added to or withdrawn from the soil solution, *exchange takes place.* Those ions that adhere more firmly are attracted more strongly by the colloid and displace other ions from its surface. The tendency for adsorption decreases in the order Al^{3+}, Ca^{2+}, Mg^{2+}, NH_4^+, K^+, and Na^+ for cations, and for anions it decreases from PO_4^{3-} through SO_4^{3-} and NO_3^- to Cl^-. Heavy metal ions can also be adsorbed, though only in trace amounts.

The *adsorptive binding of nutrient ions* offers a number of advantages. Nutrients freed by weathering and the decomposition of humus are captured and protected from leaching.

Moreover, the concentration of the soil solution remains low and relatively constant, so that the plant roots and soil organisms are not exposed to extreme osmotic conditions ; yet when they are needed the adsorbed nutrient ions are readily available to the plants.

An equilibrium exists with respect to the soil solution, the soil colloids, and the reserves of mineral substances in the soil ; it is complex and capable of adaptation. This system controls the exchange of mineral substances and ensures a continual supply of nutrient elements. The concentration of hydrogen ions in the soil solution exerts a great influence upon this ion-exchange equilibrium.

The pH of the Soil ("Soil Reaction")

Soil pH and the Availability of Nutrients

Most soils in humid regions are weakly acid to neutral, though bog soils are markedly acidic (pH about 3) and the saline and alkali soils of arid regions are basic. Acidification takes place in a number of ways : by removal of the bases through leaching, by the withdrawal from solution of exchangeable cations, by the release of organic acids by the plant roots and microorganisms, and above all by dissociation of carbonic acid, which accumulates in the soil as a product of respiration and fermentation. Depending on the parent rock and the degree of saturation of the adsorption complexes with cations, the soil is buffered to within a certain pH range. Calcareous soils are buffered primarily by the system $CaCO_3$ $Ca(HCO_3)_2$—that is, by the salt of a strong base and a weak acid ; they are therefore weakly alkaline. The pH of the soil changes in the course of the year (especially in association with the distribution of precipitation), and there are also local differences, particularly between the different horizons in the soil Therefore to characterize a habitat the pH must be measured over the entire year and if possible for the entire depth profile, or at least in the zone most densely penetrated by roots.

The pH of the soil has an effect upon its structure, on the processes of weathering and humification, and above all on the mobilization of nutrients and the exchange of ions. The most important of these relationships are summarized. In very acid soils too many Al, Fe, and Mn ions are liberated, and Ca^{2+}, Mg^{2+}, K^{+}, PO_4^{3-} and MoO^{2-} are depleted or occur in a form difficult for plants to take up. In more alkaline soils, on the other hand, Fe and Mn ions, PO_4^{3-}, and certain trace elements are fixed in relatively insoluble compounds, so that the plants are more poorly supplied with these nutrients.

Soil pH and the Plant

Soil pH has a direct effect on the viability of plants, in addition to its effect on the nutrient supply. Below pH 3 and above pH 9, the protoplasm of the root cells of most vascular plants is severely damaged. Moreover, the increased concentrations of Al^{3+} in very acid soils, and of borates in alkaline soils, act to poison the roots.

Different species display characteristic tolerance limits and requirements in their physiological behaviour with respect to soil pH. Some *Sphagnum* species prefer a strongly acid milieu ; they are very sensitive to OH^- ions and succumb even in the neutral range. According to a classification system set up by H. Ellenberg, they are to be regarded as strongly acidophilic, with a narrow tolerance span. The hair grass *Deschampsia flexuosa*, the presence of which is an indicator of acid soil, develops optimally at a pH between 4 and 5, but can also grow in the neutral range and will tolerate weakly alkaline soils. This species is "acidophilic-basitolerant". Similar behaviour is shown by *Calluna vulgaris* and *Sarothamnus scoparius*. Converse behaviour is found in coltsfoot (*Tussilago farfara*), which has an optimum in the neutral-to-alkaline range, but good tolerance down to pH 4 ; coltsfoot is classified as basiphilic-acidotolerant. Most vascular plants are amphitolerant, having a broad optimum in the range between weak acidity and weak alkalinity, and are able to exist between pH 3.5 and 8.5. a characteristic of the physiological constitution of a plant, and its nature varies with the species.

Limits to Selectivity

Plant cells cannot entirely exclude salts that are not required or are injurious, even if the plant is damaged as a result ; the biomembranes are not very permeable to ions, but the permeability is never zero. Consequently, with pronounced differences in concentration on the two sides of a membrane, some ions do leak through. Especially when the outside concentration for example, in saline soils—the cells are flooded with ions (*e.g.*, Na^+ and Cl^-) in quantities greater than are desirable.

FUNCTIONS OF MINERALS

For a long time it was supposed that the mineral elements in the plants were superfluous. Experimental evidence has, however, established that they are an essential part of the plant and perform a number of functions, the most important of which are enumerated below.

(*i*) *Nutritive Function.* Many of the minerals enter into the composition of the cell. Carbon, hydrogen and oxygen make up the cellulose and lignin, the reserve food, and most of the protoplasm of plants. They are the most important plant nutrients and constitute the bulk of the dry weight of the plant material. They are, therefore, known as the *framework elements.* Nitrogen, phosphorus and sulphur enter into the composition of the proteins which make up the protoplasm. These elements are known as the *protoplasmic elements.* Calcium is a component of the cell wall, magnesium of chlorophyll.

(*ii*) *Catalytic Function.* Some elements act as stimulants and catalysts in many metabolic processes of the plants. Some of them act as enzyme formers. They are needed in such minute quantities that any other function would seem out of question. Thus, a tomato plant would require for its normal development only 0,000125 gm. of $ZnSO_4$. Similarly, manganese, boron, and iron are required by the plants in very small quantities. Because of the small quantities in which they are required by the plants these elements are known as *micronutrients* or *trace elements.* The nutritional elements like nitrogen, calcium, and phosphorus which are needed by the plant in large quantities are called *micronutrients.*

SPECIFIC ROLES OF MINERAL ELEMENTS IN PLANTS

As already stated, the role of different mineral elements is determined by culture experiments. The role of macronutrients is relatively easy to determine. On the other hand, the determination of the role of the necessary micronutrients is frequently a very difficult matter. As these elements are required in very very minute quantities, the nutrient salts used should be of the highest degree of purity and should de dissolved in twice distilled water. The composition of the containers should also be correctly known and all possibilities of mineral elements coming from their walls eliminated. For this purpose, the solution cultures are placed in pyrex or other hard glass containers which have been cleaned with chromic acid mixture and repeatedly washed in distilled water to remove contaminating ions. Soft glass or metal containers should be coated on the inside with shellac or paraffin to prevent the solution from coming in contact with the surface. Sand cultures are not useful in such studies on account of the impurities contained in sand.

ABSORPTION OF MINERAL SALTS

It was assumed by early workers that mineral elements were absorbed passively by the plant with the absorption of water. It was held that translocation of absorbed salts to different organs of plant was dependent on the transpiration stream. But these assumptions could not explain the absorption of minerals against concentration gradient. The salt concentration of cell sap of plant tissues is usually found to be higher than the concentration of the medium in which the plant grows. The absorption of mineral elements is, therefore, independent of the absorption of water. The regions of absorption of former lie in the meristmatic and elongation zones while that of the latter in the region of root hair. Hoagland (1944) also found that the rate of absorption of salts from culture solution depended upon kind of solute and was different from the rate of absorption of water.

A solution to the problem was given by the explanation that the entrance of mineral salts into the peripheral cells of roots occurred by the diffusion due to the process of osmosis. Although it is true that limited quantities of mineral salts pass into cells, under certain conditions, by diffusion, in general, this is a relatively insignificant mechanism of salt absorption compared with certain others. Thus the early workers considered the absorption of salts as a physical or passive process and failed to recognise the importance of metabolic energy. The works done in 1930's have clearly shown that absorption of salt is predominantly an active process using metabolic energy rather than a passive one as was assumed earlier.

Formerly it was assumed that inorganic salts were passively carried away into the plant with the absorption of water. It was also suggested that the salts could diffuse into the root tissue along concentration gradients by the osmotic process. Thus the early workers took into consideration only the physical aspects of the absorption of salts and neglected the role of metabolic energy in the process. The works done in the last four decades have, however, clearly shown that mineral salt absorption is predominantly an active process rather than a passive process as was thought earlier.

PASSIVE ABSORPTION

Diffusion Mechanism

As the cell walls of roots are completely permeable, therefore, from the soil solutions ions may diffuse into the cells depending upon the concentration gradient. There is always rapid absorption of ions initially when roots are placed from a solution of lower concentration to that of a higher concentration. However, this rate soon becomes constant. The rapid rate of absorption is believed to be due to diffusion while further absorption at constant rate requires metabolic energy. In fact, the cells or tissues have been supposed to possess *outer spaces* through which free diffusion is possible and *inner spaces* where the ions become fixed. In the case of roots outer space probably extends up to endodermis. In the cells it is supposed to extend up to middle of cytoplasm.

It has been observed that an initial passive or nonmetabolic uptake of ions takes place when a plant cell or tissue is kept in a medium of relatively high salt concentration. If the above tissue is returned to a lower salt solution, some of the ions taken up will diffuse out into the external medium. The whole process is unaffected by temperature or metabolic inhibitors. It is, therefore, assumed that this free diffusion of solutes follows simple laws of diffusion.

It has, however, been found in certain cases that ions have accumulated against a concentration gradient. Surprisingly, this happened without the participation of metabolic energy. This type of passive absorption of salts is possible either by *ion exchange mechanism* or by the establishment of *Donnan equilibria.*

Ionic Exchange Mechanism

Basically the mechanism consists in the exchange of anions and cations from inside the cells for ions of the same sign and equivalent charge in the surrounding of the absorbing cells.

Jenny and Overstreet (1938) found that if excised roots of barley in which radioactive potassium (K^+) ions have been allowed to accumulate from a solution are placed in distilled water, almost none of the radioactive ions come out of the roots into the water. The cytoplasmic membranes of root cells behave as if they are impermeable to K^+ ions which readily entered the cells before their transference to distilled water and when such excised roots arc transferred to a weak solution of non-radioactive KBr, some of the radioactive K^+ ions come out of the root cells into the solution and at the same time non-radioactive K^+ ions move into the root cells from the solution. In this case cytoplasmic membranes permit the movement of K^+ ions in both the directions. An analysis of the solution around the roots does not show any increase in the total potassium. It therefore, becomes evident that radioactive K^+ ions have exchanged places with non-radioactive K^+ ions in the solution. A similar exchange process operates between soil solution and roots in nature.

The ionic exchange process has been explained by two theories, (*i*) the carbonic acid exchange theory ; and (*ii*) the contact exchange theory.

(*i*) *The carbonic acid exchange theory.* Under favourable conditions of growth, root tips are centres of high metabolic activity. CO_2 is released in their respiration. Most of this CO_2 reacts with water forming carbonic acid, a large part of which escapes into the soil. Around each root tip there is developed a zone of high carbonic acid content. Carbonic acid (H_2CO_3) ionises to form H^+ and HCO_3^- ions.

Such hydrogen ions are presumed to exchange places with other cations in the soil solution or cations on clay particles.

(*ii*) *The contact exchange theory.* An exchange of cations can also take place directly between roots and clay micelles without any role of soil solution as an intermediate in the process. An ion adsorbed electrostatically to root or clay micelles can oscillate within a certain volume of space. An exchange of ions occurs when the oscillation volume of one ion overlaps the oscillation volume of another (Fig. 2.2).

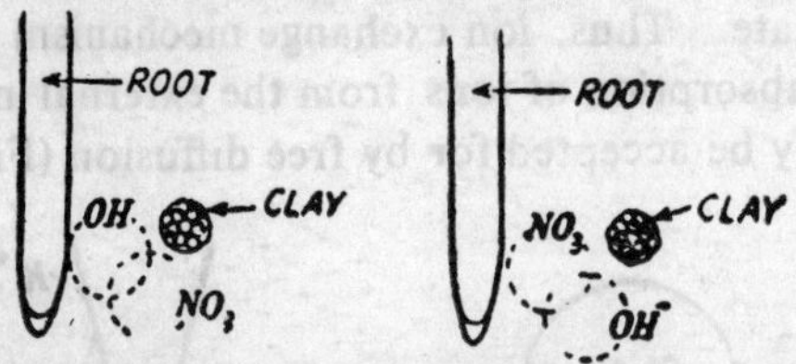

Fig. 2.2. Exchange of Ions.

In ionic exchange mechanism, anions or cations from within the cells are exchanged for anions or cations of equivalent charge of the external solution in which the tissue is immersed.

The phenomenon has been experimentally confirmed in excised barley roots in which radioactive K^+ ions exchange place with the non-radioactive K^+ ions. A similar exchange mechanism operates between soil solution and clay micelles. The ions get accumulated against a concentration gradient without the participation of metabolic energy because cations and anions of the external medium get exchanged with hydrogen ion, H^+ and hydroxyl ion, OH^-, which always remain absorbed on the surface of the membrane. H^+ and OH^- ions are readily available from water.

The process of ionic exchange has been explained by two theories, (1) the *contact exchange theory*, and (2) the *carbonic acid exchange theory* (as stated above). According to the contact exchange theory an ion may be absorbed by the plant root without being first dissolved in the soil solution. An ion absorbed electrostatically to a solid particle, such as a plant root or clay micelle, is not held too tightly, but oscillates within a certain small volume of space. An exchange of ions takes place when the oscillation volume of one ion overlaps the oscillation volume of another ion. The soil solution, however, plays an important part in the carbonic acid theory in that it provides the medium for the exchange of ions between the root and the clay micelles. Carbon dioxide released in respiration combines with water to form carbonic acid in the soil solution. Carbonic acid dissociates into (H^+) and (HCO_3^-) ions. A cation absorbed to the clay surface may be exchanged with H^+ of the soil solution. This cation then may diffuse to the root surface in exchange for H^+. The cation may also be absorbed as ion pairs with bicarbonate. Thus, ion exchange mechanism would allow for a greater absorption of ions from the external medium than could normally be accepted for by free diffusion (Fig. 2.3).

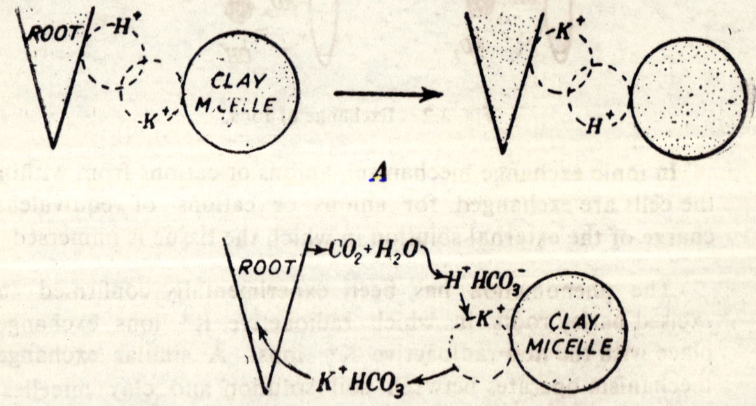

Fig, 2.3. Diagrammatic representation of (A) the contact exchange theory and (B) the carbonic acid exchange theory.

Donnan Equilibrium Theory

According to this theory the presence of a concentration of certain (fixed) or indiffusible ions within cell would require t o be balanced by ions of the other charge. Assuming that a concentration of 'fixed' anions is present on the inner side of the membrane, more cations would be absorbed in addition to the normal exchange to maintain the equilibrium. Therefore, the cation concentration would be greater in the internal solution than in the external solution.

The large, negatively charged ions (R^-) cannot pass through the plasma membrane. These anions are electrically balanced by the entry of cations $(K^+)_0$ from outside. The external anions $(Cl^-)_0$ enter into the cell because of a diffusion gradient. Each anion, however, takes along with it a cation of equal charge. Under the conditions the entry of the cation $(K^+)_0$ will be against its concentration gradient. The cation will reach a concentration inside the cell which will be greater than that of the outside of the cell. The diffusing anion concentration will be less inside than outside because of the equal number of ion pairs moving in and out of the cell. This electri-

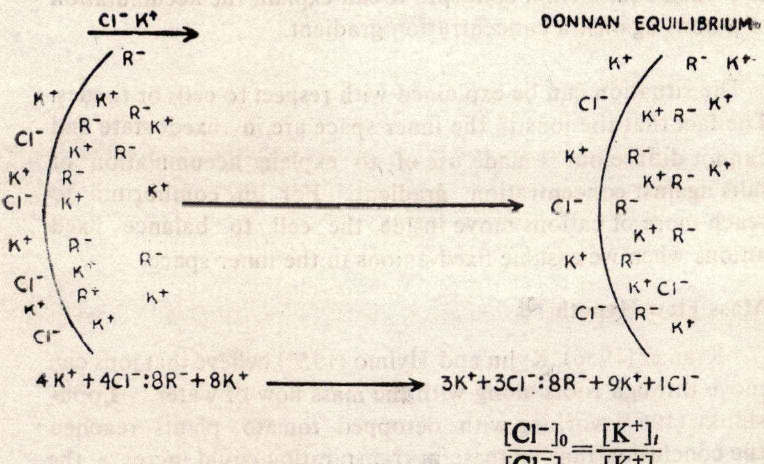

$$\frac{[Cl^-]_0}{[Cl^-]_i} = \frac{[K^+]_i}{[K^+]_0}$$

Fig. 2.4. Donnan equilibrium. At equilibrium the ratio of Cl− concentration outside to that inside the cell $\frac{(Cl^-)_0}{(Cl^-)i}$ equals the ratio of K+ concentration inside to that outside the cell $\frac{(K^+)_i}{(K^+)_0}$.

cal equilibrium which is controlled by electrical as well as diffusion phenomena is known as the *Donnan equilibrium.*

The Donnan equilibrium thus results in accumulation within the cell of an ion in a much greater concentration than that of the external medium. The electrical equilibrium is, however, maintained. The Donnan equilibrium theory can also explain the accumulation of anions against a concentration gradient without the participation of metabolic energy.

Suppose on the inner side of a membrane there is a concentration of anions to which membrane is impermeable (fixed anions). Now, if this membrane is put in a solution containing cations and anions to which the membrane in permeable, an equal number of cations and anions will diffuse across the membrane until an equilibrium is established. However, for this equilibrium to reach, more of cations move inside to balance the fixed anions on the inner side of the membrane. This is known as Donan equilibrium concept. It can explain the accumulation of anions against a concentration gradient.

The situation can be explained with respect to cells or tissues. The fact that the ions in the inner space are in fixed state and cannot diffuse out is made use of to explain accumulation of salts against concentration gradient. For an equilibrium to reach more of cations move inside the cell to balance fixed anions when we assume fixed anions in the inner space.

Mass Flow Hypothesis

Kramer (1956), Kylin and Hylmo (1957) believe that ions can move through roots along with the mass flow of water. Lopushinky (1964) working with detopped tomato plants reached the conclusion that increase in transpiration could increase the absorption of salts. Whether transpiration pull is directly responsible for absorption of salts along with water or it only decreases the concentration of salts in xylem elements and helps in the absorption of salts along with water or it only decreases the concentration of salts in xylem elements and helps in the absorption of more salts due to concentration gradient is a disputed question. However, in the light of our present knowledge, mass flow hypothesis is very insignificant for explaining the absorption of mineral salts.

The absorption of salts by simple diffusion or mass flow or accumulation of salts under ion exchange and Donan equilibrium as described above do not require any metabolic energy and, therefore, their absorption is known as Passive Absorption.

ACTIVE ABSORPTION

The passive absorption fails to give satisfactory explanation of the following observations :

(*a*) Green algae *Nitella* and *Valonia* have in their cell saps concentrations of salts much higher than in outer medium. These cannot be explained by ion exchange or Donan equilibrium.

(*b*) Absorption of ions by plants is also selective whereas concentrations of K^+ and Cl^- ions in cell sap of *valonia* were found to be much higher than in sea water, on the other hand, that of Na^+, Ca^{++} and Mg^{++} were found to be much lower.

(*c*) The region of maximum absorption of salts are meristematic and cell elongation zones of root where the cells show high metabolic activity.

(*d*) The effect of temperature on the process shows that the temperature quotient (Q_{10}) is equal to 2. Had it been purely a physical process it should have been less than 1.3.

(*e*) The absorption of ions is markedly affected by the presence of oxygen. In young barley root tips the absorption increased ten times approximately in the presence of air than in nitrogen.

(*f*) The rate of respiration is increased when roots are fed on salts. This increase in rate over the normal rate of respiration is called as SALT RESPIRATION. This shows that two processes are somehow interconnected.

(*g*) With the help of radioactive ions it has been possible to show that movement of radioactive ions is slow into the inner space, although they can diffuse easily in the outer space.

(*h*) More rapid the rate of growth, more is the absorption of salts. (Growth is the result of metabolic activity.)

All the facts enumerated above can be satisfactorily explained by the active absorption, which requires the use of metabolic energy.

There are a number of evidences for the existence of an active absorption of minerals by plants *e.g.*,

1. The rate of absorption of minerals is too rapid to be explained by passive absorption.
2. The diffusing substance is at a higher concentration within the cell and is at lower concentration outside.
3. The rate and amount of absorption of the substance is directly related to the expenditure of metabolic energy.
4. The cell activity affects the mechanism of absorption.

The absorption and accumulation of salts or their ions within the plant cells against the concentration gradient, *i.e.*, against the laws of diffusion is a characteristic feature of the salt uptake mechanism. *Hober* (1945) reported that freshwater alga *Nitella* accumulated K^+ ions to a concentration, 1,000 times greater than the concentration of K^+ in the surrounding medium. Obviously, the ions continued to be absorbed and retained even after the concentration level of the ion inside the cell had gone far above the concentration of this ion in the external medium.

The penetration of ions and their retention within the cell at a higher concentration would therefore, require expenditure of energy by the plant.

Several experimental findings have pointed out a direct relationship between metabolic energy and salt uptake or accumulation. They are as follows :

(*i*) Relation between salt accumulation and respiration.
(*ii*) Relation between inhibitors and salt uptake, and
(*iii*) Salt induced respiration.

(*i*) *Steward* (1932), *Hopkins* (19 6), and many others found a parallel curve for salt accumulation and respiration as a function of *oxygen tension.* Salt accumulation is slowed, and even prevented completely, with decrease in the oxygen content of the nutrient medium.

(*ii*) Studies with certain *metabolic inhibitors* in relation with salt uptake have also confirmed the view that the process is an energy consuming one. Workers like *Robertson* and *Turner* (1945) and *Lundegardh* (1955) reported that oxidase inhibitors, azides, and cyanides strongly inhibited salt uptake.

(*iii*) An additional evidence in support of the existence of an active process of salt uptake is provided by the phenomena of *salt induced respiration.* Salt uptake has been shown to stimulate and increase the rate of respiration. The salt-induced respiration is prevented by cyanide and carbon monoxide. This led *Lundegardh* (1955, 1958) to assume that a respiratory component (cytochrome system) may be involved in the actual process of accumulation.

THE MECHANISM OF ACTIVE SALT UPTAKE

The lipoprotein membrane of a cell is largely impermeable to free ions. In the recent past several mechanisms have been given to explain the active transport of ions across the impermeable membrane. All these mechanisms try to explain this seemingly difficult task of uptake of ions across the membrane against the concentration gradient, by visualising the presence of a carrier compound within the membrane.

Various mechanisms have been suggested to explain active absorption they are described below.

The Carrier Concept

Van den Honert (1937) was the first to propound the role of a carrier compound in the transport of ions across the membrane. The concept has derived great support from the following three observations, *viz.*,

(*i*) Isotopic exchange.

(*ii*) Saturation effects.

(*iii*) Specificity in the absorption of ions.

(*i*) Using radioactive ions, several investigators found that the ions *failed to diffuse* through a cell membrane. This led them to believe that the diffusion of ions through the impermeable membrane might be accomplished by the intervention of carriers.

(*ii*) It has been observed that the rate of the active uptake of ion does not increase beyond a maximum level, even if the salt concentration is increased very much. This is supposed to be due to the occupation of all the active sites on the carriers, a saturation effect.

(*iii*) The fact that ions are absorbed at different rates and have different levels of accumulation in the root tissue, suggests

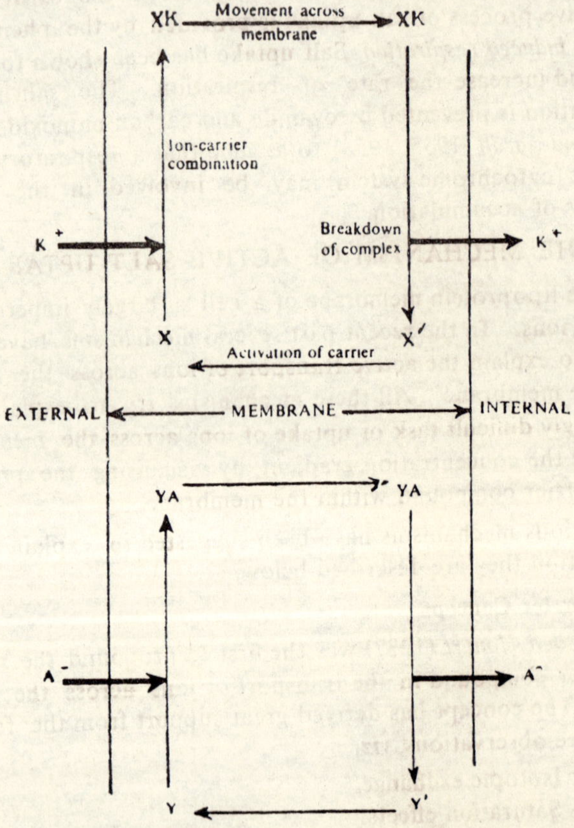

Fig. 2.5. The general concept of the operation of a carrier mechanism in ion uptake by cells. X and Y=carriers. X' and Y'=precursors of the carriers. XK and YA=carriers-ion complex, K+= cation. A−=anion,

the presence of ***specific carriers.*** Like the enzymes, the carrier would select the substrate, which will bind itself on its active site. *Devlin* (1966) believes that the similarities in the activities of carriers and enzymes is a strong evidence in support of the carrier concept in active salt absorption.

Many workers have visualized two different carriers for cations and anions. A general scheme involving the carriers is represented in Fig. 2.5. In the scheme the precursors of the carriers (X′ and Y′) are first activated. The activated carriers (X and Y) then combine with the ions to form *ion-carrier complex*, which moves across the membrane. On reaching the inner side of the membrane the complex breaks down and releases the ion. The metabolic energy is supposed to be used in one or more steps of the process. Three important hypotheses explaining the salt uptake are given below.

After diffusion in outer space the ions have to pass through inner cell membranes which are made up of lipoproteins which resist the passive absorption of ions through them. The movement of ions is supposed to be through carrier molecules. It is presumed that these form carrier ion complex. These can move across the membrane and this movement is unidirectional. On the inner side of the tonoplast the complex again breaks up into ion and carrier. Whereas the ions accumulate inside, the carrier is free to move outside.

The uptake of minerals into both cytoplasm and vacuoles is investigated by solution culture technique. It is found that K^+ ions and nitrate ions are strongly absorbed and other elements relatively less. Evidently, some elements are taken up more readily than others. The question arises how are they absorbed and why are some selectively absorbed in preference to others ?

Work with large fresh water algae *Nitella* and *Valonia* has proved to be enough useful in these investigations. The cell sap of these algae can be extracted easily and the concentration of minerals inside can be compared with that of pond water in which the alga is grown. In these species it has been found that K^+, Na^+, Ca^{++}, Mg^{++} and chloride ions are at a much higher concentration in the cell sap than in the pond water, but all these ions are not absorbed to the same extent. These results clearly show that the cytoplasmic membrane has some cytoplasm which selects the ions to be absorbed and, secondly, if this diffusion against a concentration gradient is to take place, then some energy will be utilised to drive the ions into relatively concentrated solution of the vacuole.

It was held that principally this energy came from the thermal energy of the ions themselves. Once the ions reached the vacuole they were in some way taken out of free solution, perhaps by absorption, thus lowering the internal concentration and allowing more ions to enter. However, experiments indicate that the ions are in free solution and not absorbed in any way. Another evidence that the ions are in free solution comes from the fact that the osmotic pressure of cell sap rises after a period of salt uptake.

Uptake of minerals is a property of actively metabolizing cells ; this indicates that the energy required to drive the ions into vacuole comes from some metabolic process such as respiration. This suggestion is supported by an experimental evidence in which temperature coefficient (Q_{10}) for the absorption of potassium by young barley roots is nearly 2, pointing out to the conclusion that the process is an enzymatically controlled one, and not a physical process (such as diffusion) in which case Q_{10} is about 1.3.

If respiration is involved, then the supply of oxygen should markedly affect the absorption. It was found that absorption of bromide ions by young barley root tips was nearly ten times more in the atmosphere of air than in nitrogen.

Another evidence of the close relationship between respiration and absorption of salts comes from the observation that the rate of respiration is increased on the addition of a salt to respiring tissue. This is known as '*salt respiration effect*'.

Regarding the means of uptake, evidence has been obtained by the use of radioactive ions. Radioactive Rubidium has been found to diffuse into the cytoplasm ; but the ions are passed only relatively slowly into the vacuole. It seems that cytoplasm vacuole membrane or tonoplast layer is the zone which is difficult to cross and requires the expenditure of energy. The entry into the cytoplasm through plasma membrane from the external solution appears to be relatively easy.

It is suggested that some process coupled with respiration is responsible for driving ions across the tonoplast and into the vacuole. *It is possible that a carrier is produced in the cytoplasm which can pick up ion which has diffused into the cytoplasm and transfer it to the tonoplast in such a manner that it is released into the cell sap.* The formation and possibly the regeneration of the carrier are probably the stages where the process is linked with respiratory system. The *differential accumulation* of ions in the vacuole indicates that there must be *more than one carrier* involved in the process. Although no carrier has been definitely identified, it is possible that at least one carrier for cations and one for anions is involved. It has been proposed that the production and regeneration of the carriers is connected to respiration. Two of the well known theories based on the involvement of carriers are described below :

Cytochrome Pump Hypothesis

The theory was put forward by Lundegardh (1933, 1954). According to this theory absorption of anions is linked with respiration and involves cytochrome system. It was found that there was a quantitative relationship between anion absorption and salt respiration. However, such a relationship was not observed between cation absorption and salt respiration. Lundegardh proposed that only anions were actively transported. Since both the salt respiration and anion absorption were inhibited by cyanide poisoning, it was proposed that anions were absorbed through a cytochrome system in which the cytochromes acted as anion carriers. According to this theory as the electrons pass out of the electron transport chain, a corresponding inward passage of anions occurs. The anions are picked up by the oxidised cytochrome oxidase and are transferred to the electron transport chain. The cations were considered to move passively along the electrical gradient created by the accumulation of anions at the inner surface of cytoplasmic membrane.

The theory raised considerable interest but experimental evidences go against it. All anions do not stimulate respiration to the same extent. Maximum salt respiration was recorded due to NH_4^+ ion—a cation. The theory assumes that there is carrier only for anions and not for cations. This has been criticised by different workers.

To summarise the discussion on whatever is known on the question, it may be said that there are four main stages in the mineral uptake. In the first stage mineral ions diffuse from the soil solution into the environment of root apex. In the second stage they may get absorbed on the cells, In the third stage they diffuse through the cell wall and plasma membrane into the cytoplasm. Finally, these ions accumulate in the vacuole of that cell, which is at the expense of metabolic energy—Active absorption, but if the vacuole already contains a high concentration of ions, then they may be transferred to the cytoplasm of next cell, probably through the cell-to-cell protoplasmic connections, the plasmodesmata

Lundegardh and *Burstrom* (1935), *Lundegardh* (1950, 1954) observed that there was a quantitative correlation between anion absorption and salt respiration. Such a correlation between cation absorption and salt respiration was, however, not observed. This led Lundegardh to assume that only anions were actively transported. Further since both the solt respiration and the anion absorption were inhibited by cyanide or carbon monoxide, Lundegardh proposed that the anions were absorbed via a *cytochrome pump* in which the *cytochromes* acted as anion carriers.

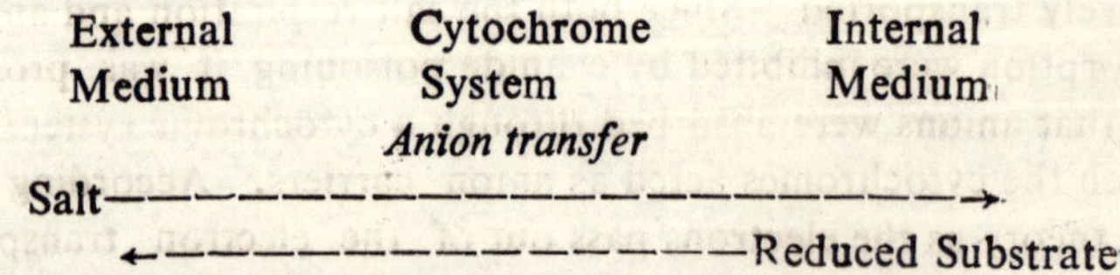

According to Lundegardh, as the electrons pass outward through the electron transport chain there is a corresponding inward passage of anions. The anions are picked up by the oxidized cytochrome oxidase and are transferred to the other

members of the chain as they transfer the electron to the next component (Fig. 2.6).

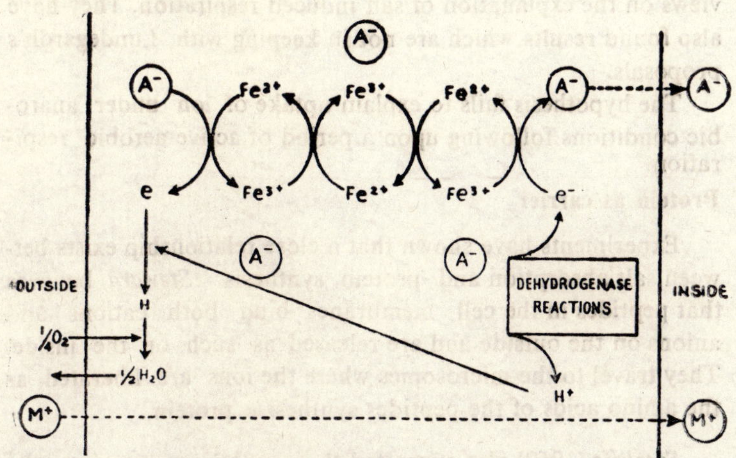

Fig. 2.6. A diagrammatic representation of Lundegardh's cytochrome theory of salt absorption.

The theory assumes that the cations (M^+) moved passively along the electrical gradient created by accumulation of the anion (A^-) at the inner surface of the membrane.

The following evidences can be cited in support of the cytochrome pump hypothesis ;

(*a*) The similarities in the action of oxidase inhibitors on salt accumulation and respiration implies a common role for cytochrome system in both the processes.

(*b*) The spectrophotometric changes, showing rapid oxidation of reduced respiratory components when salts are added to living wheat roots in distilled water.

(*c*) The stimulation of respiration by salt uptake is assumed to be due to the increased rate of electron flow which is facilitated by the inward movement of anions in exchange for electrons.

Though *Lundegardh's* hypothesis explains the role of metabolic energy in the absorption of ion, it has been adversely criticised for assuming that there is carrier only for anions and not for cations. Besides, many workers hold very different views on the explanation of salt induced respiration. They have also found results which are not in keeping with Lundegardh's proposals.

The hypothesis fails to explain uptake of ion under anarobic conditions following upon a period of active aerobic respiration.

Protein as carrier

Experiments have shown that a close relationship exists between salt absorption and protein synthesis. *Steward* believes that peptides in the cell membranes bind both cations and anions on the outside and are released as such on the inside. They travel to the microsomes where the ions are liberated as the amino acids of the peptides synthesise protein.

Sutcliffe (1962) also suggested that proteins enter into the cell membranes and carry back the ions to the ribonucleic acid templates of the endoplasmic reticulum and where the "clusters" of ions are released when the protein combines with the template. The ions combine with water to form ion rich vesicles which move in the cytoplasm and burst at the vacuolar surface releasing ions to the central pool.

Goldacre (1952) proposed that contractile proteins present within the membrane are present in folded or unfolded condition. When they are in unfolded condition they bind ions by free valencies exposed at the membrane surface. When the protein contracts the ions are drawn and released inside as the free valencies of the protein become satisfied amongst themselves in the folded molecule. The Goldacre hypothesis attaches importance to the protein content of the membrane and gives it the role of a carrier which is interconverted by the utilisation of ATP. The hypothesis, however, lacks experimental support.

Protein-lecithin as Carrier

Bennet Clark (1956) proposed that the carrier could be a protein associated with the phosphatide, *lecithin*. The carrier is amphoteric and hence both cations and anions combine with it. In the transport of ion, the lecithin, a phospholipid is synthesized and hydrolyzed in a cyclic manner. It picks up ions on the outer surface during synthesis and releases them on hydrolysis into the inner space. ATP is believed to be concerned in the synthesis of at least one of the components of this *phosphatide cycle* (Fig. 2.7).

The passive as well as active absorption of ions are affected by a host of factors, *viz.*, temperature, hydrogen ion concentration of the medium, presence or absence of light, etc.

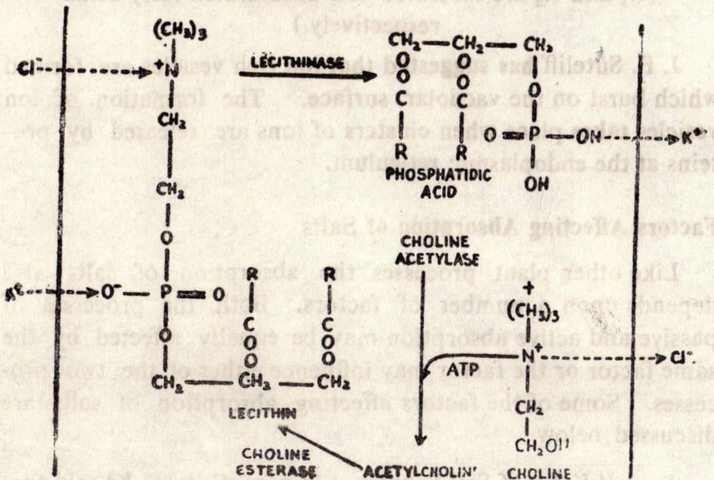

Fig. 2.7. Diagrammatic representation of the phosphatide cycle. At left, ions from outer space are picked up by the lecithin. On hydrolysis of the lecithin-ion complex, the ions are released to the inner surface. Lecithin is then resynthesized.

According to this hypothesis Bennet-Clark's hypothesis) that a phospholipid *Lecithin* and *ATP* are involved in salt absorption, Bennet Clark (1956) put forward the hypothesis that Lecithin has phosphate group to which cations can be linked and choline group to which anions can be linked; this phospholipid is hydrolysed by enzyme Lecithinase and ions are liberated. The end products, *i.e.*, phosphatidic acid and choline again give rise to Lecithine. This synthesis takes place in two steps, through enzyme choline acetylase in first and esterase in the second step. In the first step Acetyl choline is formed which requires ATP.

Lecithin

$$
\begin{array}{l}
R_1COOCH_2 \\
\quad\quad | \\
R_2COOCH \quad\quad O \\
\quad\quad | \quad\quad\quad || \quad \overline{|HOH|} \\
H_2C-O-P-OCH_2CH_2N^+\equiv(CH_3)_3 \\
\quad\quad\quad\quad\quad | \\
\quad\quad\quad\quad\quad O^-
\end{array}
$$

Phosphatidic Acid Choline

(R_1 and R_2 are saturated and unsaturated fatty acids respectively.)

J. F. Sutcliff has suggested that ion rich vesicles are formed which burst on the vacuolar surface. The formation of ion vesicles takes place when clusters of ions are released by proteins at the endoplasmic reticulum.

Factors Affecting Absorption of Salts

Like other plant processes the absorption of salts also depends upon a number of factors. Both the processes of passive and active absorption may be equally affected by the same factor or the factor may influence either of the two processes. Some of the factors affecting absorption of salts are discussed below :

1. *pH Value of Soil Solution*. In general it can be said that increase in *pH* value decreases the absorption of anions and reverse is true for cations. For example, absorption of phosphorus ($H_2PO_4^-$ and HPO_4^-) is decreased in alkaline *pH*.

2. *Temperature*. The rate of absorption of salts is increased with increase in temperature. However, temperatures of 0°C or below and 40°C or above have adverse effect. The process of diffusion is adversely affected by low temperature. An increase in temperature increases the kinetic activity of the molecules and hence diffusion. Higher temperatures and lower temperatures also adversely affect the enzymatic reactions and thus active absorption of salts.

3. *Oxygen.* Active absorption is increased in the presence of oxygen and decrease in its absence.

4. *Interaction of Salts.* Presence of calcium in soil helps in the absorption of potassium. There is also some competition amongst the various elements for absorption, *e.g.*, Barium, Strontium and Calcium compete with each other. This has been explained due to the fact that these have common binding sites of the carriers.

5. *Growth.* Increase in the rate of growth increases the absorption in early stages but with secondary growth in roots bark is formed which may be impermeable to salts.

Elongation of cells is due to hydration and this results in lowering of concentration of cell sap. The diffusion is thus increased.

6. *Light.* It increases the absorption of salts indirectly by its effect on processes of transpiration, photosynthesis and respiration. In the presence of light stomata open, transpiration increases and hence more flow of salts takes place. During photosynthesis there is more utilisation of salts and formation of food. Respiration releases the energy by oxidation of food and thus active absorption is increased.

TRANSLOCATION OF ORGANIC AND INORGANIC SOLUTES

In lower plants each cell is a self-sufficient unit. The movement of dissolved substances from one cell to the other is negligible and occurs by diffusion but in higher plants roots, the organs of absorption of water and nutrients and leaves, the organs of photosynthesis, are quite separated. The water and nutrients which are taken from soil must reach the leaves and the carbohydrates which have been synthesised in leaves must be transported not only to the aerial parts but also to the roots. This circulation or Translocation involves :

Upward Translocation. That is mainly from underground to aerial parts but also from lower branches to upper branches.

Downward Translocation. That is from aerial to underground parts or from upper to lower organs.

Lateral Translocation. This may take place within stems from one side to the other or from leaves to the stem or other lateral organs.

The upward translocation involves both of organic and inorganic solutes whereas downward translocation is mainly of organic solutes from leaves to other parts.

Translocation of Organic Solutes or Assimilates

Nature and Path of Solutes Translocated. On the basis of experiments with tracer techniques and analysis of exudates it has been observed that in most of the plants sucrose is the solute which is transported. The other sugars found to be there are Oligosaccharides, *e.g.*, raffinose, stachyose and verbascose. Mannitol, Sorbitol are other substances which have been found. Besides carbohydrates which constitute 80-90 per cent of assimilates amino acids and amides are also translocated. It is a strange fact that hexoses, glucose and fructose are not translocated.

Whereas there is general agreement that the channels of conduction of water are tracheids and vessels of the xylem, there have been different views expressed regarding the conducting channels for translocation of food material. Radial or lateral translocation mostly occurs through medullary rays. However, the important question is that of tissues concerned with longitudinal movement of food material. Mason and Maskel (1928) found that Phloem is the tissue through which downward translocation of organic solutes takes place. This has been shown experimentally by ringing or girdling experiments. When a ring of bark up to phloem is removed the carbohydrate starts accumulating above the ring.

Entry of Electrolytes into plant cells

It is now firmly believed that electrolytes penetrate into the cells as ions. Entry of one type of ion into the cell must be accompanied by entry of another type of ion of equal electrostatic charge.

It has been generally observed that salts of monovalent cations such as K^+ are taken up and accumulated at a more rapid rate than salts of di- or polyvalent cations such as Ca^{++} or Mg^{++}. Similarly, plant cells are selective regarding certain anions.

Sometimes the plants absorb the anions and cations to different extents. In such cases the electrical balance of anions and cations both inside and outside of the cells is maintained by certain changes in the ionic composition of the cell and the nutrient solution. If cation uptake is more than that of the anion, the cell produces organic acid—anions of which remain within the cell to balance the excess cations and the hydrogen ions of the organic acids pass out into the external medium to compensate the cations which have been removed.

In some cases the shortage of cation or anion within the cell is compensated by H^+ or OH^- ions which are readily available from water. In such a case, the external medium becomes either acidic or alkaline due to the differential absorption of ions.

Alkaline and acidic soils

In highly alkaline soils any added iron, manganese, copper or zinc is promptly precipitated and rendered unavailable to

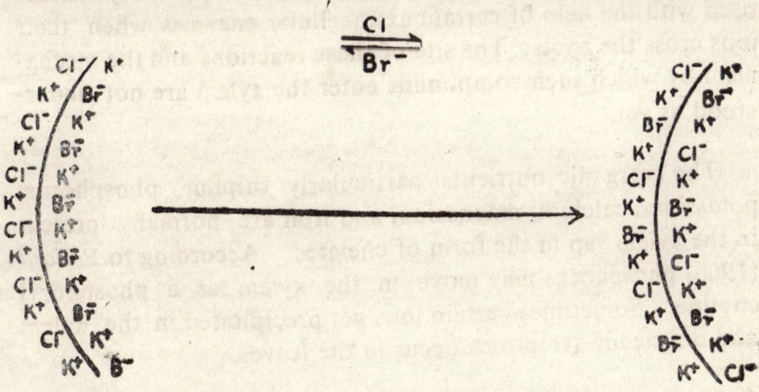

Fig. 2.8. Ion Exchange. In the figures Cl^- diffuse into the cell without disturbing the electrical neutrality because of a simultaneous and equal diffusion of Br^- out of the cell.

the plant. For this reason deficiencies of these four elements are sometimes corrected by spraying the foliage of the plant directly with a soluble salt of the nutrient rather than by soil application.

Excess alkalinity can by corrected by addition of an acidic radical like-sulphur which is slowly oxidized to sulphuric acid by microorganisms of the soil.

INORGANIC TRANSLOCATION

Transport in the xylem

Studies with tracer elements have shown translocation of inorganic substances to be largely through xylem elements. It has already been explained that solutes enter into the xylem elements of the root by an active process. Analysis of xylem sap shows minerals to constitute about one third of the solutes. Two-thirds are organic substances like sugars, organic acids, alkaloids, lactones and large amounts of organic nitrogenous materials. It should be noted that some elements are translocated as organic compounds. In many species, most or all of the nitrogen is translocated as organic compounds such as amides, amino acids and ureides. It is believed that organic nitrogen, sulphur, phosphorus compounds are probably synthesised with the help of certain extracellular enzymes when their ions cross the roots. The site of these reactions and the mechanism by which such compounds enter the xylem are not understood as yet.

The inorganic nutrients particularly sulphur, phosphorus, potassium, calcium, magnesium and iron are normally present in the xylem sap in the form of chelates. According to *Bollard* (1960) phosphorus may move in the xylem as a phosphoryl choline. Sometimes certain ions get precipitated in the xylem and deficiency symptoms occur in the leaves.

Lateral transport

Some amount of minerals are transported laterally from xylem to phloem across the cambium.

Transport in the phloem

Some minerals are also transported through phloem tissues of plants. These minerals are those which are exported from leaves immediately before abscission. The movement of ions in the phloem has been shown by radioactive tracers to be both upward and downward. *Hall* and *Baker* (1972) analysed the exudate of the bark of *Ricinus* plants and found a number of minerals present in it. Nitrogen was found to be transported in the form of amino acids.

MECHANISM OF TRANSLOCATION

Many workers believe that ions move upward chiefly by mass flow in the transpiration steam. *Biddulph* (1951), *Biddulph et. al.*, (1961), however, have performed experiments, the results of which do not agree with this. Studies with isotopes have shown accumulation of ions in various leaves at different rates which is not possible if nutrients were transported by a mechanical process. The translocation and distribution patterns of different nutrients in the leaves show many variations. Elements like phosphorus etc., collect in young leaves ; calcium and others collect in older leaves ; iron collects along leaf veins, and cobalt collects in hydathodes. Old leaves do not accumulate phosphorus ; in fact they export it in large amounts. *Biswas* and *Sen* (1959) found equal distribution or labelled phosphorus and sulphate in all leaves if the stem tip was decapitated. Obviously some sorting mechanism is involved in the transport of nutrients. *Bell* and *Biddulph* (1963) reported calcium transport to be by exchange along the walls of the conducting cells. Several workers like *Stout* and *Hoagland* (1939) and others are of the view that there is considerable exchange of materials between the xylem and phloem resulting in a gradual upward decrease in the amount of the nutrients in the xylem sap. The cambium is believed to regulate the amount of ions carried upwards in the transpiration stream by transferring some amount of ions into the phloem.

Translocation is affected by a number of factors *e.g.*, nature of the vascular connections, metabolic involvement as shown by diurnal variations, temperature, etc.

Retranslocation of ions

Considerable amount of retranslocation of certain more mobile ions from the older leaves to younger leaves, reproductive parts and to the growing parts has been observed in many cases. Redistribution of salts is probably determined by the relative metabolic activities of various parts of the plant. The movement occurs through the phloem because it is prevented by girdling or killing the phloem. While upward translocation occurs mostly in the xylem, downward as well as upward movement can occur in phloem elements. In addition to these lateral translocation between xylem and phloem has also been reported to take place. Minerals are believed to be in continuous circulation as shown in Fig. 2.9.

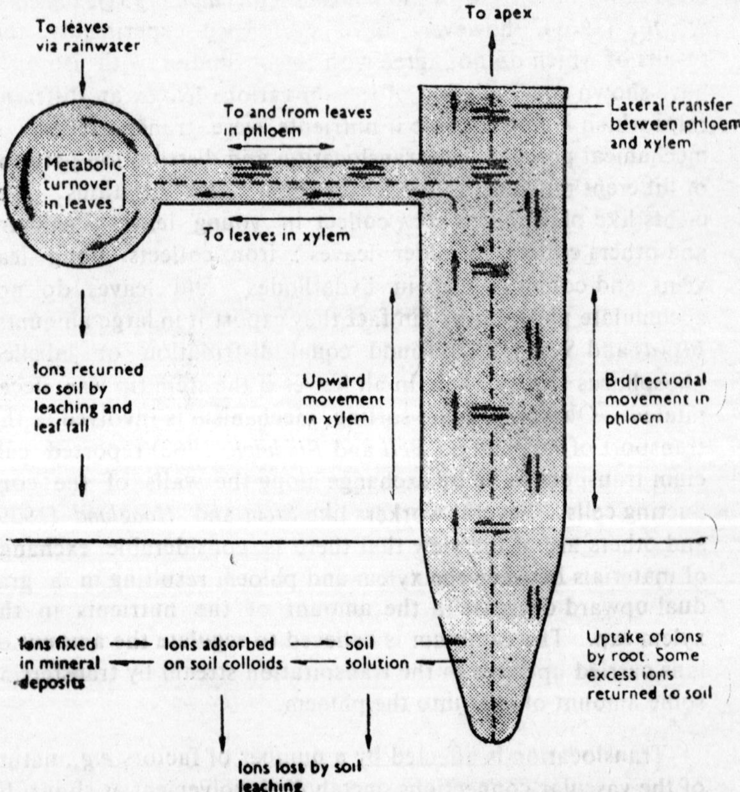

Fig. 2.9. Diagram showing the circulation of minerals within a plant and their return to the soil by leaching and lf fall.

THE MACRONUTRIENT ELEMENTS

Normally the macronutrients form an integral part of complex organic molecules. Some of them assist in the functioning of enzyme systems. A description of the role of such elements in the plants is given below.

Carbon, Hydrogen and Oxygen

Function of Carbon, Hydrogen and Oxygen. Carbon is absorbed by the plant in the form of CO_2 from the air. It is absolutely essential for the synthesis of carbohydrates and fat which build the cell wall, protoplasm etc. Hydrogen and oxygen are present in water which is absorbed by the plant from the soil. Water is indispensable for the various vital activities of the plants.

Nitrogen

Function of nitrogen. The sources of nitrogen are soil and atmosphere. Nitrogen is absorbed from the soil in the form of nitrates, nitrites and ammonium salts. The chief sources of nitrate are sodium nitrate, potassium nitrate, ammonium nitrate and calcium nitrate. Atmospheric nitrogen can be fixed and made available to the plant by certain bacteria.

Nitrogen is absolutely essential for the synthesis of protein. In addition, nitrogen is found in such important molecules as *amino acids, purines, pyrimidines, porphyrins*, and *coenzymes.* Purines and pyrimidines are found in nucleic acids, RNA and DNA which are essential for protein synthesis. The prophyrin structure is present in the structures of chlorophylls and the cytochrome enzymes. These two compounds play a very important role in photosynthesis and respiration respectively.

Deficiency of Nitrogen in the soil leads to the yellowing of leaves called as *chlorosis* due to lack of chlorophyll synthesis. The symptoms first appear in older leaves and then in younger leaves. There may be the appearance of purple colour on leaves and stems due to the formation. This fact is made use of by orchardists in developing gaudy colour of apples by decreasing nitrogen supply. There is more of vegetative growth as compared to reproductive. Characteristically in nitrogen deficiency leaves become more erect than normal and the angle between stem and leaf decreases, especially in cereals, potato, tomato and flax. The lateral buds do not sprout thus resulting into absence of tillering. The flower, fruit and seeds formation is adversely affected. Dormancy is often prolonged and the normal opening of buds is delayed.

Lutman (1934) observed that nitrogen deficiency produced different cytological symptoms in different plants. In the root tips of *setarica italica,* (Japanese millet) the size of the nuclei is reduced in normal sized cells whereas in *Fagophyrum esculentum,* buck wheat the size of the cells of root tip is reduced. In *Brassica napus* a premature vacuolation of the cells occurred. In the deficient supply of nitrogen the leaf palisade cells in potato became thin and slender and the intercellular spaces got enlarged.

Physiological role. Due to the lack of supply of nitrogen to the plant, synthesis of proteins is decreased, formation of new meristems, division and enlargement of cells is impaired. During the early stages of deficiency carbohydrates accumulate, therefore, the chloroplasts are heavily packed with starch grains. The osmotic pressure of cell sap is often high due to high sugar contents.

In excess of nitrogen, there will be more protein synthesis and hence more of cells and enlarged cells. More nitrogen means more synthesis of chlorophyll and the leaves become dark green in colour. There is more utilisation of carbohydrates.

Sulphur

Function of sulphur. It is absorbed from the soil as sulphate ion. It is used in the formation of sulphur-bearing amino acid *e.g.*; cystine, cysteine and methionine. Besides, sulphur is essential for the synthesis of sulphur-bearing vitamins *e.g.*, biotin, thiamine, and coenzyme A.

Sulphur aids in the synthesis of oils and appears to be associated with chlorophyll synthesis. Sulphur is a constituent of ferredoxin and of some lipids found in chloroplasts. Sulphur bridges (-S-S-) have an important role in determining protein structure and sulphydryl groups (-SH) are often part of the active centre of enzymes.

Sulphur deficiency symptoms. Sulphur starvation results in shortage of protein. As in nitrogen-deficient plant, there is reduced leaf, stunted growth, a general chlorosis, followed by the production of anthocyanin pigments in some species.

Sulphur is relatively immobile and therefore, deficiency symptoms appear first in young leaves.

An extensive root system, a hard woody stem and some accumulation of nitrates and carbohydrates are the other effects of sulphur deficiency. Sulphur deficiency is more likely on acid soids.

Phosphorus

Function of phosphorus. It is absorbed by the plants from the soil in the form of phosphate ions $H_2PO_4^-$ and HPO_4^{2-}. Phosphorus is one of the most important elements for the plants.

Phosphorus is a vital structural component of the nucleic acids, nucleoproteins, phytin, phospholipids, sugar phosphates, ATP, NADP and numerous phosphorylated compounds. Highest amount of phosphorus is found in meristematic parts, maturing seeds and fruits.

Phosphorus is actively involved in the synthesis of nucleoprotein—firstly by being present in the nucleic acids and secondly by activating the aminoacids for protein synthesis. The role of the coenzymes NAD and NADP and of the ATP in processes like photosynthesis and respiration is well known.

The *deficiency of phosphorus* produces symptoms similar to those of Nitrogen. There is stunted growth, leaves, fruits may develop *necrotic areas* in addition to *dark to blue green colouration of the leaves.* There is less utilisation of carbohydrates. There is poor development of vascular tissues and pith region is enlarged. Decrease in phosphorus content of the soil often results in increased nitrogen uptake which leads to the formation of free amino acids which are converted into amides instead of proteins. Deficiency of phosphorus affects the formation of ATP, DNA, RNA and other phosphorylated compounds.

Excess of phosphorus in soil decreases the availability of nitrogen. It has toxic effects in the absence of calcium.

Calcium

Function of calcium. Calcium has many roles in the structure and metabolism of plants. One important function of calcium in the plant is to give strength. It is a constituent of the middle lamella, in which it is present in the form of calcium pectate. It is essential for the continued growth of the apical meristems. Maintenance of Na^+, K^+, and Ca^{++} ions within fairly narrow limits would appear to be essential for maintaining the differential permeability of the cell membranes. If this is the case then it is possible to trace a correlation between Ca^{++} level, turgor and consequently rigidity of the stem.

Calcium affects the hydration of colloids. Toxic soluble organic acids like oxalic acid are converted into harmless insoluble calcium oxalate. There is evidence to suggest that calcium helps in the translocation of carbohydrates and amino acids and encourages root development. Calcium reduces the toxicity of such inorganic elements as sodium and magnesium.

Calcium is believed to be important in the formation of cell membranes and lipid structures. *Hewitt* (1963) is of the opinion that the structure and stability of the chromosomes is dependent on calcium. Some workers have suggested that calcium may be an activator for the enzymes phospholipase, arginine kinase, and adenosinc triphosphate, etc.

The *deficiency of calcium* causes poor growth of roots, stems and leaves. The susceptibility of different tissues differs in various species. The first indication of calcium deficiency in many plants is paling of leaf margin behind the leaf apex. The younger leaves show more pronounced chlorosis and curvature of the leaf margin. Ultimately leaf apex is also included. The chlorosis of leaves lead to Necrosis. In some species the chlorotic areas may arise in the intervenal regions thus resulting in a mottled appearance. These areas may ultimately die.

In *Beta vulgaris*, *Brassica oleracea*, tobacco and cocoa "hooking" of leaf tip occurs. In potato, leaflets remain folded and stem apex dies. In severe deficiency of calcium the shoots produced from tubers may die soon after they begin to develop. In cereals such as wheat, oats, barley leaves are tinted gray followed by white chlorosis of tips of young leaves.

Calcium deficiency may have a deleterious effect on the development of roots, which may become translucent, stunted in growth as in cereals, die back from root apex as in tomato, apple and peach or the apical growth may stop as in mustard, pea and citrus.

Calcium dificiency may not prevent flowering, however, premature dropping of flowers may occur as in members of family Leguminosae and mustard. In cereals, the flowering is usually normal but fruiting may fail. Blossom End Rot disease of tomato is attributed to calcium shortage.

Physiological role. In some plants calcium serves to neutralise toxic organic acids like oxalic acid produced during certain phases of plant metabolism. It is held that calcium forms insoluble oxalate thus removing the toxic effect of oxalic acid.

Calcium forms a constituent of certain plant proteins. It forms an essential part of the protoplasm. In plants deprived of calcium new cell walls are either not formed or are abnormal as calcium enters into the composition of middle lamella which is made of calcium pectate. Burstrom (1952) found that calcium makes the cell walls more plastic.

Calcium acts as an activator of several enzymes such as succinate dehydrogenase which is operative in Kreb's cycle, some kinases and phospholipids and Adenosin triphosphatase.

Potassium

Function of potassium. Potassium in the only monovalent cation essential for plant growth. It is found in large amounts in rapidly growing tissues. Potassium is not known to enter into the composition of important organic compounds found in plants. It acts as an activator of over 40 enzymes including the enzymes of phosphorylations and protein synthesis. It is needed by the enzyme DNA polymerase and by the mitochondrion. It is the commonest free ion in the cell, and helps to keep the cell electrically neutral and perhaps stabilises emulsions.

According to *Webster* (1953) potassium is essential for activating the enzymes concerned in the synthesis of polypeptides from amino acids. Potassium is essential for the formation of sugars and starch and also for their translocation throughout the plant. It is also needed in cell division, reduction of nitrate, development of chlorophyll, stomatal movements etc. Potassium is believed to be essential in maintaining the cellular organization, permeability and hydration.

Its deficiency causes lack of apical dominance. The leaves show patches of chlorosis which appear first in older leaves. The leaves then become necrotic from apices and margins. There may be rolling of leaves towards the dorsal side. The internodes remain small and plants show dwarfism.

In apple and some other fruits the condition is known as 'Leaf Scorch' which in appearance resembles withering due to water deficiency. To begin with the scorch is marginal but later on it may be extended to other regions. These symptoms are often associated with a tendency for the leaves to roll inwards towards the upper surface along the margins.

Chapman and Brown (1941) studied the effects of potassium deficiency in species of citrus. The symptoms produced are lesser growth, lesser production of lustreless leaves which later exhibit twisting and crinkling. The marginal leaf scorch found in potassium deficient apple trees is characteristically absent.

In the potassium deficient lemon trees boron accumulated to produce typical symptoms of boron toxicity. Gregory and Richards (1929) reached the conclusion that the variety of symptoms produced in potassium deficient plant is due to accumulation of other ions in quantities sufficient to exert a toxic effect.

Physiological role. It is held that potassium acts as a catalyst in the plants. Shortage of potassium results in decrease in rate of photosynthesis and synthesis of proteins. The amount of carbohydrates is first increased and then decreased. The concentration of soluble nitrogenous compounds first increases accumulation of carbohydrates. It has been attributed to retardation of nitrogen metabolism in early stages of potassium starvation with the result that carbohydrates which would have been involved in the synthesis of organic nitrogenous compounds remained unutilised. Potassium deficiency leads to higher rate of respiration.

Potassium activates a number of enzymes such as fructokinase and pyruvate kinase, etc. Chemical analysis of tissues indicates that potassium is present in cell sap in a soluble inorganic form. In some plants deficiency symptoms due to potassium can be reduced by the presence of sodium.

The next three major elements are S, Ca and Mg.

Sulphur

It is available as SO_4^{--} ion in the soil. The soil micro-organisms indirectly make SO_4 available. Rain and snow also increase sulphur content of soil specially in industrial areas. There may be direct absorption of sulphur dioxide also.

Sulphur is present in amino acids, *e.g.*, cysteine, cystine and methionine, in vitamins, *e g.*, biotin, thiamine and coenzyme A, as sulphydryl group in many enzymes. The sulphuretted smell in plants belonging to Sarson family and in onion and garlic is due to the presence of glucosides such as sinigrin, a condensation product of glucose, Allyl thiocyanate (mustard oil) and potassium hydrogen sulphate.

The *deficiency of sulphur* causes chlorosis and formation of anthocyanin as in nitrogen deficiency, but in contrast to that here the symptoms first appear in younger leaves. The size of the leaves is decreased. There is development of orange, red or purple pigments. Its deficiency causes "Tea Yellow" disease of tea in which stem apex is killed, decrease in nodule formation in legumes and poor formation of fruits.

Physiological role. In sulphur deficiency the total amount of reducing sugars in the plant is decreased and total amount of carbohydrates is increased. The synthesis of proteins is impaired and the amount of free amino acids and amides in plant (cotton) is increased.

Iron

Function of Iron. Iron is normally absorbed in the ferrous form, though it can be absorbed in the ferric form as well. In the latter case it is rapidly reduced within the cell into the ferrous form, the physiologically active state of the element.

In a neutral or alkaline soil, iron is in the insoluble form. Excess of liming, therefore, often results in the iron deficiency symptoms even though the soil is rich is iron. Iron is always present in the soluble form in an acidic soil and is therefore readily absorbed by the plants.

Iron is essential for the synthesis of chlorophyll. Iron is found in ferredoxin, FRS various, flavoproteins and the iron-porphyrin proteins, which include cytochromes, peroxidases and catalases. It, therefore, plays an important role in respiratory mechanism.

According to *Gauch* (1957) iron is primarily concerned in the formation of chloroplast protein in the leaves.

Its *deficiency* causes chlorosis of leaves which appears in the intervenal region so that leaf lamina shows a reticulate yellow pigmentation. The new leaves are the most affected. The general metabolism specially the processes of Photosynthesis and Respiration are affected. This results in the poor growth of plants.

Physiological role. Iron plays an important role in the oxidation and reduction phenomena in plants. Enzymes catalases, peroxidases and cytochrome system which bring about terminal oxidation contain iron as a cofactor. It also activates succinate dehydrogenase which affects the transformation of succinic acid to fumeric acid in Kreb's cycle.

Boron

As boric acid it is exchangeable while its Ca and Mg salts are soluble. With increase in *p*H value its availability decreases, but according to some it is not the *p*H value but the amount of Ca present in soil which affects the availability. It may be remembered here that acidity of soil is generally removed by liming of soil.

It is generally agreed that its primary function is in translocation of sugars. Its deficiency effects due to the lack of sugar supply, are widespread in plant processes. It is believed to play a role in the absorption and utilisation of calcium and other cations.

Its *deficiency* first of all leads to hypertrophy and then the death of shoot apices. The leaves curl and become brittle. There is poor growth of roots and flowers fail to develop. It also affects the internal tissues of organs which are fleshy. In apples internal tissues become corky. Its deficiency also causes heart rot of beet roots.

Chlorine

The evidences in favour of its being an essential clement are increasing. It has been found to be required for normal growth of tomato and sugar beet. It appears to be present as chloride in plants.

It is required during the first part of photosynthesis, *i.e.*, in photophosphorylation both cyclic and non-cyclic.

Its *deficiency* affects photosynthesis and has been found to affect the yield in tomatoes and wheat.

Magnesium

Function of magnesium. Magnesium is indispensable for the formation of chlorophyll, since it is the one and the only mineral constituent of the chlorophyll molecule.

Magnesium is intimately associated with the phosphorus metabolism and is particularly essential for the enzymes involved in the phosphate transfer reaction. It plays an important role in the synthesis of ATP from ADP and inorganic phosphate. Magnesium is an activator for many of the enzyme involved in *carbohydrate metabolism*. Such enzymes and the reactions with which they are concerned are indicated in Table 2.1. Magnesium acts as an activator for the enzymes in the synthesis of the nucleic acids (DNA, RNA). In all these reactions ATP is also required. Nason and McElroy (1963) have suggested that a chelate complex is formed in which

Table 2.1. Some of thc enzymes involved in carbohydrate metabolism which require Mg^{2+} as an activator.

Enzyme	*Reaction*
Glucokinase	glucose+ATP —→ glucose 6-P
Fructokinase	fructose+ATP—→fructose-1-P
Galactokinase	galactose+ATP—→galactose 1-P
Hexokinase	hexose+ATP—→hexose-6-P
Triosekinase	gylceraldehyde+ATP—→phospho-glyceraldehyde
Gluconolactonase	6-phosphogluconolacton →6-phosphogluconate
6-Phosphogluconic dehydrogenase	6-phosphogluconate—→ribulose-5-P
Phosphopentokinase	ribulose-5-p+ATP—→ribulose-1, 5-diphosphate
Enolase	2-phosphoglycerate—→phospho-enolpyruvate
Pyruvic kinase	phosphoenolpyruvate—→pyruvate +ATP
Carboxylase	pyruvate—→acetaldehyde
Phosphoglyceric kinase	1, 3-diphosphoglycerate+ADP —→3-Phosphoglycerate+ATP

the magnesium and the pyrophosphate group of the coenzymes ATP or ADP are linked to the concerned enzyme. It is also believed to be an important binding agent in microsomal particles where protein synthesis takes place.

Deficiency Symptoms. As magnesium is a constituent of chlorophyll '*a*' and '*b*' its deficiency produces some sort of chlorosis followed by the appearance of some other pigments like anthocyanine. The symptoms are practically always seen first in the oldest leaves and then in the younger ones. In many members of family cruciferae, the symptoms appear in the form of yellow green, mottled areas that are uniformly distributed over most of the older leaves while vascular tissues remain green. This is followed by bright yellow or purple pigmentation which is generally marked along the margins of the leaf. Chlorosis and pigmentation is often followed by necrosis. In the shortage of magnesium primary phloem is usually poorly formed and pith region becomes narrow.

Physiological role. It has been suggested that magnesium is involved in some way with phosphorus metabolism, since a higher concentration of magnesium is found in those parts of the plant which contain a higher content of phosphorus as for example, seeds, apices of roots and shoots, etc. Hewitt held that this is probably due to activating action of magnesium in enzymes connected with phosphorus metabolism.

Magnesium is an activator of many enzymes such as fructo-kinase, gluco-kinase, hexo-kinase and numerous other kinases concerned with carbohydrate metabolism. It is also an activator or cofactor of most of the enzymes involved in Kreb's cycle. There is probability of its involvement as a binding agent in protein synthesis.

Copper

It is required by plants in very small doses and becomes toxic in higher concentration. It is available as cuprous and cupric ions, while some of it forms stable complexes with organic matter and becomes non-available. In the presence of phosphate its uptake is decreased due to the formation of insoluble copper phosphate.

It is required for normal metabolism. It is a component of enzymes phenolases, ascorbic acid oxidase. The element is highly toxic in higher concentrations and, therefore, it is used as a fungicide.

Its *deficiency* results into necrosis of apices of young leaves which proceeds along margins and ultimately the leaves are lost. Its deficiency also produces a disease called *exanthema*, *i.e.*, trees start producing gum through slits on bark or produce blisters.

Zinc

It is available in divalent form. Its availability is affected by *p*H value of soil. In acidic soils more zinc is available because of action of acids on its salts ZnS and $ZnCO_3$. In the presence of calcium hydrogen phosphate its behaviour is similar to that of copper described above, *i.e.*, its uptake is decreased due to formation of its insoluble phosphates.

It is an activator for several enzymes. It is required in the synthesis of IAA. It takes part in protein synthesis for which there is an indirect evidence that amino acids and amides accumulate in its deficiency. It enters into the composition of carbonic anhydrase which catalyse the reaction $H_2CO_3 \rightleftharpoons CO_3 + H_2O$. This enzyme is present in chloroplasts.

Zinc deficiency symptoms. Zinc deficiency cause reduced stem growth due to decreased synthesis of auxin. It results in stunted vegetative growth and distorted leaves. Chlorosis in the interveinal regions of the leaves is another effect of the element's deficiency. The internodes are reduced in size and the effect is sometimes referred to as 'little leaf' disease. Other deficiency diseases are rosettes of pecan, white bud of maize, "mottle leaf" disease of apples and walnut. The absence of zinc also suppresses seed formation and causes malformations in fruiting trees.

Molybdenum

The need of this element was first demonstrated in tomatoes by Arnon and Stout (1939). It is present in soil as molybdate ions. It is also found as its di-, tri- and pentaoxides. Out of these it is made available only in the form of its trioxide. Its absorption is maximum in the presence of phosphates and least in the presence of sulphates.

It is required for nitrogen and phosphorus metabolism. Though, the mechanism of its involvement in the latter is not yet completely known. In the nitrogen metabolism it is required for reduction of nitrates and fixation of nitrogen. It plays a role in the formation of ascorbic acid. The activity of enzyme nitrate reductase depends on molybdenum.

Molybdenum deficiency symptoms. Its deficiency results first in the chlorotic interveinal mottling of the lower leaves. It also inhibits the flower formation. "Whiptail" disease of cauliflower has been reported to be due to deficiency of the element.

Supply of Mineral Elements to Crops

We have so far discussed about the role of various elements and symptoms produced in their deficiency. We shall now briefly discuss their practical uses. When crops are grown in the soil, the amount of the elements absorbed by the crop decrease in soil. If these are not supplied to the soil before sowing another crop, the result will be poor yield or sometimes even no yield at all if the soil has been completely depleted of elements by repeated crops. To keep the soil fertile, its proper management is required. The various methods to keep the fertility of the soil so as to get proper yield of crops are given below :

1. *Addition of Manures and Fertilizers.* The common practice is to add Farm Yard Manure to the soil. If it is properly converted into compost and then supplied to the field the results are much better. Compost is made in pits which are made of sizes depending upon the number of cattle in the farm. Entire refuse is put along with the dung. Here proper decomposition takes place which fails when the dung is just placed in the open in heaps. Further improvement in the compost is done by adding required quantities of fertilizer, *e.g.*, phosphates in the pit itself. The dung used for burning

purposes is indeed a huge national wastage. Compost is also prepared from the town refuse and underground sewage system.

To meet the increasing demands of soil and to have the maximum yield from the limited space, manures are supplemented with chemicals or artificial manures called as *Fertilizers.* The common fertilizers which supply N, P, and K are ammonium sulphate, ammonium nitrate, calcium ammonium nitrate (CAN or Kisan Khad), urea, calcium superphosphate, and muriate of potassium. A fertilizer in which all these three critical elements are present is called as *complete fertilizer.* Such packed fertilizers are now being marketed by various firms. Before applying these, the soil must be got tested by soil chemists and the nutritional requirements got evaluated for the crop to be grown. Otherwise over-dosing is as bad as under-dosing.

2. *Green Manuring*. This is the practice of growing crops and ploughing them into the soil when their foliage is quite developed. Usually plants belonging to leguminosae are taken for this purpose. When ploughed back their decomposition by microflora increases the nitrogen content of the soil.

3. *Crop Rotation*. It is another sound practice to get good yields. It means to grow different crops alternately in the field. Here again one of the crop taken is legume. Leaving the land fallow (uncultivated) also helps in its recuperation. Extensive research has been done on the effect of fertilizers on the yield of different crops. The agriculture departments of different States issue instructions from time to time on the use of fertilizers. Their method of application, the time of application, the quantity of different fertilizers required, etc. are given for different crops. One must make the judicious use of all these.

THE MICRONUTRIENT (TRACE) ELEMENTS

In general, it may be stated that the micronutrients and a few of the macronutrients play their part by participating in

Table 2.2. A Summary of Minerals used by Plants.

Element	*Form in which Absorbed*	*Functions*	*Deficiency Symptoms*
Major Elements			
1. Nitrogen	NO_3^- or NH_4^+	1. Enters into the composition of amino acids, proteins and nucleotides. 2. Photosynthesis and respiration.	(1) Yellowing of leaves (Chlorosis). (2) Production of pigments other than chlorophyll. (3) Small sized plants.
2. Potassium	K^+	1. Amino acid and protein synthesis. 2. Activator for enzymes. 3. It is responsible for apical dominance.	(1) Marginal leaf scorch. (2) Hooking of leaf tip. (3) Stunted growth. (4) Shortening of Internodes. (5) Narrow pith.
3. Calcium	Ca^{++}	1. Constituent of cell walls as calcium pectate. 2. Essential for cell division (Mitosis). 3. Formation of cell membranes. 4. Translocation of carbohydrates. 5. Acts as an activator of Kinases (enzymes). 6. Neutralises acidity of soil.	(1) Chlorosis of leaves. (2) Poor growth of roots stem and leaves. (3) Dropping of flowers. (4) Fruit formation may fail.
4. Phosphorus	$H_2PO_4^-$ (orthophosphate)	1. Formation of high energy phosphate ATP and other phosphate bonds. 2. Synthesis of nucleo proteins.	(1) Leaves dull, development of purple or red colour. (2) Development of Necrotic areas on leaves. (3) Dark to blue green colouration of the leaves.

(*Contd.*)

5. Magnesium	Mg^{++}	1. Part of the Chlorophyll molecule. 2. Acts as an activator of enzymes involved in phosphate and carbohydrate metabolism.	(1) Chlorosis of older leaves (2) Veins of leaves remain green. (3) Phloem poorly formed, pith narrow.
6. Sulphur	SO_4^{--}	1. Enters into the structure of proteins containing thiol (–SH) group. 2. Present in glucosides like Sinigrin.	(1) Chlorosis of younger leaves. (2) "Tea Yellow" disease of tea.
Minor Elements			
7. Iron	F^{++}	1. Chlorophyll synthesis. 2. Constituent of cytochromes.	(1) Young leaves turn yellow white, veins remain green. (2) Poor growth.
8. Copper	Cu^{++}	Activator of phenolases and ascorbic oxidase enzyme systems.	(1) Die back of shoots. (2) Exanthema. (3) Necrotic spots.
9. Manganese	Mn^{++}	1. Activator of enzymes like carboxylases. 2. Plays a role in nitrate reduction.	Chlorotic and grey specks in intervenal areas of leaf.
10. *Zinc*	Zn^{++}	1. Activator of some enzymes. 2. Involved in synthesis of IAA.	(1) Leaf malformation. (2) White necrotic spots.
11. Boron	BO_3^{---} or BO_4^{--} (Borate or Tetraborate)	1. Involved in carbohydrate transport. 2. Influences uptake and utilisation of Ca.	(1) Leaves become brittle. (2) Death of short apex.
12. Molybdenum	Mo^{+++} or Mo^{++++}	1. Plays a role in phosphorus and nitrogen metabolism. 2. Activates enzyme nitrate reductase.	(1) Reduction in the size of plant. (2) Necrosis. (3) Inhibition of flowering.

enzyme systems. It is believed that there are two basic ways in which this may be brought about. Micronutrients like copper and molybdenum and the macronutrient iron acts as electron carriers by virtue of valency changes. The other important method by which micronutrients (and also the macronutrient, magnesium) act is by combining with the enzyme—substrate complex. The special role of the micronutrient elements is described below.

Manganese

Function of manganese. Primarily manganese functions as an enzyme activator is several reactions of respiration and nitrogen metabolism. The enzymes—malic dehydrogenase and oxalosuccinic decarboxylase in respiration and nitrite reductase and hydroxylamine reductase in nitrogen metabolism require manganese as activator. Manganese also plays some role in the synthesis of chlorophyll and in the transfer of electrons from H_2O to photoxidized chlorophyll in photosynthesis.

Its *deficiency* produces chlorosis of the leaves which is followed by necrosis. The younger parts are affected first in some species while in other species the older may be first affected. The chlorosis appears in the form of patches between the veins of the leaves. The grey streak disease of oats and other cereals in which internal chlorotic streaks appear in the leaves is due to deficiency of Mn.

Excess of Mn may become injurious although sensitivity of different species varies. Chlorosis of marginal and then intervenal regions may occur due to Mn toxicity. Chlorosis may be followed by necrosis and ultimately death of the plant.

Boron

Function of boron. Boron differs from the other micronutrients in that there is no evidence to suggest its connection with the enzyme systems and it also differs in that it is absorbed as an anion, *i.e.*, borate or tetraborate, rather than as a cation like the other metallic nutrients. Although many roles for boron in plant metabolism have been suggested, its function in the translocation of sugar is the only one, which has been accepted by all. It is believed that boron facilitates the translocation of sugars in plants. It has been seen that places of high metabolic activity in a plant requiring high quantities of sugar, are the first to be affected in the absence of boron.

Boron is also believed to affect flowering and fruiting, cell division, metabolism, active salt absorption, photosynthesis etc.

Boron deficiency symptoms. Boron deficiency results in the death of shoot tip. The root tips die and the shoot growth is stunted. Flowers are not formed. The leaves have a coppery texture. Internally, carbohydrates and aminoacids accumulate.

Boron deficiency tends to shift metabolism from glycolysis to pentose phosphate pathway.

Physiological diseases like internal cork of apples, top rot of tobacco, cracked stem of celery, browning of cauliflower and heart rot of sugarbeet are developed as a result of boron deficiency.

Role of other Elements

Several other elements, over and above the 16 elements, which are essential, are present in plants in large or small amounts. It is supposed that most of these non-essential elements are quite important for the mormal growth and development of the plants. Among them the most common are sodium, silicon, iodine, flourine, cobalt, selenium, aluminium etc.

Sodium is essential for blue green algae and at least for some of the higher plants. It regulates the transport of amino acids to the nucleus and therefore, controls the synthesis of nucleoprotein.

Silica is essential for cell walls of diatoms but its function in other plants is not known. It also occurs in cell walls of some grasses and sedges.

Cobalt is part of vitamin B_{12} which is itself a part of an enzyme. It is known to be essential for blue green algae only.

Table 2.3 shows the occurence, uptake, distribution, incorporation and function of macronutrienls.

Table 2.3. Occurence, uptake, distribution, incorporation and function of macronutrients.

Bio-element	*Bound form in soil*	*Accessible form in soil*	*Taken up as*	*Incorporation in plant*	*Function in plant*	*Sites of accumulation*	*Transportability*	*Deficiency symptoms*
1	2	3	4	5	6	7	8	9
N	Organically bound, nitrate	supplied by microbial decomposition ; NH_4^+ absorbed on clay minerals and humus ; NO_3^- in solution	NO_3^-, NH_4^+, (urea)	Free as NO_3^- ion (vacuoles) in organic compounds, in protein, nucleic acids sec. plant substances	Essential component of protoplasm and enzymes	Young shoots, leaves, buds, seeds, storage organs	Good, primarily in organically bound form	Stunting or dwarfism ; spindly appearance ; shoot ; root ratio shifted toward roots ; premature yellowing of old leaves, sometimes reddening

Table 2.3. (*Contd.*)

1	2	3	4	5	6	7	8	9
P	Organically bound, phosphates of Ca, Fe, Al	As PO_4^{-3}, HPO_4^{2-}, rel. insoluble, absorbed and in chelated complexes, Microbial release slight	HPO_4^{2-}/ HPO_4^{-}	Free as ion, in esteric compounds, nucleotides, phosphatides, phytin	Basal metabolism and synthesis (phosphorylation)	More in reproductive organs than in vegetative (pollen granules)	Good, in organically bound form	Disturbance of reproductive processes (delayed flowering), stunting, dark green or bronze-violet discoloration of leaves and needle-tip drying in conifers

S	Organically bound, sulfur-containing minerals, sulfates of Ca, Mg andNa	SO_4^{2-} readily soluble, little adsorbed	SO_4^{2-} from soil (SO_2 from air)	Free as ion, bound as SH-or SS- group and as ester, in protein, coenzymes, sec. plant substances	Component of protoplasm and enzymes	Leaves, seeds	Good in organic form, poor as ion	Similar to N-deficiency, intercostal chlorosis of young leaves
K	Feldspar, mica, clay minerals	Adsorbed > dis-solved	K^+	Dissolved as ion (primarily in cell sap) and adsorbed	Colloidal effect (promotes hydration). Synergists : NH_4^+, NA^+. Antagonist : Ca^{2+}. Enzyme activation (photosynthesis, nitrate reductase), osmoregulation (stomata)	Division zones, young tissue, bark parenchyma, sites of intense metabolism	Good	Disturbed water balance (tip drying), curling of edges of older leaves, old conifer needles dropped prematurely, root rot

(Contd.)

Table 2.3. (*Contd.*)

1	2	3	4	5	6	7	8	9
Mg	Carbonate (dolomite), silicate (augite, hornblende, olivine), sulfate chloride	Dissolved > > adsorbed ; deficient in acid soils	Mg^{2+}	As ion dissolved and adsorbed, bound in complexes, organically bound in chlorophyll and pectates, component of enzymes and ribosomes	Regulation of hydration (antagonist to Ca^{2+}) ; basal metabolism (photosynthesis, phosphate transfer ; synergists ; Mn, Zn	Leaves	Good in part	Stunted growth, interveinal chloroses of old leaves
Ca	Carbonate, gypsum, phosphate, silicate (feldspar, augite)	Adsorbed > > dissolved ; deficient in very acid soils	Ca^{2+}	As ion ; as salt dissolved ; crystallized and encrusted as chelate ; organically bound in pectates	Regulation of hydration (antagonists : K^+, Mg^{2+}) ; enzyme activator (amylase, ATPase), regulator of growth in length	Leaves, tree bark	Very poor	Disturbance in growth by division (small cells), tip drying, leaf deformation, impaired root growth

Fe	Sulfides, oxides, phosphates, silicates (augite, hornblende, biotite)	Adsorbed > > mobilized ; fixed in chalk soils	Fe^{2+}, Fe (III)-chelate	In metal-organic compounds ; component of enzymes (heme, cytochrome, ferredoxin)	Basal metabolism (redox reactions), nitrogen metabolism, chlorophyll synthesis	Leaves	Poor	Straw-yellow interveinal chloroses, in extreme case white coloration of young leaves (veins green) apical-bud formation suppressed
Mn	Amorphous oxide (MnO_2), carbonates, in silicates	Adsorbed> >dissolved ; better available in acid soils ; accumulates under reducing conditions	Mn^{2+} Mn-chelate	In metal-organic compounds and complexes ; component of enzymes	Basal metabolism (oxidases, photosythesis, phosphate transfer), stabilizes chloroplast structure, nitrogen metabolism ; nucleic-acid synthesis synergists : Mg, Zn	Leaves	Poor in part	Inhibition of growth, chloroses and necroses on young leaves, leaf abscission

(Contd.)

Table 2.3. *(Contd.)*

1	2	3	4	5	6	7	8	9
Zn	Phosphates, carbonates, sulfides, oxides, in silicates	Adsorbed ⊳ ⊳ soluble ; mobilization acid > basic	Zn^{2+} Zn-chelates	Bound in complexes	Chlorophyll formation, enzyme activator, basal metabolism (dehydrogenases), protein breakdown, biosynthesis of growth regulators (IAA)	Roots, shoots	Poor	Stunted growth, white-green discoloration of older leaves, disturbances in fructification
Cu	Sulfides sulfates, carbonates	Adsorbed, mobilization acid > basic, strong fixation of humus	Cu^{2+} and Cu-chelates	Bound as complexes, component of enzymes (plastocyanin, phenol oxidases)	Basal metabolism (photosynthesis, oxidases) ; nitrogen metabolism ; sec. metabolism	Woody axes of shoots	Poor	Tip drying, leaf curl, spotty chloroses of young leaves

Mo	Molybdates, in silicates	Adsorbed, mobilization basic > acid	MoO_4^{2-}	In metal-organic compounds ; component of enzymes	Nitrogen fixation (reductases), phosphorus metabolism, iron absorption and translocation		Poor	Disturbance of growth and deformation of shoots, browning of leaf edges
B	Tourmaline, borates	Adsorbed > ⊳ soluble, availability acid > basic	HBO_3^{2-} $H_2BO_3^-$	Bound to carbohydrates as complexes ; esteric binding	Carbohydrate transport and metabolism ; phenol metabolism activation of growth regulators (growth of pollen tubes)	Leaves, tips of shoots	Poor	Disturbance of growth (meristem necroses), diminished branching of roots, phloem necroses, disturbances of fructification, excessive cork formation

(*Contd.*)

Table 2.3. (*Contd.*)

1	2	3	4	5	6	7	8	9
Cl	Salt, silicates	Soluble > ≫ adsorbed	Cl^-	Free as ion, mostly stored in cell sap	Colloidal effect (increases hydration); enzyme activation (photosynthesis)	Leaves	Good	Leaf curl, root thickening

Interaction of nutrients

The principle of plant demand implies that the rate of absorption of a nutrient which is supplied in a constant external concentration will be influenced by variations in the concentration of other nutrients which affect its utilization. This type of interaction is well illustrated in the work of Williams (1948) already cited ; increased growth caused by supplying additional phosphate led to a greatly enhanced uptake of nitrate, the external concentration of which was maintained constant. Such effects were familiar to earlier students of plant nutrition ; Gregory (1937) drew attention to the fact that when the supply of nitrate was limiting, plants were unable to respond to an additional supply of phosphate which, however, caused increased growth when more nitrate was provided.

This is not the only way in which ions can interact in the process of absorption. Since the net uptake of anions equals that of cations—otherwise the electrostatic balance would not be maintained—an increased supply of a readily absorbed cation may reduce the uptake of other cations even though their concentration remains unchanged. From the work of Hoagland and Martin (1933) shows that increased absorption of potassium brought about by adding that ion to the soil caused a concomitant reduction in the uptake of calcium and magnesium Fig. 2.10. In other situations the increased supply of a mobile ion of one sign can enhance the uptake of ions of the opposite sign.

Not all ionic interactions are so simply explained ; the 'lime induced chlorosis' observed in soils containing large concentrations of calcium can be primarily due to the high pH reducing the solubility of manganese and iron. There are other ways also in which the absorption of one ion can exert an influence. For example, the presence of calcium may lead to a general enhancement of absorption ; this 'Viets' effect is attributed to the influence of this ion on the permeability of membranes. The interaction between calcium and hydrogen ions can be particularly important ; when the pH is low but still within the physiological range the addition of calcium can confer a

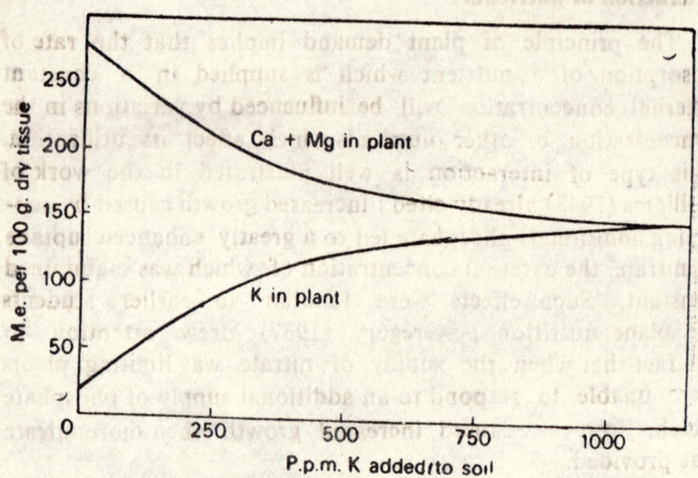

Fig. 2.10. Cationic competition in nutrient uptake. Increasing applications of potassium to soil increased the uptake of that ion and reduced uptake of magnesium and calcium concomitantly.

particular benefit on plant growth. The loss of potassium from roots which can be induced by a lowering of the external pH is likewise restricted by an abundant supply of calcium. The effects of low pH cannot, however, be entirely explained, or offset, by interactions with other ions. If the pH falls below about 3, root growth of many species ceases and there is a considerable efflux of ions. These effects have been attributed to interference with the integrity of membranes.

Contribution of different parts of the root system to the nutrition of the intact plant

In natural conditions different parts of the root system often experience very contrasting environments depending on their distance below to gain entry to the stele by passing across the plasmalemma and cytoplasm of the endodermal cells. The deposition of suberin lamellae in the endodermis is considered to bar the apoplastic pathway by interposing a layer of hydrophobic material between the cell walls and the plasmalemma.

The observations of Marschner and Richter (1973) who studied the uptake of potassium, sodium and calcium by different parts of root systems are broadly compatible with the above findings. Other nutrient ions have been studied in less detail. The absence of a suitable radioactive tracer is the main reason why information comparable to that has not hitherto been available for ammonium or nitrate ions ; however, it was recently demonstrated that ammonium ion can be absorbed by the entire root systems of cereal plants in a broadly similar manner to phosphate and potassium ions ; information on nitrate is not yet available. Kashirad *et. al.* (1973) have found that the absorption of ferric iron by different parts of root systems of *Zea mays* can be much influenced by the amount of phosphate which is simultaneously supplied.

The significance, from the viewpoint of plant growth, of the contrasting extent to which different nutrients are absorbed by tissues of differing age becomes more apparent when the mobility of ions within the plant is also considered. Potassium is readily redistributed within plants as well as being able to enter virtually the entire root system. Calcium in contrast not only enters solely into the younger tissues but it is translocated within roots toward their apices to a negligible extent. Thus, with calcium but not potassium a continuing supply in the immediate neighbourhood of growing root apices is a prime requirement for growth. The situation with other nutrients, for example phosphate and nitrate, is more complex because of localized responses to nutrient enrichment in the rooting zone.

Relationships between the length, volume and surface area of roots and their ability to absorb nutrients

From the viewpoint of studying the function of root systems in the soil, it is important to know whether their ability to absorb ions is related more closely to their length, to their surface area, to their volume or to their weight. These are the four ways in which root distribution can be expressed in practice. It has commonly been assumed that the surface area of roots is the most relevant parameter ; this seemed a logical view so long as the primary barrier to nutrient uptake was assumed to be the external surface of the root, or another cylindrical interface.

However, it is now evident that when the external nutrient concentration is similar over the entire root surface—as in solution culture—the ability of different parts of the root systems of cereals to absorb nutrients varies more closely with the volume of tissues than with their length or surface area. The same appears to be true of the dicotyledonous root before secondary thickening. Since the dry matter content varies little between diffent parts of root systems similar relationships between the age of roots and nutrient uptake would be found if absorption were expressed per unit weight of roots, either fresh or dry. These relationships can be greatly modified in the soil especially with those ions such as phosphate which diffuse so slowly that their rate of arrival at the soil/root interface rather than subsequent physiological processes may largely govern absorption.

Root absorbing power. In studies of the movement of nutrients to roots through the soil it can be important to consider the flux across the root surface, that is to say their rate of arrival per unit surface area. For this purpose the concept of 'root absorbing power' which has been denoted as 'α' has been used by a number of workers.

Experiments under controlled laboratory conditions permit a more detailed analysis of the manner in which root systems respond to localized enrichment of fertilizers. A strictly localized proliferation of lateral roots can occur if one part of the root system is provided with a more favourable supply of nitrate or phosphate though the eventual length of the root axes themselves is little affected ; in contrast, local enrichments with potassium can affect the whole root system equally.

The provision of favourable concentrations of either phosphate or nitrate to only a limited part of the root system can largely offset a limited supply to other parts of the root system. Fig. 2.11. The entire root system of some plants received 1.0 mM nitrate ; in others only a 4 cm zone of a single seminal root

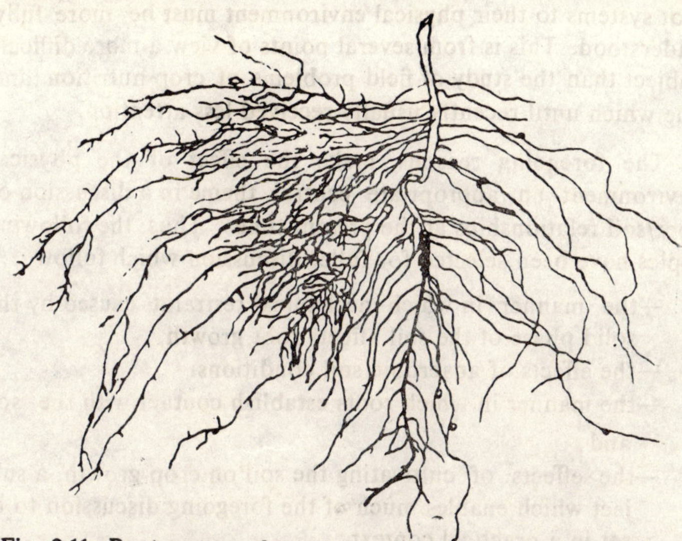

Fig. 2.11. Root system of pea plant (Pisum arvense) showing proliferation of roots due to a fertilizer containing phosphate and potassium being placed 5 cm to the left of the seed and 7.5 cm below the soil surface,

axis received this concentration, the remainder of the root system being exposed to 0.01 mM nitrate. In these latter plants the weight of shoots, their concentration of nitrogen and their growth rate were appreciably reduced after ten days. But nine days later, although the plants were still smaller, the relative growth rate of the shoots and the concentration of nitrogen in them had become similar to that of plants in which the whole root system received the high concentration. This compensatory response was due to both the greater size and the greater uptake per unit weight of roots in the enriched zone.

Modern constraints have thus tended both to extend arable farming to soils in which physical problems are more likely to occur and at the same time to make it less easy to employ some of the traditional methods for offsetting any deleterious effects. It will be shown that the new and simpler methods of cultivation can sometimes provide important opportunities for maintaining the fertility of the soil, but if the maximum advantage is to be taken of them, without undue risk, the response of root systems to their physical environment must be more fully understood. This is from several points of view a more difficult subject than the study of field problems of crop nutrition and one which until recently usually received less attention.

The foregoing reasons make the effect of the physical environment an appropriate central theme in a discussion of root/soil relationships at the present time. Thus, the following topics have been selected for fuller discussion which follow :

—the manner in which mechanical restraint caused by the solid phase of the soil affects root growth.
—the effects of anaerobic soil conditions.
—the manner in which roots establish contact with the soil and
—the effects of cultivating the soil on crop growth, a subject which enables much of the foregoing discussion to be set in a practical context.

The Elimination of Minerals

The ascending sap carries minerals into the shoot, where they accumulate in the cell walls or in the vacuoles. Small quantities of minerals are removed as components of the various materials eliminated by the plant. Most of the minerals incorporated are released, however, only when *parts of the plant die and are shed.* Loss and replacement of leaves and bark are thus a necessary and regular process of elimination in perennial plants.

Frey-Wyssling (1949) has proposed a scheme in which three processes of direct elimination are distinguished. *Recretion* is the elimination of salts in the same form in which they were taken up. Recretion takes place over the entire surface of a plant, from which salts are washed away by rain. K^+, Na^+, Mg^{2+}, and Mn ions are leached out relatively easily. Many species growing in saline habitats have glands specialized for the elimination of salt, and species of saxifrage eliminate calcium through the hydathodes. *Secretion* is the release of assimilation products such as the sugar in nectar and the carbonic and amino acids given off by the roots. Planktonic algae secrete into the water quite large amounts of soluble carbohydrates, C_4 acids, and amino acids. *Excretion* is the elimination of products of intermediate metabolism and end products of catabolism.

The ecological significance of secretions and excretions lies not so much in regulating the composition of the plant as in their role in biotic interference: for example, they act as attractants of pollinators, as deterrents to herbivores, and as inhibitors in intraspecific and interspecific competition (allelopathy and antibiosis). Substances with *allelopathic* effects are organic compounds given off by plants which injure other plants or prevent them from becoming established in the vicinity. In most cases these are ethylene, etheric oils, phenol compounds, alkaloids, glycosides, and cumarin derivatives released into the air, given off by the roots, or washed out of the shoot and into the ground by rain.

3

Water in Relation to Plants

Life evolved in water, and water remains the essential medium in which biochemical processes take place. Protoplasm displays the signs of life only when provided with water—if it dries out, it does not necessarily die, but it must at least enter the anabiotic state, in which vital processes are suspended.

Plants are composed mainly of water. On the average, protoplasm contains 85—90% water, and even the lipid-rich cell organelles such as chloroplasts and mitochondria are 50% water. The water content of fruits is particularly high (85-95% of the fresh weight), as is that of soft leaves (80—90%) and roots (70—95%). Freshly cut wood contains about 50% water. The parts of plants having the least water are ripe seeds (usually 10—15%) ; some seeds with large stores of fat contain only 5—7% water.

Of all the environmental factors that affect plant life, none is more important than water. Water is essential for all plants, from aquatic species to those living in deserts. The geranium and rubber plant in the living room utilize water in exactly the same ways as do the plants in mile after mile of irrigated Western croplands. In the West, more than 21 million acres are productive only because irrigation provides water not available through rainfall. At least another 10 million acres could be converted to agriculture if water could be obtained. Not only in our own country but also in many other parts of the world the quantity of available water is a limiting factor in crop production and so directly affects man's food supply.

Scientists have long recognized the significant role played by water in plant life and for more than three centuries have conducted an immense amount of research on plant-water relations. So complex is this relationship, however, and so intimately connected is it with other plant processes that much remains to be investigated in this important field of plant physiology.

Uses of water

Virtually all physiological processes in plants take place in the presence of water. Water is necessary for such essential chemical reactions as those in digestion, in which water combines with the material digested, and in photosynthesis, in which the hydrogen of the water joins with the carbon and oxygen of carbon dioxide to form sugar. The minerals from the soil, and the foods manufactured by the leaves, move, after being dissolved in water, from the regions of absorption or manufacture to other parts of the plant. Water in the cell vacuoles holds the cytoplasm in position against the cell wall and maintains the turgor of the cell. This turgor, in turn, contributes to the rigidity and support of the entire plant or parts of the plant. Water, through its effect upon cell division and enlargement, is necessary for growth. It is an essential constituent of protoplasm and is particularly abundant in young and growing tissues. Fresh fruits are composed largely of water: apple, 84 per cent; banana, 75 percent; peach, 87 per cent; and grape, 82 per cent. The percentage of water in physiologically active plants ranges from 30 to 95, although relatively inactive parts, such as seeds, and grains, such as corn, contain only 12 to 15 per cent.

WATER RELATIONS OF THE PLANT CELL

The Water in the Cell

The water in plant cells occurs in several forms: It is a chemically bound constituent of protoplasm; as water of hydration, it is associated with ions, dissolved organic substances and macromolecules, filling the gaps between fine structures of the protoplasm and of the cell wall ; it is stored as a reserve supply in cell compartments and vacuoles ; finally, as interstitial water, it serves as a transport medium in the spaces between cells and in the conducting elements of the vessel and sieve-tube systems.

Water of hydration

In accordance with their dipole nature, water molecules aggregate and accumulate at charged surfaces in flexible arrays (known as structured water). Strongly charged ions of about the same size as water molecules bind the latter more firmly, the greater their charge and the smaller the radius of the ion. The sodium ion, with the same charge but half the surface area of the potassium ion, has twice the "density" of charge and accordingly almost four times as thick an envelope of hydration. Water molecules are bound tightly at the surfaces of the ions by electrostatic forces. A similar situation occurs at the surfaces of protein molecules and polysaccharides. Water molecules become associated with polar groups (hydroxyl, carboxyl and amino groups) and form several layers of structured water ; the water molecules are more readily displaceable the further they are from the polar group.

Water of hydration accounts for only 5—10% of the total cell water, but this amount is absolutely essential to life. Only a slight decrease in the content of water of hydration results in severe alterations in protoplasmic structure and in the death of the cell. Most of the water in the protoplasm fills the spaces in the cytoplasm and cell wall, but it is not entirely free to move ; its mobility is limited by capillary forces, and it is retained by dissolved substances. The cell walls of plants hold water with a pressure of 15—150 bar, depending on the density with which the fibers are packed. The surface forces holding water to structural elements in a matrix (cell wall, plasma colloids) can be expressed in terms of the "matric" pressure τ.

Stored water

The most easily translocated water is that in those cell compartments specialized as reservoirs for solutions. But even this water is not completely mobile ; in addition to hydration and matric effects, dissolved substances such as sugars, glycosides, organic acids and ions affect the net diffusion of water, through semipermeable membranes, from one solution to another. This effect is described in terms of the *osmotic pressure* of each solution.

If two solutions (for example, cell sap and pure water) having different concentrations of various solutes are separated by a membrane permeable only to water, there is a tendency for net flow of water to occur into the solution having the greater solute concentration. This is not paradoxical, and can be understood in the sense that the water diffuses *down* its *own* concentration gradient. If now the more concentrated fluid is confined (as by the cell wall enclosing the protoplast), the inflow of water causes the pressure in the cell to rise ; it rises until the pressure difference just counterbalances the tendency of water to flow in. Although this equilibrium pressure difference depends upon the concentrations of the *two* solutions, one speaks of the osmotic pressure of a single solution, implying the pressure that will just counterbalance the water influx which would occur were it in communication with pure water, through a membrane permeable to water.

The osmotic pressure of a solution increases with absolute temperature and with the concentration n of dissolved particles (the molarity in terms of molecules and ions that do not further dissociate). Van't Hoff pointed out in 1855 that, at least for dilute solutions, the osmotic pressure π is given approximately by

$$\pi = nRT.$$

Here R is the universal gas constant (8.3143 J. K^{-1}. mol^{-1}), so that a 1 M solution at standard conditions has an osmotic pressure of about 22 bar.

Solutions of dissociating substances contain more independent particles than their molarity indicates, so that for electrolytes the equation includes another factor taking into account the degree of dissociation. Conversely, macromolecular substances can be prominent constituents, in terms of weight, without raising the osmotic pressure appreciably. Through the polymerization of small molecules (with marked osmotic effect) to macromolecules—for example, the conversion of sugar to starches—and the reversal of this procedure by hydrolysis, the cell can rapidly alter its osmotic pressure ; as a result, the net influx of water can be regulated.

The Water Potential of Plant Cells

The binding of water to macromolecular structures and dissolved substances results in a decreased availability for chemical reactions and for use as a solvent. It is the availability of water, rather than its total quantity, that influences the biochemical activity of protoplasm.

Table 3.1. Values of relative humidity (in the air above a solution) and the osmotic pressure (in bar) of the solution, when the two phases are enclosed and allowed to equilibrate at 20°C.

bar	% *RH*	*bar*	% *RH*
0	100	97.9	93.0
6.7	99.5	112	92.0
13.5	99.0	126	91.0
20.3	98.5	141	90.0
27.2	98.0	301	80.0
34.1	97.5	481	70.0
41.0	97.0	687	60.0
55.0	96.0	933	50.0
69.1	95.0	∞	0
83.2	94.0		

The thermodynamic state of the water in a cell can be compared with that of pure water, and the difference expressed in terms of potential energy. The water potential ψ, as defined by Slatyer and Taylor (1960), is equal to the difference in free energy per unit volume between matrically bound, or pressurized or osmotically constrained water and that of pure water. The dimensions of water potential are energy per unit mass or per unit volume (J. kg^{-1}). Because a dyne-cm is an erg, force per unit area and energy per unit volume have the same dimensions, the conversion factor being approximately 100 J. kg^{-1}=1 bar.

The increasing acceptance of a terminology in which potential terms are used consistently and exclusively has done a great deal to eliminate sources of confusion in the literature. In the past water relations have been discussed in terms of osmotic pressure, diffusion pressure deficit, suction pressure, hydrostatic pressure, the pressure exerted by an enclosing structure, and the depression of water potential—all of which are defined in such a way as to represent positive values. Although the use of positive variables has been considered an advantage, this has been more than offset by the multiplicity of terms and of the symbols used to represent them. Here we adopt a terminology based entirely on potential, in which the water potential ψ (negative) expresses the net effect of the osmotic potential ψ_π (negative), the matric potential ψ_τ (negative) and the pressure potential ψ_P (positive). The relationships between these quantities will be discussed in the following sections.

Water Potential and the Cellular Translocation of Water

Between sites differing in water potential there is a *potential difference* $\Delta\psi$. This situation is analogous to that in an electrical system, in which different points are at different voltages. Just as there is a tendency for current to flow from the higher-voltage to the lower-voltage points, in living cells there is a tendency toward the reduction of water-potential differences. In the cell this occurs by the transport of water or other substances ; *water transport* always proceeds in the direction toward the system at lower water potential. A state of equilibrium within the cell, and between the cells and their surroundings, is usually rapidly achieved (as long as there is no obstacle to diffusion).

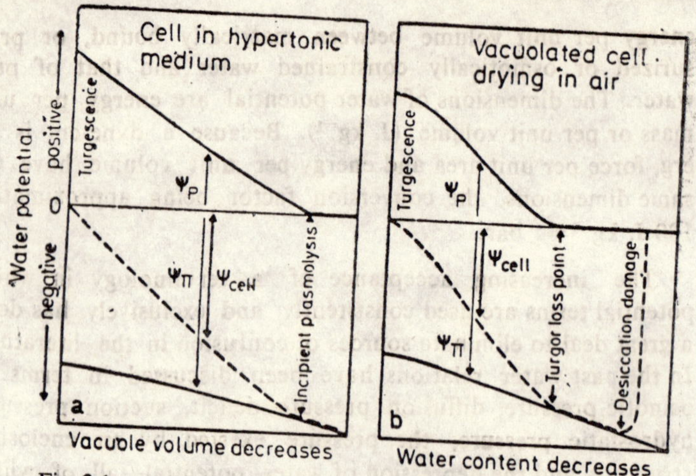

Fig. 3.1. a and b. Water-potential diagrams for vacuolated plant cells in a hypertonic medium (a) and for desiccating vacuolated cells in air (b).

The water balance of cells is of course intimately related to the availability of water in their environment. For a cell in air, as Table 3.1 indicates, there is an equilibrium between the humidity and the water potential of the cell ; if the relative humidity of the air is less than the given equilibrium value, water will be lost and the water potential will become more negative. Within a cell, too, water shifts from one cell compartment to another until an equilibrium is reached, at which point the water potentials of the different compartments are equal. The processes underlying osmotic water transport can be most simply explained in terms of the model of the single cell in hypotonic and hypertonic media that has been derived from *experiments* in plant physiology (Fig. 3.1). Note, however, that this situation occurs *in nature* only in the case of marine plants in tide pools that are drying up or becoming diluted with fresh water.

The Single Cell in a Hypotonic Medium

If a cell with a non-rigid wall is placed in a hypotonic medium, water enters the cell ; it flows through the protoplasm into the cell sap, and as a result the volume of the vacuole increases considerably. The expansion of the vacuole presses the plasma against the cell wall, which stretches elastically and exerts a counterpressure (the turgor pressure P). The tendency for water to enter the cell is determined by the difference in water potential between the cell contents and the external medium, the water potential of the cell being given by

$$\psi_{cell}=\psi_{\pi}+\psi_{P}.$$

As water enters the cell, ψ_{π} rises (becomes less negative, as the cell solution becomes diluted) and ψ_{P}, due to the restoring force of the cell wall, also increases. ψ_{cell} thus becomes increasingly less negative, until it equals the water potential of the hypotonic medium.

If the cell is in pure water, equilibrium will be reached only when ψ_{cell} is zero. Under these conditions, the positive term in Eq. above ψ_{P}, must just balance the negative term ; *i.e.*,

$$\psi_{\pi}+\psi_{P}=0, \text{ and } \pi=P.$$

The water potential equation holds for the cell as a whole, the protoplasm behaving like a semipermeable membrane, so that the entire system functions as an osmometer. Separate potentials can also be identified for the components of the system—the vacuole, the protoplasm and the cell wall. Whenever the water content changes, these components immediately adjust to a new state of equilibrium. That is,

$$\psi_{vacuole}=\psi_{\pi_{vac}}+\psi_{P}$$

$$\psi_{protoplasm}=\psi_{\tau_{ppl}}+\psi_{\pi_{ppl}}+\psi_{P}$$

$$\psi_{cell\ wall}=\psi_{\tau_{cw}}$$

at equilibrium, $\psi_{vacuole}=\psi_{protoplasm}=\psi_{cell\ wall}$.

The Single Cell in a Hypertonic Medium ; Incipient Plasmolysis. If a cell is placed in a solution more concentrated than that in the vacuole, water is drawn out of the cell. As a result, the volume of the vacuole is reduced and the elastic cell wall is less dis tended, exerting less pressure on the protoplasts. Finally the volume of the cell shrinks to a limiting value, beyond which the cell wall can no longer follow. At that point the protoplast begins to pull away from the cell wall ; this stage is called *incipient plasmolysis*. In the state of incipient plasmolysis, the osmotic pressure is unopposed by the tension of the cell wall :

$$P=0, \text{ so that } \psi_{cell}=\psi_{\pi} .$$

Tissues in Aqueous Media

When cells are not isolated, but form a tissue, their expansion is facilitated or opposed by that of their neighbours ; *i.e.*, P in the equation for a given cell is modified by external factors. If the adjacent cells are pushing against the cell considered, P is increased and equilibrium is reached when the water content is still low (and the volume of the vacuole is relatively small). This is important in tissues with delicate walls, for under such conditions they can maintain suitable turgidity without excessive filling with water. If, on the other hand, the cell walls form a rigid surface, they would not readily follow the shrinking protoplasts. Under these conditions P ceases to counteract ψ_{cell} when only a moderate amount of water has been withdrawn.

Cells of Land Plants in their Natural Surroundings

The cells of land plants lose water by evaporation from the surfaces of the cell walls. As the cell wall dries out, water flows into it from the protoplasm and, in homoiohydric plants, is replaced by water from the vacuole ; the cells begin to wilt. But plasmolysis does not occur, since the cell walls are impermeable to air. The walls follow the protoplasts ; they are drawn inward and can even fold up in the process. As turgor diminishes there is a rapidly increasing tendency for water to be drawn into the cell until the turgor loss point is reached ; from then on this tendency increases in proportion to the increase in cell-sap concentration. Finally, if desiccation is so extreme that the biomembranes are damaged, the osmotic system breaks down and no more water can be drawn in.

Anatomy of Roots

If a longitudinal median section be cut through a root tip such as has just been described, and examined under a microscope, the anatomy of the several regions of the root can be observed. The root cap is a more or less thimble-shaped assemblage of cells covering the distal end of the meristematic region. It apparently minimizes mechanical abrasion of the root tip as it grows through the soil. Such abrasion gradually tears off the outer terminal cells of the root cap, but these are replaced by new root-cap cells which are formed by cell divisions occurring in the lowermost layers of the cells of the meristematic regions.

The meristematic tissue of a root tip is composed of small, thin-walled cells with prominent nuclei. As new cells are formed by cell division they begin to enlarge, principally in the direction of the long axis of the root. The division of meristematic cells and their subsequent elongation result in projecting the growing region and root cap forward through the soil and account for the growth in length of the roots. The region of cell elongation is seldom more than a few millimeters in length. This contrasts markedly with the corresponding region of a stem tip which may be as much as 10 cm. in length or even longer in some species. Only a small part of the root tip—a few millimeters in length at most—is pushed through the soil. Since, in a growing root tip, the elongation of cells ensues as soon as they are formed by cell division, a cell which is one day in the region of cell division is the next day in the region of cell elongation, subsequent divisions in the meristematic tissue having produced additional layers of cells beyond it.

The anatomy of a representative root, as shown in cross section through the root-hair zone, is illustrated in Fig. 3.2. The structure of a young root shows a number of distinctive features. The cortex is relatively thicker than that of stems. This characteristic is especially noticeable in fleshy roots in which the thickness of the cortex is often many times the radius of the stele. The intercellular spaces of the root cortex are also more prominent than those in the stem cortex in which the cells are rather densely packed. An endodermis is almost invariably present in roots; this is generally considered to represent the innermost layer of the cortex. While an endodermis is found in the stems of many species, it is by

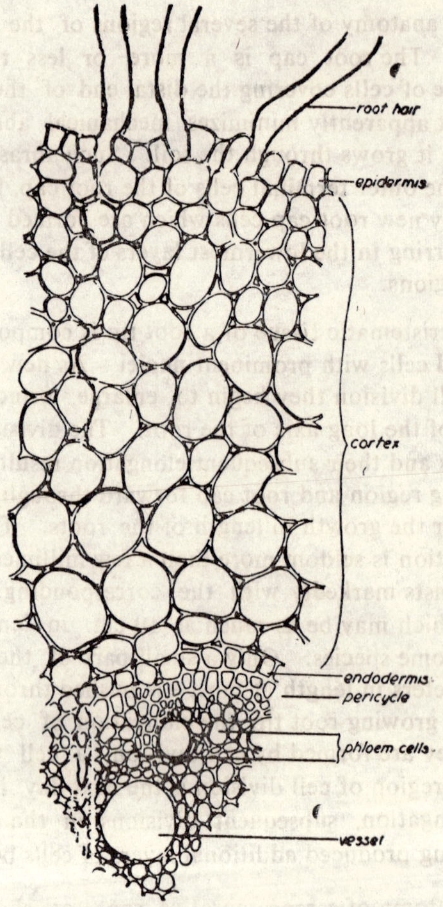

Fig 3.2. Cross section of a segment of a young maize root through the root-hair zone.

no means of universal occurrence. Just within the endodermis is present a narrow zone of parenchymatous pericyclic tissue. Usually this is continuous, but in some species, as described below, it may be discontinuous. Lateral roots most commonly originate in the pericycle.

In roots the primary xylem and the primary phloem are present in a radial pattern. The primary xylem as seen in cross section appears as a number (usually 2 to 5, although sometimes as many as 20) of radially situated strands. In many roots the center of the stele is composed of xylem ; in some, especially monocots, it is composed of pith. Usually the xylem strands terminate radially in contact with the pericycle, but in some species they abut directly on the endodermis, breaking up the pericycle, as seen in cross section, into a discontinuous series of arcs. The primary phloem of roots occurs as patches of tissue (as seen in cross section) which alternate with the strands of xylem.

The structure of the individual types of cells occurring in the root tissues is essentially similar to the structure of corresponding types of cells occurring in the stem.

With few exceptions the roots of all perennials and many annuals grow in diameter as they increase in age by means of a cambium layer much as do most stems. The cambium layer is initiated in young roots in such a way that it lies inside of the strands of phloem tissues, and outside of the xylem strands. In cross section the original cambium layer appears as a wavy band, passing inside of each phloem strand, and outside of each xylem strand. Once differentiated this cambium layer produces secondary xylem on its inside face, and secondary phloem on its outside, just as the cambium of stems does. The initial formation of secondary tissues by a root cambium is usually more rapid in the segments of the cambium internal to the primary phloem strands. Because of this differential growth rate the cambium of a root rapidly attains a circular aspect in cross section. The further differentiation of secondary phloem and secondary xylem proceeds in essentially the same manner that it does in stems.

Most perennial roots sooner or later become encased in layers of cork cells. The initial cork cambium often originates in the pericycle. As layers of cork cells are produced by the cork cambium the cortex of the root, including the endodermis, is ruptured and the cells of these tissues die and decay away. Older roots therefore have a characteristic smooth, brownish, corky covering that is pierced only by lenticels. With increasing age, secondary cork cambiums may arise progressively more and more deeply in the phloem tissues. This results in the gradual loss of the pericycle and older phloem tissues. The bark of older roots, therefore, is essentially similar to that of the trunks or larger branches of trees. Thick layers of bark do not as a rule accumulate on roots as they do on the trunks of some species of trees because of the rapid decay of all dead underground tissues.

In the roots of species in which no secondary thickening occurs, as in many monocots, the epidermal layer of cells may persist intact, usually becoming suberized. In other such species the epidermis may die and decay, but when this occurs an underlying layer of the cortex cells in turn becomes suberized.

Lateral branches of the primary root system usually originate in the pericycle, most of them being formed in the region just above the root-hair zone. Usually the locus of origin of a lateral root is opposite one of the primary xylem strands. The first step in the formation of a lateral root in most species is the development of a group of meristematic cells by the division of several adjacent pericycle cells in the layer just inside of the endodermis (Fig. 3.3). By successive divisions these cells rapidly form an apical root meristem with its characteristic root cap, region of cell division, etc. As this develops the endodermis and tissues exterior to it are first stretched and later ruptured. The elongating lateral root penetrates through the tissues exter-

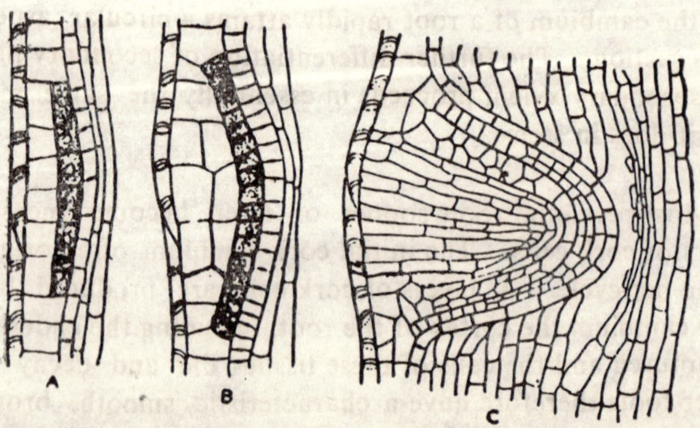

Fig. 3.3. Three stages in the formation of a lateral root.

nal to it, partly by mechanical pressure and perhaps partly by digesting the tissues through which it passes. Eventually the lateral rootlet emerges from the root of which it forms a branch and becomes an externally visible part of the root system.

Root Hairs

These structures are confined to the root-hair zone, which may be from a few millimeters to many centimeters in length, depending on the species and the conditions under which the root develops. Since the root-hair zone lies back of the region of cell enlargement, there is no forward progression through the soil of the individual root-hairs along the axis of the root. In any rapidly growing root tip new hairs are continually developing just back of the zone of elongation. New root hairs are thus constantly developing in contact with different portions of the soil, a fact which is of fundamental significance in the absorption of water and perhaps of mineral salts. Many root hairs are short-lived structures and often die within a few weeks or even less after they develop. In certain species, at least, some of the root hairs may remain alive for an entire growing season, and in a few species they may become suberized or lignified and persist for a year or longer.

Very little information is available regarding the abundance and distribution of root hairs on the root systems of mature plants. Dittmer (1937) found root hairs to be present on all of the roots of a four-months-old rye plant. This plant bore a total of about 14 billion root hairs. On the other hand, some tree species, especially certain conifers, bear few or no root hairs. The roots of some species, such as maize, produce an abundance of root hairs when they develop in the soil or moist air, but few or none when they develop under water as in a solution culture. The roots of other species, on the contrary, produce root hairs in abundance whether they develop in the soil or in water. Under field conditions maximum development of root hairs on most terrestrial plants appears to occur in soils with a moisture content between the field capacity and permanent wilting percentage.

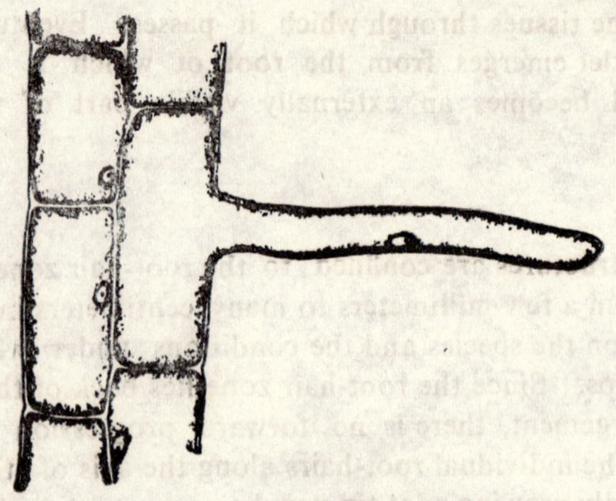

Fig. 3.4. A young root hair.

A root hair is essentially a tubular outgrowth of the peripheral wall of an epidermal cell, closed at its distal extremity, projecting more or less at right angles from the long axis of the epidermal cell of which it is an integral part. Root hairs develop only from certain of the epidermal cells. They range in length from less than a millimeter to about a centimeter and are usually about 10 μ in diameter. On some roots as many as several hundred root hairs may be borne on a square millimeter of root surface, although the number per unit area is usually less. The presence of root hairs on a given area of a root often increases the exposed surface of that area from three-to tenfold.

The cell wall of a root hair is composed principally of cellulose and pectic compounds. The outer lamella of the wall seems to be composed entirely of pectic compounds, largely calcium pectate. The tenacity with which root hairs and soil particles adhere is accounted for by this pectic coating. Because of this intimate contact between the root hairs and soil particles it is difficult to separate the two by washing or in any other way without injuring or destroying most of the root hairs. The inner wall of a root hair is lined with a thin layer of cytoplasm which is continuous with the cytoplasm of the epidermal cell of which the root hair is a part. In water or in moist air the root hairs are usually straight, but in soil they are more or less contorted, and sometimes even branched, conforming to the shape and distribution of the soil particles among which they penetrate. Detailed investigations of the development of root hairs have been made by Sinnott and Bloch (1939) and Cormack (1949).

Movement of water into the plant

Water moves into the plant chiefly through the *root hairs,* which are hairlike outgrowths from the epidermal cells near the root tip. The mechanisms which account for the movement of soil water into the root hairs are not thoroughly understood, although much progress has been made in recent years. Osmotic forces alone have been credited as being sufficient to account for water absorption by the root hairs. Under certain conditions, chiefly when the soil moisture content is high and the transpiration rate low, water may be absorbed by osmosis. Because the concentration of solutes in the root hairs is usually higher than that of the surrounding soil solution, the relative concentration of water in the root hairs is lower than that of the soil. Water therefore moves from the soil into the root hairs. It continues to move by osmosis from the epidermal cells into the vacuoles of the adjoining cells, and finally into the conducting cells of the root. This water movement into the plant by means of osmotic forces is called *active absorption* and is the main cause of root pressure.

Considerable experimental evidence, however, now shows that active absorption is not the chief method involved in the movement of water from the soil into the root. This active method appears to prevail only when the soil moisture content is high and the transpiration rate is low. When transpiration is high, another process, termed *passive absorption*, predominates over active absorption or, more likely, replaces it. Water movement by osmosis is probably too slow to replace the water lost from the top of a plant that is actively transpiring. The forces set up in the transpiring leaves—the same forces involved in the ascent of water just described—then come into play. Water loss from the surfaces of the mesophyll cells sets up a tension in the water of the conducting cells. This tension is transmitted down to the smallest roots. A water deficit in the conducting cells causes water to move from adjacent living cells. A gradient of water deficit is thus set up, extending across the root to the root hairs. A deficit in the root hairs, in turn, results in the movement of water into them from the soil particles and soil spaces with which they come in contact.

Passive absorption is, then, a kind of mass flow of water into the plant, and the root hairs merely act as surfaces for absorption. Water is absorbed not *by* the root hairs but *through* them. The passive nature of water absorption by rapidly transpiring plants is further indicated by experiments in which the roots of small potted sunflower, tomato, tobacco, and other plants were killed by heat. The plants nevertheless continued to absorb sufficient water to remain alive for several days after the roots had been killed.

The Relation between Roots and Soil Water

From the standpoint of the absorption of water by plants, a clear distinction should be made between conditions under which capillary movement of water occurs readily in a soil and those under which such movement of water is slow or nonexistent. If capillary movement occurs readily, translocation of water may take place toward the young roots whenever they are absorbing water. There are two principal conditions under which capillary translocation of water can occur in soils at appreciable rates : (1) in any zone of soil which is not more than a few feet above a water table, and (2) in the upper layers of any soil after a heavy rain or irrigation, but before the water content of the soil has decreased to its field capacity. As the water in the films surrounding the soil particles with which the root tips are in contact becomes depleted, more water moves toward those particles by capillarity. The actual rate of such capillary movement of water through the soil may become a factor influencing the rate of absorption. Root systems, however, are not static, but are more or less continually growing through the soil. The rate of root growth of most species decreases, as a general rule, with increasing wetness of the soil.

above the field capacity, because of the corresponding reduction in soil aeration. Hence in soils in which capillary movement of water occurs—which are necessarily relatively wet—the rate of elongation of roots is generally less than in otherwise similar but somewhat dryer soils. This continued growth of root tips through the soil brings them into contact with other portions of the soil water, so that, even in capillary movement to certain parts of the soil ceases, capillary movement of water to the roots may be re-established by the extension of the root tips themselves into zones of the soil that have not yet been depleted of water which can move by capillarity.

Many plants, much of the time, grow in soils at water contents between the permanent wilting percentage and the field capacity. In this range of soil-water contents capillary movement of water is very slow. Once most as a result of passive absorption Kramer, (1939). The effectiveness of active absorption in reducing the water content of a soil also appears to be considerably less than that of passive absorption. Detopped herbaceous plants cease exuding water when the soil-water content has been reduced to about halfway between the permanent wilting percentage and the field capacity Kramer, (1941). The active absorption mechanism apparently does not absorb water against a diffusion-pressure deficit of more than 1 or 2 atm.; whereas the passive absorption mechanism results in absorption of water at relatively rapid rates until the diffusion-pressure deficit of the substrate approaches the permanent wilting percentage value (approximately 15 atm.).

4

Movement of Water in Plants

In many parts of the world an inadequate supply of water is the most conspicuous and continuing limitation to crop growth. Water relations thus occupy a central place in any survey of developmental physiology. But much of this subject lies outside the scope of a discussion of root function in the soil. Meteorological factors and the characteristics of plant shoots can greatly influence the effects on crops of variations in water supply.

The driving force for the movement of water

If the water economy of plants is to be seen in perspective, account much be taken of their evolutionary history. They originated in an aquatic environment and photosynthesis, the unique metabolic process on which plant growth depends, require the absorption of carbon dioxide from the environment with the simultaneous release of oxygen. In submerged plants this exchange occurs in solution but when terrestrial plants evolved, the site of photosynthesis being necessarily in the above-ground parts, the exchange of carbon dioxide and oxygen with an unsaturated atmosphere had the inescapable consequence that water vapour was simultaneously lost. No membrane system was evolved—nor has synthetic chemistry yet produced one—which permits the free exchange of oxygen and carbon dioxide, between an aqueous phase in plant cells and an unsaturated atmosphere without the loss of water.

Transpiration is thus an inescapable physical consequence of the development of terrestrial vegetation.

Studies of the utilization of water by crops entered a new phase with the work of Penman (1948). Earlier discussions had frequently been concerned with the 'transpiration coefficient', that is to say the quantity of water used per unit of dry matter produced, a physiological relationship between growth and water loss often being implied. Penman, however, showed that, during summer in temperate regions, the quantity of water lost from a complete ground cover of short grass, which was well supplied with water, is about 0.8 of the water loss from the same area of a free water surface; in spring or autumn the corresponding figure was somewhat lower, *c.* 0.7. Somewhat different values apply to other types of ground cover and Penman's work provided a valuable and widely used basis for estimating the water requirements of growing crops.

Transpiration by leaves is thus the driving force for the movement of water through plants. The simplicity of this relationship should not, however, obscure the complexity of processes which can regulate the movement of water within plants, especially when they are subject to a degree of water stress. This matter can be meaningfully discussed only with a thermodynamic background which is here considered only briefly; Slatyer (1967); Kramer (1969); Kozlowski (1968, 1972) and Weatherley (1970) are among the standard sources from which fuller information can be obtained.

Water potential

The movement of water depends on the existence of gradients of decreasing water potential—that is to say its free energy—which bears no direct relationship to the total quantity of water present. The transfer of water from the soil through plants to the atmosphere is thus appropriately considered in terms of potential gradients. None the less the quantity of water present exerts an important indirect effect; it largely determines how the potential changes when water is introduced into the system or withdrawn.

The same basic thermodynamic principles apply to the movement of nutrient ions into plants. However, in the majority of circumstances it is adequate to consider only their concentration in the solution with which roots are in contact, since it is usually sufficiently dilute for the chemical activity of an ion to vary closely with its concentration.

The following general relationship between the components which determine water potential apply equally to plant tissues and to the soil :

$$\psi_w = \psi_s + \psi_m + \psi_p$$

Where ψ_w is the water potential in plant cells or in the soil, and ψ_s ψ_m and ψ_p are the contributions due to solutes, matric forces and pressure. These quantities are often expressed in bars (1 bar is approximately equal to a pressure of one atmosphere) and ψ_s is numerically equal to the negative of the osmotic pressure expressed in the same units.

The relative magnitudes of ψ_s and ψ_m and ψ_p differ greatly between plant tissues and the soil. In plants ψ_m is usually small ; ψ_s is negative and in fully turgid cells it is balanced by ψ_p a positive quantity due to wall pressure, so that ψ_w is zero. When cells are not fully turgid ψ_s is numerically greater than ψ_p so that ψ_w is negative.

In soil the pressure component ψ_p is relatively insignificant, so also is ψ_s when the concentration of the soil solution is low. The matric potential ψ_m which results from the interaction of water with the solid phase assumes an increasingly large negative value as the water content of the soil decreases. It largely determines ψ_w for soil except when loss of water results in a high concentration of soil solution so that ψ_s achieves an appreciable negative value and enhances the negative potential. For some purposes it is appropriate to take account also of the gravitational component ψ_g of soil water potential. Because of its major importance in determining soil water potential in many circumstances, the matric potential ψ_m is sometimes alone referred to in discussions of plant/soil water relations.

Though the foregoing is the most usual terminology it has been cogently argued by Spanner (1973) that the state of water both in plants and soil should be considered in terms of the older concept 'suction potential' which is numerically equal to water potential but opposite in sign and differently derived. This eliminates the ambiguity that a lower 'water potential' has in fact a higher numerical value but with a negative sign. However, to conform to the more common practice, suction potential is not used in this book except when it is convenient to define the forces required to withdraw from the soil.

The further term *pF* is also frequently used to describe the state of water in soil. By definition the *pF* is the logarithm of the free energy of water, measured as the head of water in centimetres. This notation has the convenience of facilitating the graphic representation of relationships between suction, or water potential and the water content of soil over a wide range.

PASSAGE OF WATER THROUGH THE PLANT

Resistances to flow

The smaller loss of water from a complete wegetation cover than from a free water surface, as shown by Penman, reflects resistances to the flow of water along its pathway from the soil to the transpiring leaves. Ohm's law provides a convenient analogy which has been widely used since van den Honert (1948) drew attention to its usefulness ; if water moves at a constant rate through a system, the ratio of the drop in water potential to the resistance to water flow at each step in the transfer process will be constant. Thus, the relative magnitudes of the resistances to water movement along the soil-plant-atmosphere pathway, under constant conditions, can be derived from the change in potential. To describe the passage of water realistically the simple Ohm's law analogy must, however, be amplified in two respects. *First*, the major resistances to water flow in the plant can vary in a complex manner. *Second*, account must be taken of the 'capacity' of the system, to use another electrical analogy ; if increased transpiration leads to a lowering of the water potential in any part of the transpiration stream, water will be withdrawn from the surrounding cells with a corresponding drop in their water potential.

Fig. 4.1 in simplified form aspects of the system which are of major relevance in the present discussion. It does not, however, indicate all processes which cause water to move in plants. Although transpiration is dominant under normal circumstances it is not the sole cause of water movement. If plants are maintained in a humid atmosphere the flow of water through them does not usually cease completely; in some species droplets of water may be released by guttation through hydathodes at leaf tips. This process is attributed to gradients in water potential created by osmotic forces; the bleeding sap which can pass out of cut stems or roots, which is sometimes regarded as evidence of the 'active' uptake of water is similarly explained. These processes are not considered further as they account for only a very small fraction of the total water which moves through transpiring plants.

Movement of water through plant shoots to the atmosphere

When plants are transpiring in a relatively dry atmosphere the decrease in water potential between the interior of the leaf and air exceeds by a considerable factor that in any other part of the system. Between the soil and the transpiring leaf the decrease in water potential is unlikely to exceed about 50 bar in an unwilted plant and is often very considerably less, but the drop in water potential from the interior of the leaf to the atmosphere may approach 1000 bar. It follows, therefore, that the highest resistance to water movement in a plant system can occur in the leaf.

The movement of water across the cuticle on the leaf surface is small except in plants habituated to constantly humid environments and the major gas exchange occurs through the stomata. The finely balanced mechanisms which control the movement of the guard cells have been the subject of considerable research. The two main circumstances which usually lead to stomatal closure are an increase in the concen-

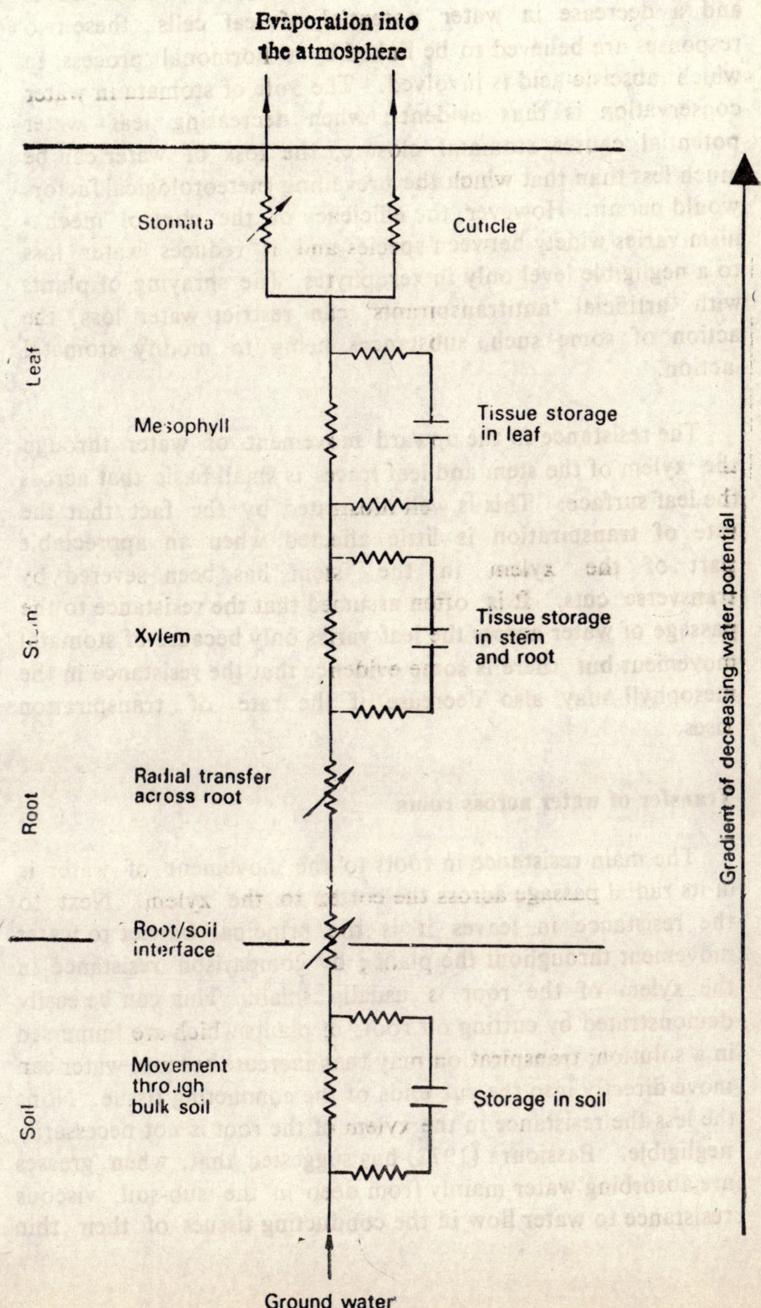

Fig. 4.1. Diagrammatic representation of passage of water from the soil to the atmosphere through plants.

tration of carbon dioxide in the gaseous phase within leaves and a decrease in water potential of leaf cells ; these two responses are believed to be linked by a hormonal process in which abscisic acid is involved. The role of stomata in water conservation is thus evident ; when decreasing leaf water potential causes stomatal closure, the loss of water can be much less than that which the prevailing meteorological factors would permit. However, the efficiency of the control mechanism varies widely between species and it reduces water loss to a negligible level only in xerophytes. The spraying of plants with artificial 'antitranspirants' can restrict water loss, the action of some such substances being to modify stomatal action.

The resistance to the upward movement of water through the xylem of the stem and leaf traces is small basic that across the leaf surface. This is well illustrated by the fact that the rate of transpiration is little affected when an appreciable part of the xylem in the stem has been severed by transverse cuts. It is often assumed that the resistance to the passage of water across the leaf varies only because of stomatal movement but there is some evidence that the resistance in the mesophyll may also decrease if the rate of transpiration rises.

Transfer of water across roots

The main resistance in roots to the movement of water is in its radial passage across the cortex to the xylem. Next to the resistance in leaves it is the principal barrier to water movement throughout the plant ; by comparison resistance in the xylem of the root is usually small. This can be easily demonstrated by cutting off roots of plants which are immersed in a solution; transpiration may then increase because water can move directly into the cut ends of the conducting tissue. None the less the resistance in the xylem of the root is not necessarily negligible. Passioura (1972) has suggested that, when grasses are absorbing water mainly from deep in the sub-soil, viscous resistance to water flow in the conducting tissues of their thin roots can be important. Newman (1974) cites evidence which points to the same conclusion if roots exceed *c.* 1 m in length.

It is usually considered that water crosses the cortex of roots to the endodermis mainly, if not entirely, in the free space, but some workers consider that cytoplasmic movement, or transfer between vacuoles in cortical cells, may make some contribution.

When plants of different species are grown in the same environment considerable contrasts in the permeability of roots to water have been detected and there is much evidence that the resistance to the radial transfer of water across a root can vary with environmental factors. Decreased transpiration, caused by anaerobiosis, or unfavourable temperature in the rooting medium, has been attributed to the reduction in the permeability of membranes; the application of respiratory inhibitors can have a similar effect. A review of the effects of temperature by Kramer (1969) indicates that the restriction of absorption caused by low root temperatures is greatest in plants adapted to warm habitats.

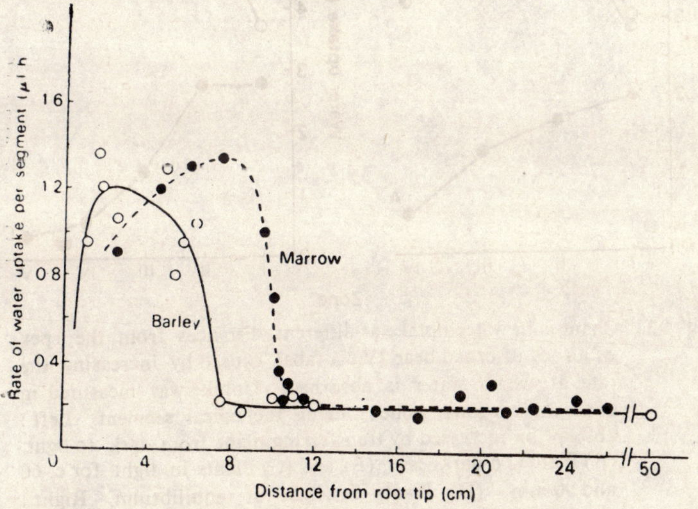

Fig. 4.2. Rate of water uptake by short segments (3.5 mm) of intact seminal axes of barley (Hordeum vulgare) and marrow (Cucurbita pepo) (Graham et al., 1974).

Variations in the rate of transpiration may also modify the resistance to the radial transfer of water across roots. Weatherley and his co-workers found in several species that if transpiration is increased by a reduction in the humidity of the atmosphere, the resistance to the movement of water across the root decreased. This effect has not, however, been found by all investigators and some observations suggest that it may occur only at a limited range of transpiration rates. Differences in experimental methods may perhaps help to explain discrepancies between investigations.

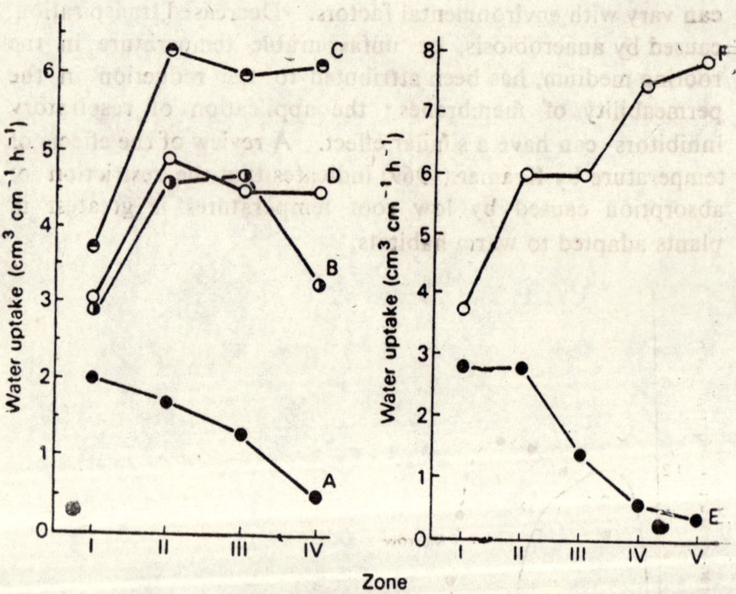

Fig. 4.3. Changes in water uptake at different distances from the apex of a root of broad bean (Vicia faba) caused by increasing the rate at which water is absorbed. Uptake was measured in zones 2.5 cm long, zone 1 being the apical segment. Left: absorption increased by transferring plant from dark to light. (A) Plants in darkness. (B) and (C) Plants in light for c. 60 and 90 min. (D) Plants in light at equilibrium. Right: absorption of test root increased by placing remainder of root system in a salt solution of osmotic potential of c.—2 bar. (E) Beginning of experiment—all roots in tap water. (F) Remainder of the root system in salt solution for c. 4 h, test root in tap water (Brouwer, 1953).

Comparative studies of the entry of water into different parts of root systems provide additional evidence that the radial movement of water across roots is influenced by the rate of transpiration. When the supply of water is ample it enters more rapidly into the apical zones of roots where the endodermis is not suberized (Fig. 4.2). However, experiments by Brouwer (195?) indicate that changes in the rate of which water is absorbed can modify the extent to which it enters different parts of young roots. Some of his results are summarized in Fig. 4.3. The uptake of water by a single root of broad bean (Vicia faba) was enhanced either by transferring the plant from dark to light or by placing the other roots in an osmolyte of osmotic potential *c.* 2.0 bar. Both treatments led to considerably more rapid uptake of water, especially 7.5—12.5 cm from the apex; absorption by them then exceeded that in the apical 2.5 cm in which the rate of uptake had been greater when transpiration was low. Newman (1974) has reviewed other evidence that the entry of water into different parts of young roots can be varied by external factors but this subject is still very incompletely understood even when the entire root system experiences the same water potential. As will be shown further it is possible that, when plants receive a restricted water supply in soil, further complications may be introduced as a result of discontinuous contact between roots and the soil.

None the less, all investigations of the uptake of water by different parts of the root system indicate that water can enter most readily into the young unsuberized zones of roots, but absorption is not confined to them (Fig. 4.2). In woody perennial plants, of which over 99 per cent of the root system may be suberized, slow absorption through the older tissues is likely to make a major contribution to the total water uptake.

Sources of more detailed information on the water relations of plants include Kramer (1969); Hsiao (1973) ; Newman (1974) and Monteith and Weatherley (1976).

EFFECTS OF THE EXTERNAL SUPPLY OF WATER ON THE FUNCTION AND GROWTH OF ROOT SYSTEMS

We should learn first the effects of the supply of nutrients on absorption by roots and on the form of root systems primarily on the basis of observations when roots were grown in solution culture and no other factors were varied. There is no corresponding simple and direct way for studying the effects of differences in the external water potential ; it cannot be varied in controlled artificial systems with the same facility as the concentration of nutrients. Evidence on the effects of water potential on root function comes mainly from experiments in soil. Thus there is some overlap between this chapter, which are more directly concerned with soil conditions.

General relationships between water potential and root function

The terms 'permanent wilting point' and 'wilting percentage' are used to describe the maximum water potential, or water content, is soil which causes plants to wilt. Slatyer (1967) has pointed out that the permanent wilting point should not be regarded as a soil constant since wilting depends on the loss of turgor from leaves and it is influenced by the rate of transpiration as well as by the osmotic pressure of the vacuolar sap of cells. Beyond this the practical usefulness of the 'wilting point' concept in field situations is often limited because ample water in part of the rooting zone may permit growth to continue although the soil adjacent to much of the root system is close to the wilting point.

It is generally considered that a water potential of —15 bar throughout the rooting zone would lead to the permanent wilting of the majority of crop plants, but the rate of root extension starts to decrease at much higher water potential, *e.g.*, often about *c.* —0.5 bar, though root extension may continue slowly until the water potential falls to —10 bar or lower. Some experiments with maize (corn, Zea mays) and tomatoes (*Lycopersicon esculentum*) indeed indicate that if part of a root system is well supplied with water, some growth can still occur in roots, which are surrounded by soil in which the water potential is less than −40 bar ; they may remain alive at appreciably lower water potentials.

When osmotic forcess contribute appreciably to the water potential, for example in saline soils, or after large additions of fertilizer, the restriction of growth may be greater than if low soil water potential arises from matric forces alone. Osmotic potentials *c.* −2 to −4 bar, can be injurious to many plants, but halophytes are more tolerant. The injury caused by low osmotic potential may be associated with interference with hormonal processes.

Because of the experimental problems of varying the water potential uniformly through the rooting zone, other factors being maintained constant, the effects of water potential on nutrient uptake have been little studied independently of those on growth. However, if low water potential reduces root growth a restriction of nutrient uptake is to be expected because of the reduced metabolic activity of the roots ; moreover the metabolic demand of the whole plant for nutrients can be reduced. If osmotic as opposed to matric forces are responsible for low water potential, nutrient uptake may also be affected by competitive processes ; for example, in saline soils a high concentration of sodium can depress the absorption of other cations.

Effects of gradients in water potential within the rooting zone

Except when the soil is close to or above field capacity appreciable gradients in water potential usually occur. When water is lost in dry weather its potential usually decreases more rapidly near the surface than in the deeper soil layers. This is due not only to the downward movement of water, but also to its evaporation into the atmosphere and absorption by roots which are often more abundant near the surface. The gradient can be enhanced if gravitational drainage is slower in the deeper layers and sometimes also because their capacity for water retention is higher.

Climate, the distribution of roots and soil characteristics all interact in determining the magnitude of gradients of water potential in the soil under growing plants. Thus, the relationship observed in any set of circumstances cannot be regarded as widely representative. None the less (Fig. 4.4), construction

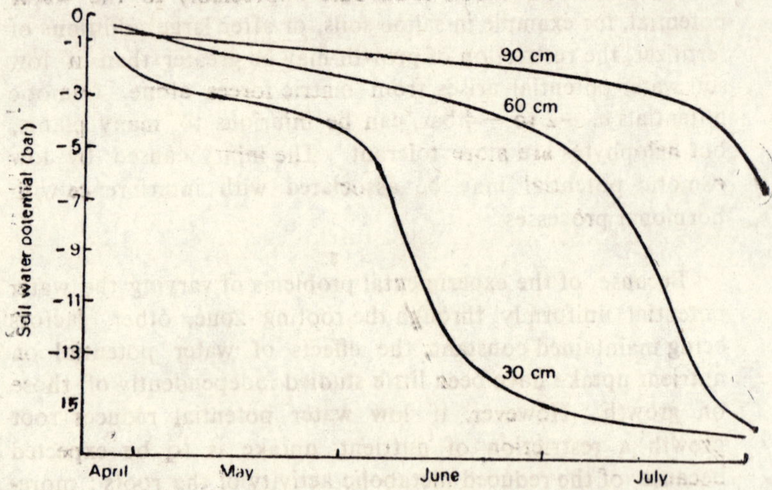

Fig. 4.4. Changes in water potential at three depths in a light alluvial soil underlying an established pasture during three summer months in which the pasture was sheltered from rain. The water potential at 15 cm differed little from that at 30 cm during the major part of the period.

from the work of Garwood and Williams (1967), illustrates the type of situation which can develop in dry summer weather in a temperate locality ; during three summer months a sward of perennial ryegrass (*Lolium perenne*) in southern England was protected from rain by a transparent cover at sufficient height to allow free air circulation. A period of drought was thus reasonable simulated. The water potential fell considerably more rapidly at and above 30 cm than at greater depths, and as time passed the uptake of water became increasingly dependent on roots which had penetrated deeply. The general manner in which these conditions affect the distribution of roots and the absorption of nutrients will now be considered.

MECHANISM OF WATER ABSORPTION

The mechanism of absorption of water is considered to be of two types namely *Active Absorption* and *Passive Absorption.* Under conditions of slow transpiration and well irrigated soil the forces responsible for absorption of water develop in the roots. This type of absorption is called *Active absorption.* Under conditions of high transpiration absorption is brought about by the forces which are developed in the shoots and transmitted to the roots, the latter serving only as channels. This type of absorption is called *Passive absorption.*

ACTIVE ABSORPTION

Two types of theories are put forth to explain active absorption of water, namely, secretory or non-osmotic theories and second osmotic theories.

Secretory or Non-osmotic Theories of Active Absorption

Advocates of secretory theories assume that living tissues of the root secrete water into the xylem. The process is believed to be dependent on the energy released by respiration. A large number of such theories were given from time to time. Two of them are discussed here.

(*i*) De Candole (1832) held that root tips are contractile organs which show rhythmic movements of contraction and expansion. On account of this alternate contraction and expansion the water is taken in from outside and forced in.

(*ii*) Bose (1927) assumed that due to the friction of the soil there is developed a pulsation in the root cells and due to these pulsations roots show alternate contraction and expansion. These cause in-take of water.

These non-osmotic theories were not much accepted as they did not explain absorption clearly. However, during the last two decades some facts have been noticed which have revived these vital theories. These are :

(*i*) Bennet Clark et al (1944) showed that osmotic pressure of cell sap determined by plasmolytic method is always higher than osmotic pressure of same cell sap determined by cryoscopic method (See Osmosis) which means that in cryoscopic method cell sap has been diluted by the activity of the cells. The difference in two values is a measure of non-osmotic absorption of water.

A similar observation was made by Van Over Beek (1942) who found that the osmotic pressure of the exudate from xylem of detopped tomato plant was somewhat lower than the osmotic pressure of solution required to stop exudation. He regarded this differenec as a measure of non-osmotic water absorption.

(*ii*) A number of experiments have shown that on treating roots with auxins the absorption of water increases and also respiratory rate increases. It means more water is taken in under increased metabolic activity (respiration) or in other words, the rate of absorption increases due to energy released in respiration. Thus if absorption has to take place against diffusion gradient (Physical forces) it has to utilize respiratory energy showing the involvement of non-osmotic factor in absorption.

(*iii*) When roots are treated with poisons like KCN the rate of respiration and absorption of water decreases. If the process of water absorption is dependent upon only physical forces any amount of poison should not affect the rate of absorption. Rosene (1944, 47) reported that KCN partially inhibited water uptake and exudation from onion roots. She suggested that perhaps all water uptake is non-osmotic and inhibition by KCN is incomplete only because it does not completely inhibit respiration in the concentration used in her experiments.

So the secretory theories may not be adequate but these facts indicate that some absorption is brought about by vital activity of the cells.

Osmotic Theories

As described earlier the cell wall is made up of pectic compounds. The cell wall imbibes large quantities of water due to hydrophilic properties of pectic compounds. From cell wall water passes into root hair by osmosis because of higher osmotic pressure of sap of root hair. Cytoplasmic layer of the root hair acts as the differentially permeable membrane.

On absorbing water a root hair becomes turgid and its suction pressure falls. The adjoining cortical cells have much higher suction pressure and, therefore, absorb water from root hair. In this way due to an increasing gradient of suction pressure (DPD) water moves from cell to cell in the cortex, reaches the endodermis, from the passage cells of the latter it diffuses into xylem of the root. Atkin (1916) considered the cortical cells of the root as a complex differentially permeable membrane separating the soil solution from the sap in xylem. He observed that so long as the cortical cells of the root are turgid, water can move across them irrespective of their osmotic pressure. Priestley (1928) regarded the endodermis as the differentially permeable membrane of the root osmometer.

A bigger problem in connection with an osmotic explanation of active absorption is how a sufficient concentration of sap is maintained in the xylem of roots to produce DPD gradient across the root from the soil. Atkins believed that parenchyma cells surrounding the xylem secrete sugars into the xylem while Priestley (1922) believed that cells which are differentiating into xylem might supply these solutes. None of these hypotheses explain the situation in those plants where the solutes are inorganic in nature, as is true of many herbs and some woody plants. It appears that normal uptake of minerals may be responsible for the maintenance of a sufficient concentration of solutes in the root xylem to cause inward movement of water by osmosis.

The best evidence to support the osmotic absorption of water comes from the phenomena of root pressure and sap exudation as observed in decapitated plants. Arsiz, Helder and Van Nie (1951) believed that root pressure depends on active transport of salts into the xylem which is controlled by metabolic processes. Water movement then occurs into the xylem because of the osmotic effects of the sap present in the xylem elements.

PASSIVE ABSORPTION

In rapidly transpiring plants absorption of water takes place due to forces set up in the leaves. In the intercellular spaces of a leaf relative humidity is almost 100% due to presence of water in the surrounding mesophyll tissue. In the outer atmosphere relative humidity is never 100% ; as a result water vapours move out through stomata from intercellular spaces of leaf to outer atmosphere, *i.e.*, from their higher concentration to their lower concentration. This outward movement of water from intercellular spaces causes an increase in their DPD and water from mesophyll cells moves into intercellular spaces thus creating an increase of DPD in the mesophyll cells which therefore receive water from xylem vessels due to lesser DPD of water in xylem vessels than that in mesophyll cells. The water in xylem of leaf is in contact with the xylem vessels of stem. The movement of water from vessels of leaf to mesophyll cells set up a tension under the impulse of which water from stem is pulled up into the leaves. Now vessels in stem are in contact with those in roots. Thus ultimately loss of water through stomata causes the upward movement of water through root, stem and then to leaves. This force of transpiration causes the movement of water as if water is being pulled at the top. As in this absorption roots play only a passive role of serving as channel only for water movement is called *passive absorption*.

The two mechanisms of water absorption, *viz.*, active and passive, work independent of each other. Passive absorption is the main method in many plants. The absorption of water by this method is 10—100 times more than by active absorption. The best evidence in this favour is that the absorption goes on even if roots are cut in transpiring plants. In gymnosperms this is the only method of absorption. However, in most of the cases this is not the only method. What happens in the absence of transpiration or low transpiration as at night ? Active absorption plays the main role under such conditions.

THE MECHANISM OF ACTIVE WATER ABSORPTION

The absorption of water from the soil takes place chiefly through the root hairs. The root hairs are in contact with the water films on the soil particles. Inside the cell wall of the root hair is a thin lining of cytoplasm which encloses a large vacuole filled with cell sap. The cytoplasm and cell sap of the root hair are continuous with those of the root cell of which it is a prolongation (Fig. 4.5). The cell wall of the root hair is a permeable membrane. Its cytoplasmic lining is a semipermeable membrane separating the cell sap from the

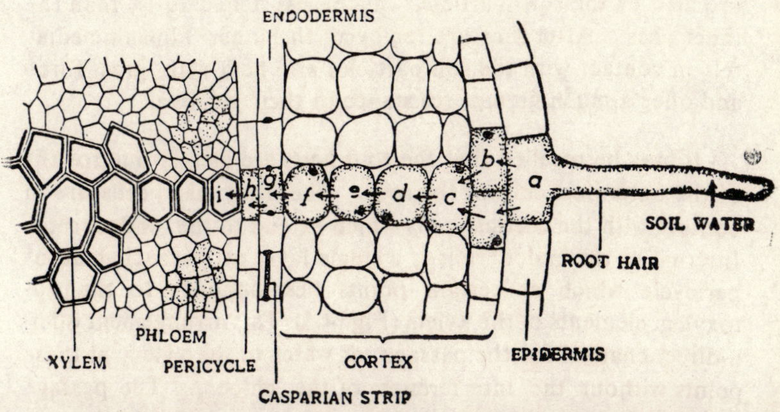

Fig. 4.5. Diagram showing the path of water from the soil into the root. The arrows indicate the direction of movement of water.

soil water. The cell sap is an aqueous solution of mineral salts, sugars and organic acids. On account of the presence of the last two classes of organic solutes, to which the cytoplasmic membrane is not permeable, the osmotic pressure of the cell sap is higher than the osmotic pressure of the soil solutions. The value of the latter in well watered soils is usually less than 1 atmos, and that of the former at least 3 to 8 atmos. The cell wall imbibes large quantities of water on account of its pectic constituents which possess pronounced hydrophilic properties. This imbibition of water is the first step in the process of water absorption. From this point water passes into the root hair across the semipermeable cytoplasmic membrane by osmosis on account of the higher osmotic pressure of the cell sap. As the films of capillary water are removed by absorption by the root hairs, the films of water from the adjacent soil particles are drawn on. These in turn draw on films of water on the particles adjacent to them and in this way water may move from considerable distances to the root hairs. It should, however, be noted that this capillary movement of water towards the root hairs is due to the cohesive power of water molecules and is not simply the result of "sucking power" of the root hairs. The outer films of water are held by the soil particles much less tenaciously than the inner ones. After they are removed, the inner films immediately in contact with the soil particles are held with great force and offer a much greater resistance to their removal.

It may be recalled that the root hairs are merely outgrowths of the outermost cells of the root. The root hair cells are in contact with the cortical cells which extend to the endodermis. Internal to the endodermis is a single layer of parenchymatous pericycle which at certain points lies opposite to the protoxylem elements of the xylem (Fig. 4.5). This arrangement offers a direct channel for the passage of water to the xylem at these points without the interference of the phloem. The passage of water through the endodermis requires some explanation. In many species the cells of the endodermis have lignified, suberised and partially cutinised radial and inner tangential walls and sometimes the outer tangential walls too are so modified. In other species, a strip of suberised thickening, called the *casparian strip,* stretches along the inner surface of the radial and tangential walls. These peculiarities of the endodermal cells obstruct the movement of water through them. In most cases, however, in the endodermis opposite the protoxylem groups are to be found specialised *passage cells* with thickenings only on their radial walls. In such cells the outer and inner tangential walls remain thin and thus facilitate the radial movement of water. In those species in which there are no passage cells there are pits in the thickened tangential walls for the passage of water.

As a result of the absorption of water from the soil, the root hair cell *a* becomes fully turgid, its osmotic pressure falls due to dilution and its turgor pressure increases. As a consequence, its suction pressure will fall below that of the adjacent cortical cell *b*, with the result that water will pass from *a* to *b* (Fig. 4.5). The diffusion of water into *b* likewise reduces its suction pressure which falls below that of the next cortical cell *c*, with the result that water passes from *b*, into *c*. In the same manner water passes from the cell *c* to *d*, from *d* to *e*, from *e* to *f* and from there into the endodermal cell *g*. From here it is passed on to the pericycle cell *h* which will eventually become turgid. It will then exert no suction pressure and hence will readily give up water to the xylem vessel *i* with which it is in contact. The walls of the xylem vessel are inelastic so that there is no turgor pressure and the whole of the osmotic pressure of the xylem sap constitutes its suction pressure. This being higher than the reduced suction pressure of the parenchyma (pericycle) cell *h*, water will be drawn into the xylem vessel. It will be clear that the factor which determines the passage of water from the soil to the xylem via the root hairs and cortical cells is the existence of what may be called a *gradient of suction pressure* from the root hair to the xylem vessel. In this process one may regard all the layers of the cortical cells as constituting a single complex semipermeable membrane on the outside of which is the soil solution, and on the inside the solution of the xylem vessels. So long as the osmotic pressure of the xylem contents is higher than that of the soil water, water will continue to pass from the soil to the xylem even if the osmotic pressure of the intervening cortical cells is higher than that of xylem. The force with which water is drawn in from the soil depends upon the difference between the osmotic pressure of xylem sap and the soil solution.

According to the mechanism of water absorption outlined above, water is forced into the xylem vessels by the surrounding cortical cells with a certain force. This induces a pressure sufficient in certain cases to raise the water to many feet in the xylem. This pressure is called *root pressure.*

The existence of root pressure may be demonstrated by cutting the stem of a well watered plant just above the soil level when sap will begin to exude from the cut stump. This phenomenon is called *bleeding*. Another manifestation of root pressure is the oozing out of drops of water from the uninjured edges or tips of leaves of such herbaceous plants as garden nasturtium, and many grasses. This phenomenon is called *guttation*.

The sap pressures in the stem may be the result of root pressures but quite often are the result of pressures arising in the tissues of the stems themselves. Thus, the same plant may show at the same time positive stem pressure and negative root pressure. The stem pressures are best known in the sugar maple. The stem sap contains sucrose, ammonia, protein, organic acids and a number of mineral substances.

Root pressure and the phenomena of guttation and bleeding that depend upon it are most common in plants that grow in moist warm and well aerated soils, with their tops surrounded by humid air. This is because it occurs only when absorption is sufficiently in excess of transpiration. Conditions which hinder water intake such as poor aeration, cold or dry soil, a high concentration of solutes in the soil, or the presence of toxic substances, and conditions which favour rapid transpiration reduce or prevent root pressure. Since the proper combination of conditions most often exists at night, root pressure (and therefore bleeding and guttation) is best, manifested in the early morning.

The magnitude of root pressure varies in different plants and under different environmental conditions. Under the most favourable conditions a root pressure of 1 atmos. or even 1.5 atmos. may be exhibited by plants. Higher magnitudes have also been reported in some cases. Root pressures upto 3 atmos. have been reported by Clark and others for birch. Molisch observed pressures up to 6.4 atmos. in *juglans regia* and White (1938) reported root pressure of 10 atmos. in plants grown in water culture.

Root pressure may be measured in a potted plant which is well watered. The shoot is cut off a few centimeters above the soil level and the stump is connected to a manometer by means of strong rubber tubing (Fig. 4.6). The rise of level in the free limb of the manometer is a measure of root pressure.

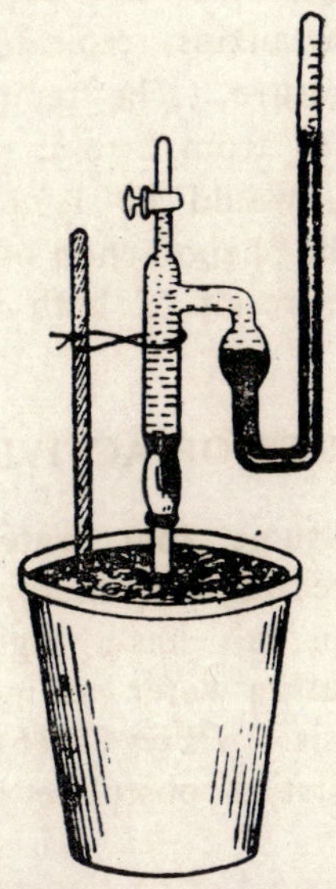

Fig. 4.6. An apparatus for the measurement of root pressure.

Root pressure is a vital phenomenon and depends upon the activity of living cells of the root. It ceases when the root cells are killed by the addition of poisonous substances and anaesthetics to the soil. There is sufficient evidence to show that root pressure is the result of osmotic phenomena going on in the roots and the necessity of living cells is due to the fact that these provide the differentially permeable membrane which are essential for the osmotic processes.

It will be clear from the foregoing discussion that water absorption is caused principally by forces developed in the roots as evidenced by root pressure and associated phenomena and that it depends upon the activity of the living cells of the root. This mechanism of water absorption is known as the *theory of active absorption* and gives a satisfactory explanation of water absorption in slowly transpiring, well watered plants. Active water absorption must not be regarded as a purely physical phenomenon resulting from osmotic movements of water. There is evidence to show that active absorption is also a physiological mechanism and depends upon metabolic energy. It is affected by auxins, respiratory inhibitors, oxygen availability, and temperature. The temperature coefficient of active absorption varies from 2 to 3, whereas for a strictly physical process the Q_{10} would be 1.2 to 1.3. It seems that active absorption and the phenomenon of root pressure which depends upon it are the result of both osmotic and physiological mechanisms.

OSMOTIC THEORIES OF ACTIVE ABSORPTION

Osmotic theories assume that water moves by diffusion along a gradient of decreasing free energy or diffusion pressure (D.P.D.). If the xylem sap has a higher osmotic potential than that of the soil solution water can move from the soil into the root xylem by osmosis. *Atkins* (1916) and *Priestley* (1920-22) were amongst the first to postulate an osmotic theory of active absorption.

According to Atkins the entire root functions as a differentially permeable membrane but *Priestley*, *Arnold* (1952) and others believe the endodermis to be the effective differentially permeable membrane in this system.

The value of osmotic pressure of the cell sap of the root hairs is generally 2 atmos. and may be as high as 3 to 8 atmospheric pressure in some cases. The osmotic pressure of the soil water is, however, much lower than that of the cell sap and is generally less than 1 atmos. The higher diffusion pressure deficit of the cell sap of the root hair causes endosmosis of soil water across the semipermeable cytoplasmic membrane. Before this the soil water is imbibed in large quantities by the cell wall which is permeable to it.

A number of investigators have supported the operation of an osmotic mechanism of absorption of water by demonstrating a close relationship between root pressure and higher osmotic pressure (accumulation of salts) of the xylem sap. The water absorption is caused by an increase in the water potential or D.P.D. of the xylem sap brought about by the accumulation of salt in the xylem.

The chief problem in connection with an osmotic explanation of root pressure is therefore, to explain as to how a sufficient concentration of solutes is maintained in the root xylem to maintain a higher D.P.D. in the xylem sap.

While *Atkins* (1916) believed that the adjoining parenchyma cells secrete sugar into the xylem vessels, *Priestley* (1922) thought that the *contents* of the cells differentiating into xylem elements provided the necessary solutes. *Anderson* and *House* (1967) reported that salts were absorbed by the xylem elements containing living protoplasts.

NON OSMOTIC SECRETION THEORIES OF ACTIVE ABSORPTION

Investigators like *Bennet Clark et al.* (1936), *Thimann* (1951) and *Bogen* and *Prell* (1953) believe that there is a non-osmotic active water uptake mechanism in plants. Water is believed to be absorbed against a concentration gradient or at an accelerated rate. This 'active' intake of water will obviously require an expenditure of energy released from the metabolic activity of the root cells such as respiration etc.

Bennet-Clark et al. (1936) observed that the apparent mean osmotic potential of cell sap determined by plasmolytic method was often lower than the cryoscopically determined value. *Lyon* (1942) and *Currier* (1944) also reported similar results. According to several workers this discrepancy indicated the existence of a *non-osmotic component* in the absorption process. They suggested that a '*secretion force*' developed due to a mechanism pumping water actively into the vacuole, with expenditure of metabolic energy. *Kramer* (1956) and others, however, believe that the discrepancies were due to experimental errors which might include dilution of vacuolar sap during expression, penetration of plasmolyticum, and neglect of matric potential caused by the presence of hydrophilic colloids *e.g.*, proteins in the sap.

Van Overbeek (1942) observed that the osmotic pressure of the exudate from tomato root systems was somewhat lower than the osmotic pressure of the solution required to stop exudation or absorption of water. He attributed this difference to be due to non-osmotic secretion of water into the xylem. *Findlay* (1966) did not find any difference between the osmotic pressure of the external solution and the xylem sap.

The observations of Van Overbeek and Findlay have, however, been criticised by *Klepper* (1967) and others on account of the fact that salt is removed from the xylem sap as it moves upward giving a comparatively lower value of the osmotic pressure than that of the absorbing region.

There is a large amount of evidence to show some relationship between respiration and water uptake and this together with the often observed diurnal periodicity in root pressure has been taken as a proof for a mechanism more directly dependent on metabolism. The evidences cited in favour of this are :

(*i*) Correlation between respiratory rate and water uptake.
(*ii*) Reduction of water uptake due to low temperature, lack of oxygen and respiratory inhibitors.
(*iii*) Active salt uptake.
(*iv*) Stimulation of water uptake by auxins etc.

Reinders (1942), *Kelly* (1947) have found aerobic conditions to be necessary for water uptake. *Rosene* (1947), *Follman* (1957) and others have reported decreased water uptake caused by respiratory inhibitors.

It is believed by many workers that low temperature, lack of oxygen and respiratory inhibitors apparently reduce the permeability of the membranes and increase the resistance of cytoplasm to water flow. According to *Lopushinsky* (1964) and some others even the passive water movement through roots caused by a pressure gradient is greatly reduced by respiratory inhibitors. *Kramer* (1969) has emphasized that accumulation of salt which is essential for water uptake is dependent on expenditure of metabolic energy. The fact that uptake of salt and water is much reduced by respiration inhibitors does not by itself prove that active transport is essential for both. The inhibition of respiration reduces the active absorption of solutes which may affect the osmotic potential of the absorbing cells. In should also be noted that cytoplasm is much more permeable to water than to dissolved substances and hence a great deal more energy would need to be spent to maintain a concentration gradient for water than for solutes. *Levitt* (1947) calculated that the whole of the energy released from respiration in a red beet root cell is insufficient to maintain a gradient of D.P.D. (water potential) of more than about one atm. (bar).

It can therefore be concluded that the relation between respiration and water uptake is indirect. Energy released in respiration appears to affect both membrane permeability of cells and the absorption and accumulation of salt in the xylem of entire roots and both of these influence water uptake.

It has been observed that auxin increases water uptake of plant tissues. It occurs only in the presence of fiee oxygen and is affected by respiration inhibitors. Since auxin induced water uptake is reduced by respiration inhibitors, *Hackett* and *Thimann* (1952) concluded that it was directly linked to respiration. Auxin induced water uptake however, appears to be brought about by its effect in reducing wall pressure, increasing cell wall extensibility, permeability and osmotic potential of plant cells.

Many earlier claims of nonosmotic water uptake can now be explained in terms of osmosis. If active water uptake does occur it is a negligible factor in the over-all picture of plant water relation *Kozlowski*, (1964).

ELECTROOSMOTIC THEORIES

Keller (1930) and others proposed that water is absorbed by the xylem by electroosmotic mechanism. The cellulose membranes in contact with water are negatively charged, water moves toward the negative pole, which is in the interior of the roots. *Fensom* (1957, 1958) reported that cycles in bielectric potentials were correlated with cycles in exudation from roots. He proposed an electrokinetic theory of water transport.

Lundegardh (1940), *Blinks* and *Airth* (1951), however, consider the observed difference in potential to be too small to produce appreciable water movement. The theories are not considered to be significant in the water uptake process.

OTHER METHODS OF WATER ABSORPTION

Any cell in an unsaturated condition may absorb water by virtue of its suction pressure when water is supplied to it. Leaves moistened with rain or dew may absorb water in this way. The amount of water thus absorbed is, however, very insignificant. Moreover, the cuticle of leaf epidermis when thus moistened becomes somewhat permeable to water and the water drops act like small lenses in sunlight focussing the sun's rays on to the leaf cells with the result that cuticular transpiration increases, thus inducing wilting. The direct watering of plants in sunlight is, therefore, injurious to the plants.

A special method of absorption of water is met with in many epiphytic flowering plants, especially certain orchids. These plants have long, shining aerial adventitious roots which develop on the outside of their cortex a special water absorbing tissue called the *velamen* (Fig. 4.7). This is several cells thick. The outermost layer of the cortex forms a special layer called the *exodermis* whose cells have thick radial and outer walls except some passage cells lying opposite the protoxytem groups. These cells remain thin-walled. Similar passage cells are also present in the endodermis. These velamen cells have holes in their walls and through them absorb water like a sponge from dew or rain water. The water is then passed on to the xylem through the passage cells. Other epiphytes absorb water through hygroscopic hairs developed on their aerial parts.

Some conifers and many forest trees and shrubs do not develop root hairs. Their absorbing surfaces are considerably modified by the presence of mycorrhizae. Many of the forest trees, like those of the Abietineae, Salicaceae, Betulaceae, and Fagaceae, have two distinct types of root—long roots and short roots. Long roots have root caps, produce root hairs near their tips, and are long-lived. Short roots lack root caps, grow slowly, are short and thick, have few root hairs and branch repeatedly in a dichotonous manner, sometimes even

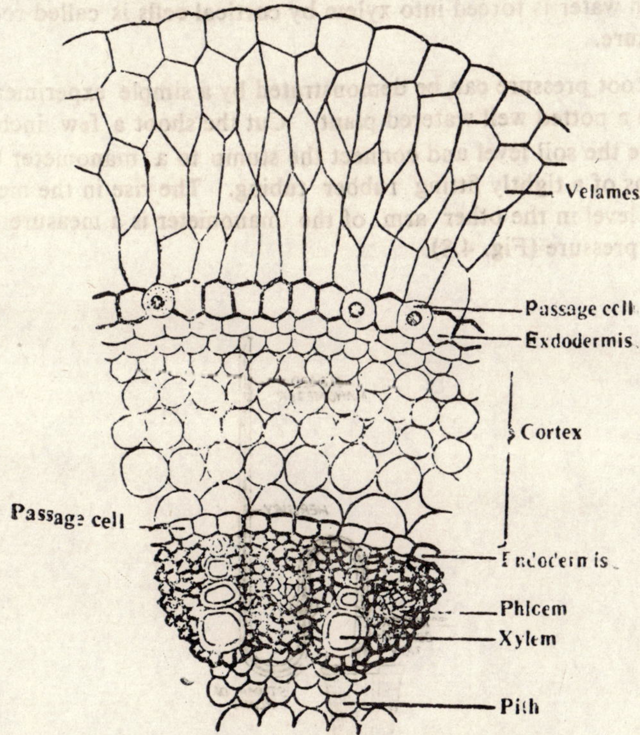

Fig. 4.7. T.S. of the aerial root of an orchid showing velamen and the passage cells in the exodermis and endodermis.

forming corolloid clusters as in pinus. The short root form mycorrhizae as a result of invasion by hyphae of various species of fungi commonly found in forest soils. These mycorrhizae are of considerable importance to forest trees as they increase enormously the absorbing surface of the root system. The fungal hyphae absorb from the soil large quantities of water and mineral substances which they pass on the roots of the trees.

ROOT PRESSURE

According to the osmotic mechanism of water absorption outlined above water is forced into xylem by the cortical cells of root. This pressure in some cases is sufficient to raise the water to many feet in the xylem. The actual pressure with which water is forced into xylem by cortical cells is called root pressure.

Root pressure can be demonstrated by a simple experiment. Take a potted well watered plant. Cut the shoot a few inches above the soil level and connect the stump to a manometer by means of a tightly fitting rubber tubing. The rise in the mercury level in the other arm of the manometer is a measure of root pressure (Fig. 4.8).

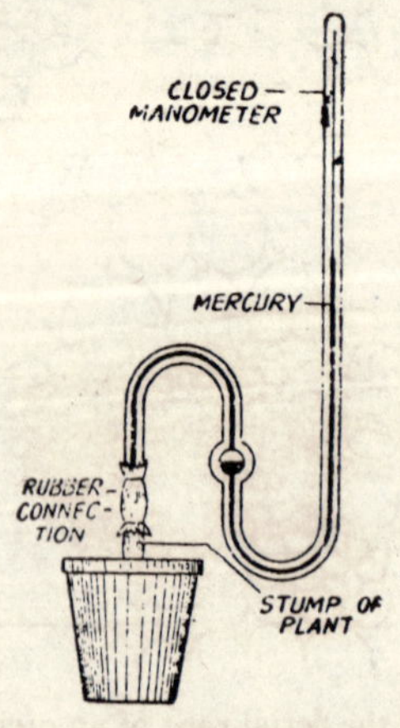

Fig. 4.8. Apparatus showing root pressure.

Root pressure depends upon the activity of living cortical cells of the root and therefore is a vital phenomenon. In roots treated with poisonous substances no root pressure is observed.

When the stem of a well watered plant is cut, sap begins to exude from the cut end. This phenomenon is called bleeding. In certain herbaceous plants such as garden nasturtium and many grasses water oozes out in the form of droplets from uninjured tips and edges of leaves. This phenomenon called Guttation is very common during early morning and is a manifestation of root pressure. Root pressure and guttation are most commonly noticed in plants growing in moist aerated soil surrounded by humid atmosphere. It is so because they can be observed only when absorption is in excess of transpiration.

FACTORS AFFECTING THE RATE OF ABSORPTION OF WATER

Any factor which affects the rate of transpiration, the DPD of root cells and of soil water will influence the rate of absorption. The following soil factors have considerable influence on the rate of absorption of water :

Available Soil Water

The rate of absorption of water is generally uniform in the range of water between the *field capacity* and *permanent wilting percentage*. Decrease in the soil water below the permanent wilting percentage causes considerable decrease in the absorption of water. If the soil-water is increased appreciably beyond the field capacity, the aeration of the soil is affected causing decrease in the absorption of water.

Concentration of the Soil Solution

The osmotic pressure of a solution depends upon its concentration. If the soil solution is highly concentrated due to the presence of salts, etc., its osmotic pressure will be greatly increased which will inhibit the absorption of water. The lack of water absorption by plants growing in saline water is an example of physiological dryness.

Soil Aeration

Absorption of water by the roots takes place at a rapid rate in well aerated soils and at a greatly retarded rate in soils, which are deficient in oxygen supply. Oxygen deficiency retards the growth and development of fresh roots and disturbs their metabolic activities. The accumulation of carbon dioxide in the soil is even more detrimental for water absorption. Accumulation of carbon dioxide increases the viscosity of the protoplasm and decreases its permeability, both of which reduce the rate of absorption of water. The type of condition causing almost total check on the absorption of water is found during water-logging of soils. This is also an example of physiological dryness.

Soil Temperature

The maximum absorption of water takes place generally between 20°C and 30°C. Increase in temperature beyond 30° retards the rate of absorption. Lowering of temperature below 20°C also causes an appreciable reduction in the rate of absorption which is almost nil at freezing temperature. The decrease in the rate of absorption with lowering of temperature, is due to several factors which are : (1) slower rate of elongation of root (2) reduced rate of diffusion of soil water into roots, (3) decreased permeability of the cell membrane, (4) increased viscosity of water, protoplasm and cell membranes and (5) reduced metabolic activity of root cells.

Extent of Root System

The amount of water that can be absorbed by roots from the soil depends upon their number, branching and number of root hairs. The total surface area of the root that comes in contact with soil water, efficiency of conducting elements, growth and metabolic state of roots control the amount of water that can be absorbed.

Other Methods of Water Absorption

In addition to the normal absorption of water by roots some special methods of absorption are met with in some plants. These include mycorrhiza, Velamen and absorption of water through epidermal cells.

Some forest trees and gymnosperms do not have root hairs. The younger parts of their roots are covered over by hyphae of some fungus—the association being known as *Mycorrhiza*. The hyphae of the mycorrhizal fungus function as absorbing organs of the root. Members of family Salicaceae, Betulaceae, Fragaceae and Abietinae possess such mycorrhizae. They have two types of roots namely long and short roots. The long roots possess root hairs whereas the short roots which branch repeatedly in a dichotomous manner lack root hairs and are invaded by mycorrhizal hyphae. These hyphae absorb water and minerals from the soil and pass them on to the roots of host.

Another special method of water absorption is met with in epiphytes and certain orchids. In these plants water is absorbed by long, shining aerial roots. These roots have a special water absorbing tissue called Velamen, outer to the *Exodermis* —the outermost layer of the cortex. The cells of velamen are thick walled, dead and perforated, function like a sponge and absorb dew or rain water. The cells of Exodermal and Endodermal layers which are thickened on their tangential and radial walls have some unthickened cells called *passage cells* lying opposite the protoxylem groups. The water absorbed by velamen is transferred to the cortex and xylem of root through these passage cells. Many epiphytes absorb water from the atmosphere through hygroscopic hairs developed on their aerial parts.

The absorption of water also occurs to a small extent through the surface cells of the leaves. The amount of absorption depends upon the DPD of leaf cells and permeability of cutin layer.

PASSIVE ABSORPTION OF WATER

Whenever transpiration is occurring the diffusion pressure deficit of the water in the cells of the leaf increases. Water, therefore, diffuses out of the xylem elements of the leaf into mesophyll cells. During rapid transpiration a tension is developed in the water column present in the xylem elements.

The tension is equivalent to the negative pressure and so the diffusion pressure deficit of the water is further increased by the amount of the tension imposed on the water column. The tension developed in the conducting elements is transmitted along their entire length to their lower ends in the root ends. In this situation the diffusion pressure deficit of the peripheral cells of young roots is greatly increased. As a matter of fact the diffusion pressure deficit of the root hairs is considerably increased even if the water in the root cells does not pass into a state of tension. A gradient of diffusion pressure deficit is developed across the root, increasing steadily from the epidermal cells of the root to the xylem elements.

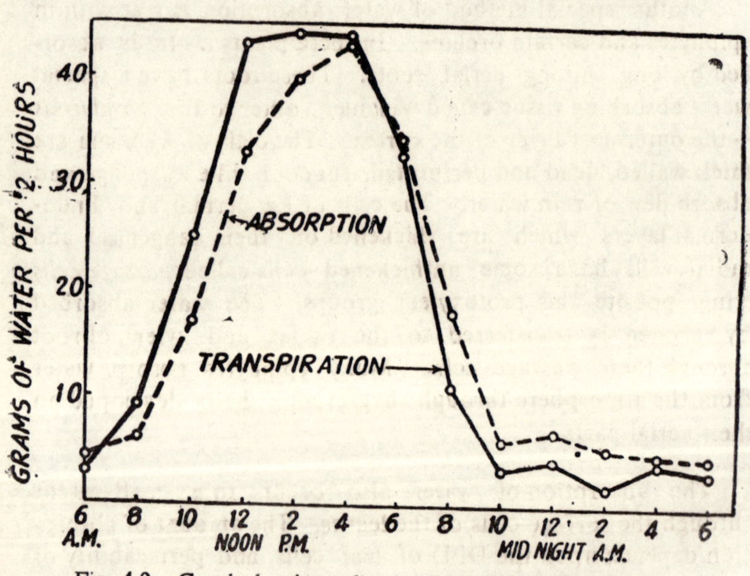

Fig. 4.9. Graph showing a direct relationship between transpiration and passive absorption of water.

Under the circumstances stated above the osmotic pressure of the cell sap of root hair increases to about 3-5 atmospheres, or even more than that, while the osmotic pressure of the soil water remains at less than one atmosphere. The higher diffusion pressure deficit of the root hair causes endosmosis of soil water.

The process of absorption in this case is due to factors originating in the upper portion of the plant, and the cells of the root play only a passive role. In passive absorption large quantities of water are absorbed at relatively rapid rates. The mechanism also ensures a continuous supply of water under conditions of high rates of transpiration.

The best evidence in support of passive type of absorption comes from the experimental finding that the rate of absorption of a plant is approximately equal to the rate of transpiration, provided water is available in the soil (Fig. 4.9).

PATHWAY OF WATER ACROSS ROOT CELLS

If there is an increasing gradient of diffusion pressure deficit water will enter the root hair, and passing through the cortical cells, will finally enter the xylem vessel. The root hair absorbs water from the soil and becomes fully turgid. Its suction pressure falls. The adjoining cell of the cortex which has a much higher suction pressure absorbs water from the root hair. In this way water moves from cell to cell in the cortex and reaches the endodermis (Fig. 4.10). Water cannot diffuse through the thickened walls of the endodermal cells due

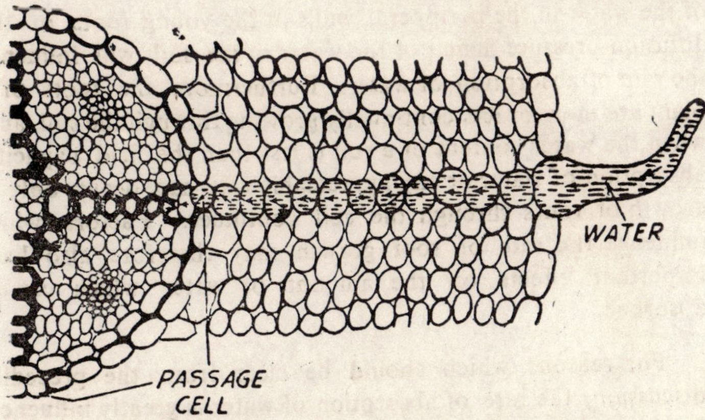

Fig. 4.10. Diagram showing the pathway of water across the cortical cells of the root.

to the *casparian strips*. However, water can easily diffuse through isolated thin walled endodermal cells called *passage cells*. The passage cells are generally opposite the protoxylem elements of the root.

The next problem and which is more difficult to explain is the mechanism of the unidirectional flow of water into the xylem ducts. Various theories have been put forward to explain the process but all of them are purely hypothetical. According to *Lepeschkin* the cell next to the xylem element has differential permeability at different sides. The cell wall towards the protoxylem is more permeable so that water from the fully turgid cell diffuses into the xylem vessel. According to *Priestley* the xylem element in the absence of the turgor pressure will always have a higher suction pressure than the next outer cell and so water from the turgid cell is absorbed into the xylem ducts.

The force with which water will be drawn by the xylem ducts from the soil will depend entirely on the difference between the osmotic pressure of the soil water and that of the xylem elements.

Environmental Factors Influencing the Rate of Absorption of Water

Any factor which influences the diffusion-pressure deficit of the water in the peripheral walls of the young roots or the diffusion-pressure deficit of the water in the soil will influence the rate of absorption of water. Furthermore the roots of a plant are more or less continually growing through the soil, and, when the water content of a soil is less than the field capacity, absorption of water at any appreciable rate can occur only if growth of roots through the soil continues. Factors which influence the rate of root growth may therefore also have important effects on the amount of water which can be absorbed.

For reasons which should be clear from the preceding discussion, the rate of absorption of water is greatly influenced by the rate of transpiration. Hence any factor which influences the rate of transpiration will indirectly influence the rate of absorption of water. Contrariwise, also we know that any factor which influences the rate of absorption will also influence the rate of transpiration. The more important soil factors which influence the rate of absorption of water will now be discussed.

1. *Available Soil Water*. In general, the term available soil water is used to refer to that fraction of the soil moisture in excess of that present at the permanent wilting percentage. Absorption of water appears to take place almost equally well over the range of soil-water contents between the field capacity and the permanent wilting capacity, especially in light and medium-textured soils. In heavier textured soils there is considerable evidence that the rate of absorption of water is somewhat retarded in the lower part of this range. A significant fact in this connection is that the osmotic pressure of the leaves increases as the soil-water content decreases (Fig. 4.11). Hence as the diffusion-pressure deficit of the soil water becomes greater, the potential diffusion-pressure deficit of the water in the plant likewise becomes greater. The effectiveness of the passive absorption mechanism undoubtedly increases accordingly. This upward adjustment in the osmotic pressures of the cells of the plant with decrease in soil-water content is one of the main factors resulting in a nearly uniform rate of absorption from many soils in the range of soil-water contents between the field capacity and the permanent wilting percentage.

At soil-water contents below the permanent wilting percentage, water intake is so slow that the turgidity of the leaf cells ordinarily cannot be maintained. Relatively high soil-water contents, in the range above the field capacity, result in a diminution in the rate of absorption by many species, because of the accompanying decrease in soil aeration.

2. *Soil Temperature*. Reduction in the absorption rate of water occurs in many kinds of plants at soil temperatures well above freezing, but the exact magnitude of this effect differs according to species. In general, plants native

to warm climates undergo a greater reduction in rate of water intake when the soil is chilled than those which are habitants of cooler climates. For example, Kramer (1942) found that watermelon and cotton, warm season crops, absorbed only 20 per cent as much water at 10°C. as at 25°C., while collards, a cool season crop, absorbed 75 per cent as

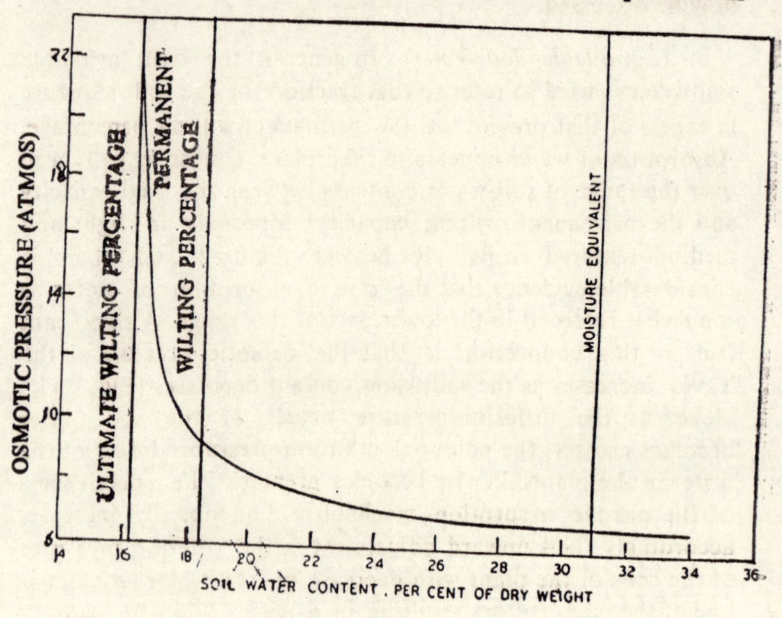

Fig. 4.11. Relation between soil-water content and osmotic pressure of leaves of sunflower plants.

much water at the lower of these two temperatures as at the higher. The relation between soil temperature and the rate of absorption of water (for which the rate of transpiration is taken as the index) by sunflower. In this species, the rate of transpiration diminishes rapidly with a decrease in soil temperature below about 55°F. (13°C.). A similar relation doubtless holds for other species, except that the soil temperature at which an appreciable diminution in the rate of water influx is initiated differs from one kind of a plant to another. For obvious reasons, virtually no water is aborbed by roots from frozen soils.

The mechanism whereby low soil temperatures cause a retardation in the rate of absorption of water is doubtless a complex one. The principal factors involved are : (1) retardation in the rate of root elongation, (2) decreased rate of movement of water from soil into roots, (3) decreased permeability of cell membranes, (4) increased viscosity of the protoplasm and cell walls, (5) increased viscosity of water, (6) decreased vapor pressure of water, and (7) decreased metabolic activity of root cells (Kramer. 1949).

Soil temperatures may sometimes become high enough to exert a retarding effect on the rate of absorption of water. For example, Haas (1936) found that absorption of water by lemons, oranges, and grapefruits decreased when soil temperatures exceeded 30° or 35°C.

3. *Aeration of the Soil.* In general, absorption of water by the roots of most species of plants proceeds more rapidly in soils which are well aerated than in those which are not. In such soils oxygen concentrations may approach a zero value, and carbon-dioxide concentrations ranging up to 10 per cent are not uncommon. The comparative effects of a maintained near zero oxygen content and a maintained relatively high (20 per cent) carbon-dioxide content in the soil atmosphere on the intake of water by tomato, tobacco, sunflower, coleus, cotton, and corn plants were investigated by Whitney (1942). Severe oxygen deficiency, obtained by passing nitrogen through the sand medium, resulted in a marked reduction in water absorption by all of these species as compared with plants whose roots were aerated with atmospheric air, although the effect was more pronounced on some than on others. When the roots of these species were exposed to excess carbon dioxide in the presence of oxygen, a condition which was obtained by passing a mixture of 20 per cent carbon dioxide, 20 per cent oxygen, and 60 per cent nitrogen through the sand, there was no significant decrease in the rate of water intake by corn or sunflower plants, and only a relatively small decrease in the rate for tobacco, tomato, coleus, and cotton plants as compared with similar plants aerated with atmospheric air.

When the roots were exposed to the combined effects of oxygen deficiency and carbon-dioxide excess (gaseous phase in the sand consisting of 80 per cent nitrogen and 20 per cent carbon dioxide), the retarding effects on water absorption by all species investigated except coleus and tomato were little if any greater than the effect of severe oxygen deficiency alone. In general, these results indicate that a near absence of oxygen has a greater influence in retarding the rate of water intake by many kinds of plants than the presence of carbon dioxide in concentrations considerably in excess of those usually found in soils. Exposure of roots to pure carbon dioxide, or to high concentrations (50 per cent or more) of this gas, at least for short periods, however, has a marked retarding effect on absorption of water by a number of species (Kramer 1940, Chang and Loomis 1945). Hoagland and Broyer (1942) interpret this result as being caused by an initial decreasing effect of carbon dioxide on the permeability of root cells to water.

Saturation or near-saturation of a soil with water is the most common cause of deficient soil aeration. Under such conditions it appears probable that the reduced rates of water absorption which are evidenced by most plants result more from a severe oxygen deficiency than from an accumulation of carbon dioxide. A drastic reduction in the supply of available oxygen in the soil reduces the rate of respiration of the root cells and this, in turn, influences other metabolic processes as well as the rate of root growth. One of the results of this sequence of disturbed physiological conditions is a lower rate of absorption of water. While the roots of most kinds of plants can survive for at least short periods in soils practically devoid of oxygen, as a result of the occurrence of anaerobic respiration, maintenance of this process for any considerable interval of time leads to death of roots of many species.

In contrast with other species, the roots of hydrophytes naturally grow in water-saturated soils and absorb water readily from such soils. Some species of hydrophytes have well-developed intercellular air spaces which are continuous from the leaves through the stems into the roots. It has been shown in some such species that oxygen moves to the roots through such channels (Conway 1937, Laing 1940) and such movement of oxygen doubtless occurs in many other kinds of hydrophytes. In other species which grow with their roots submerged, however, no such prominent system of air passages is present. Even in such species some movement may occur to the roots from aerial organs through intercellular spaces of usual dimensions. The roots of these species in which little or no oxygen becomes available through downward movement from the tops apparently are able to carry on their metabolic processes at relatively low oxygen concentrations.

4. *Concentration of the Soil Solution.* The concentration of the soil solution in most soils in humid regions is so small that it has only a slight effect on the diffusion-pressure deficit of the soil water. In alkali or saline soils, on the other hand, the concentration of the dissolved salts in the soil water is often sufficient to raise the osmotic pressure to a very considerable value—in extreme situations to 100 atm, or even higher. Copious applications of fertilizers to greenhouse or agricultural soils, especially if sandy, or irrigation with water containing dissolved salts in considerable concentration, often result in raising the osmotic pressure of the soil solution to a value of several atmospheres or more.

The diffusion-pressure deficit of the soil water, except when soils are relatively dry, *i.e.*, below the field capacity, is essentially equal to the osmotic pressure of the soil solution. In general, the rate of absorption of water decreases, often almost proportionately, with increase in the osmotic pressure of the substrate. The effect of solutes on the movement of water into roots appears to be principally an osmotic one, specific effects of ions playing only a secondary role. Plants may, within limits, become adjusted to an increase in the concentration of the substrate solution, inasmuch as the osmotic pressure of the cells of a plant may increase under such conditions. Hence a plant in which the absorption rate is markedly reduced when its roots are first brought into contact with a substrate solution of higher osmotic pressure than the one in which it had been growing may, after an interval of time, largely or entirely regain its original rate of water absorption. Most plants can develop normally only when the osmotic pressure of the substrate solution does not exceed a few atmospheres. Halophytes, *i.e.*, plants indigenous to saline or alkali soils or substrates, are the only important exception to this statement.

Absorption of Water and Water Vapour by Leaves. Leaves and other aerial parts of plants frequently become wet as a result of rain, dew, or fog. Floods also sometimes result in a temporary immersion of aerial plant organs. Wetzel (1924) found that, of a large number of species investigated, practically all absorbed some water directly through the leaves. The turgidity of the leaves was restored from the wilted condition in most species after immersion in water for 24 hr. or less. Absorption of water occurred directly through the epidermal cells rather than through the stomates. In nature, however, absorption of water through the leaves is only rarely a factor of importance in the water economy of the plant.

Prolonged immersion of the leaves of many species in water leads to opening of the stomates and, under certain conditions, also to an injection of the intercellular spaces with liquid water. Tiny droplets of water are sometimes projected through the stomates as a results of splashing during heavy rains or during the artificial spraying of leaves. Partial or complete injection of the intercellular spaces with liquid water is sometimes brought about in this way, especially in species with large stomates.

Under certain conditions leaves and other aerial organs can absorb water vapour directly from the atmosphere. Stone *et al.* (1950) have shown that seedlings of Coulter pine, (*Pimus Coulteri)* in a state of permanent wilting, absorb water-vapor from saturated or near-saturated atmospheres.

5

The Translocation of Water

Most terrestrial plants obtain the water which is necessary for their existence from the soil. An overwhelmingly large proportion of the water which is absorbed by the roots of land plants is lost in the process of transpiration. Smaller quantities are utilized in growth and in photo-synthesis, and in some species limited amounts of water may be lost by guttation. Water must therefore move through the intervening tissues and organs from the absorbing regions of the root to the tissues in which it is utilized, or from which it passes out of the plant. The process whereby water moves through the plant is termed the *conduction, transport,* or *translocation* of water.

In herbaceous species and many shrubby plants the distance through which water moves in passing from the root tips to the leaves is usually not more than a few feet. Even in such plants appearances are sometimes deceptive, as some herbaceous and shrubby species such as alfalfa may have such deep root systems that some of the absorbed water often ascends for distances as great as 20 feet or more before it reaches the level of the soil surface. It is in trees, however, that the most striking illustrations of the upward movement of water occur. The tallest tree of which we have an authentic record is a specimen of the Coast Redwood (*Sequoia sempervirens*) which has attained a height of 364 feet. Many other individuals of this species and of several others; including the Big Trees (*Sequoia gigantea*) of California, the Douglas Fir (*Pseudotsuga douglasi*) of the Pacific Northwest, and the Blue Gums (*Eucalyptus*) of Australia exceed 300 feet in height.

Trees ranging from 100 to 200 feet in height were of common occurrence in the virgin forests of eastern North America. Since the root systems of trees always penetrate at least a few feet into the ground, the actual vertical distance through which at least a part of the water absorbed is elevated is always more than the height of the tree. In trees, therefore, water must ascend to heights ranging up to nearly 400 feet above the level of water absorption.

The mechanism by which this feat is accomplished in tall trees has been the subject of much experimentation and even more speculation. The ensuing discussion of the "ascent of sap" through plants will be principally in terms of its movement through trees, because much of the experimental work on this problem has been performed on woody species. Any explanation of this phenomenon which can be shown to be adequate for tall trees should also prove satisfactory for vascular species of lesser stature.

The water which moves through plants is not pure, but almost invariably contains dissolved in it small quantities of inorganic solutes, and often organic solutes as well. This dilute solution is often called the *xylem sap*.

Path of Water Stream

The upward translocation of water in the plant takes place through the xylem. This can be readily proved by a simple experiment. A leafy shoot of balsam plant is cut under water and placed in a weak solution of Basic Fuchsine. After sometime the veins of the leaves will be found to be injected with coloured water and coloured streaks can be traced along the whole length of the semi-transparent stem. If transverse sections of the stem are cut at different levels it will be found that only the tracheids and vessels are coloured. This proves that the water moved up through these cells.

A confirmation of this may be obtained from the ringing experiment. A ring of bark about one inch in length external to the wood is removed from the stem of a woody twig without injuring the xylem (Fig. 5.1). In another twig, by careful manipulation, the wood element is removed for a short distance without injuring the phloem and the cortex. The ringing operations are conducted while the twigs are under water. In the former case the leaves will remain fresh and will not show

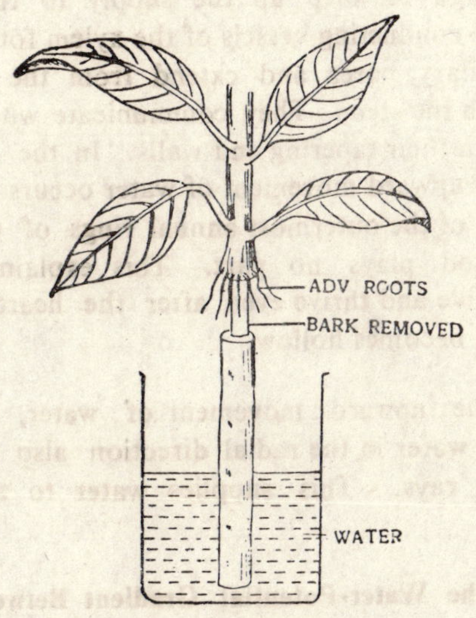

Fig. 5.1. The ringing experiment.

any sign of wilting while in the latter the leaves will wilt in a few minutes. This shows that the tissues outside the xylem are not concerned in the ascent of water because the interruption of their continuity does not interfere with the rise of water.

That the water rises upward through the lumen of the xylem ducts and not through the walls of the xylem by imbibition, as was supposed by Sach, may be demonstrated by a simple experiment. The cut end of a leafy shoot is placed for about an hour in melted paraffin which is then allowed to cool and solidify. A similar control shoot is placed in water. A fresh cut is made near the end of each shoot so as to expose the cell walls and both are placed in a solution of Aniline Blue. After sometime compare the ascent of the dye in the shoot with blocked vessels and in the normal shoot. It will be observed that the shoot treated with paraffin soon wilts. This is due to the failure of water to rise up the stem to the transpiring leaves as the lumens of the xylem ducts have been plugged by the solidified paraffin. This shows that water rises in the lumen and not in the walls of the tracheids and vessels. Some water may rise through the cell wells by imbibition but it is not enough to keep up the supply to the transpiring surfaces. The conducting vessels of the xylem form continuous tubes of capillary bores and extend from the roots to the leaves through the stem. They communicate with each other through pits in their tapering end walls. In the woody plants and trees, the upward movement of water occurs in the vessels and tracheids of the outermost annual rings of the sapwood. The heartwood plays no part. This explains why a tree continues to live and thrive even after the heartwood decays and the trunk becomes hollow.

Besides the upward movement of water, a cell to cell movement of water in the radial direction also occurs along the vascular rays. This supplies water to the cortex and epidermis.

The Plant in the Water-Potential Gradient Between Soil and Atmosphere

The plant bridges the steep *water-potential gradient between soil and air.* Because the shoot is exposed to the vapor-pressure deficit of the air (that is, to a low water potential), a flow of water through the plant is set in motion.

The rules governing *water transport* are analogous to those for the flow of electrical current, described by Ohm's Law. Analog circuit diagrams are thus a useful way to represent the relationships in the soil-plant-atmosphere system. The *potential gradient* in the continuum soil-plant-atmosphere is the *driving force for water transport* through the plant. Owing to the various resistances to hydraulic conductivity within the plant, there develops a gradient in water potential from the leaf surface through the conduction system to the root.

B. Huber, H. Gradmann and T.H. van den Honert have expressed the relationship between *water flow* (J), the *petential gradient* $\Delta\psi$, and the *hydraulic and transition resistances* (*r*) in the following form :

$$J=\frac{\Delta\psi}{\Sigma r}\cdot$$

The water potential ψ_z at a particular location in the plant is lower, the lower the *water potential in the soil*, the more effective the *force of gravity* (low ψ_g in tall trees or vines), the greater the *hydraulic resistances r* between the soil and the reference point *z* in the shoot, and the more *water flows through the plant* (sum of the partial fluxes f_i). A formula expressing these relationships has been given by Richter (1973) :

$$\psi_z=\psi_{soil}+\psi_g+\Sigma^z{}_{soil} f_i\cdot r_i.$$

From the two equations it is evident that the plant would be expected to exhibit a *steep gradient in water potential* only when large quantities of water are flowing through it—that is, when conditions demand *intensive transpiration* (Fig. 5.2). In this situation the cohesive force acting on

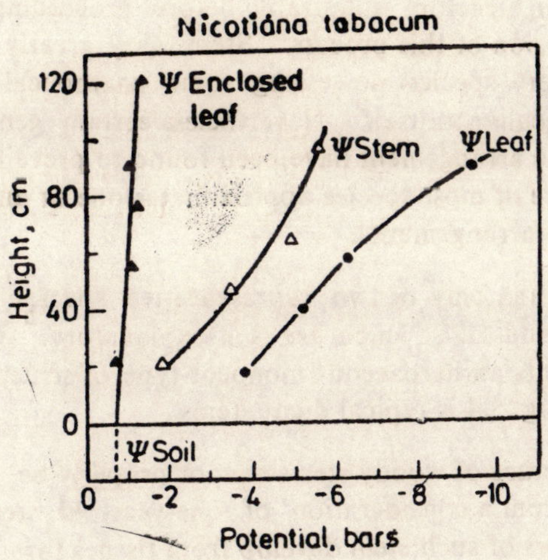

Fig. 5.2. Vertical gradients of leaf and stem water potential in tobacco plants transpiring at a high rate and in plants with transpiration restricted by enclosing leaves in polyethylene bags.

the conducting system is large. This effect is of practical value, because by *determining the xylem water potential* one can estimate the water-potential state of the entire plant : When a leaf or twig is cut off, the column of water in the xylem is broken and the water in the vessels is drawn back into the tissues. If the twig, immediately after it is cut off, is introduced into a pressure chamber like that developed by Scholander *et al.* (1965), the external pressure can gradually be increased until the water potential in the cells is compensated and the water moves back into its original position in the vessels. The extra pressure applied, which can be read off from a manometer, corresponds to the negative pressure previously existing in the twig, a function of the water potential of the cells and the cohesive tension of the water in the xylem ducts of this part of the plant.

A natomy of Stems

Since the mechanisms of the movement of water can scarcely be understood intelligently without some knowledge of the anatomy of the tissues through which it moves, a brief review of stem structure is desirable before proceeding further with a discussion of this process. Stems vary greatly in their structure, every species possessing some anatomical features which are peculiar to itself. Nevertheless certain general patterns of tissue arrangement have been found to prevail, and the stem structure of most species approximates one or another of these general arrangements.

The stem anatomy of two representative species is shown in Fig. 5.3 and 5.4, which are self-explanatory. The corn stem represents an herbaceous monocot type of structure while that of the Fig. 5.4 is typical dicot stems.

The structure of woody stems cannot properly be appreciated merely from a consideration of one-year-old stems. The *primary tissues* of such stem develop from tissues formed at the apical growing tip during growth of the stem in length. Nearly all perennial stems also grow in diameter as a result of the development of *secondary tissues* from the cambium. By

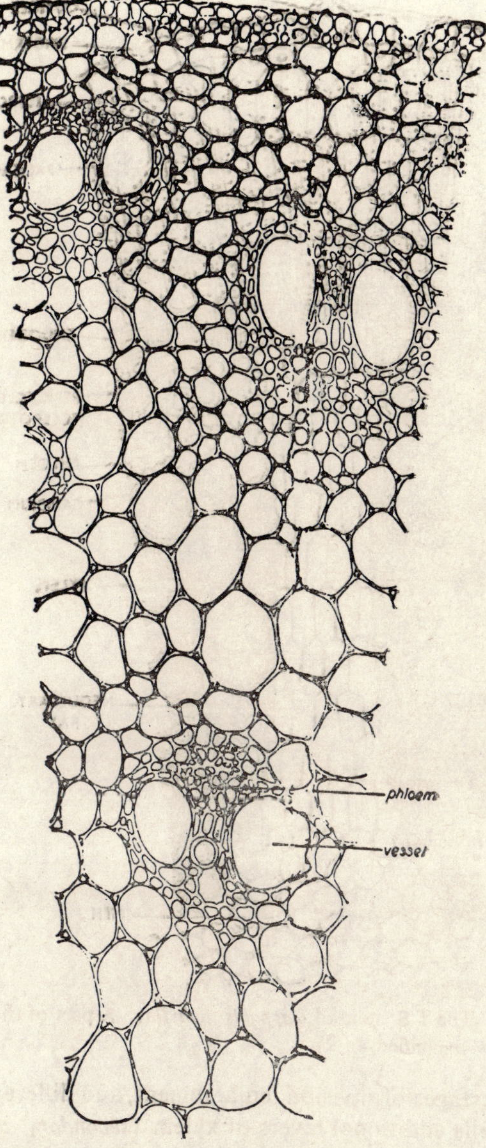

Fig. 5.3. Segment of young corn (maize) stem as seen in cross section. Three vascular bundles and parts of two others are shown. The large-celled tissue between the bundles is sometimes classed as pith ; sometimes as parenchyma.

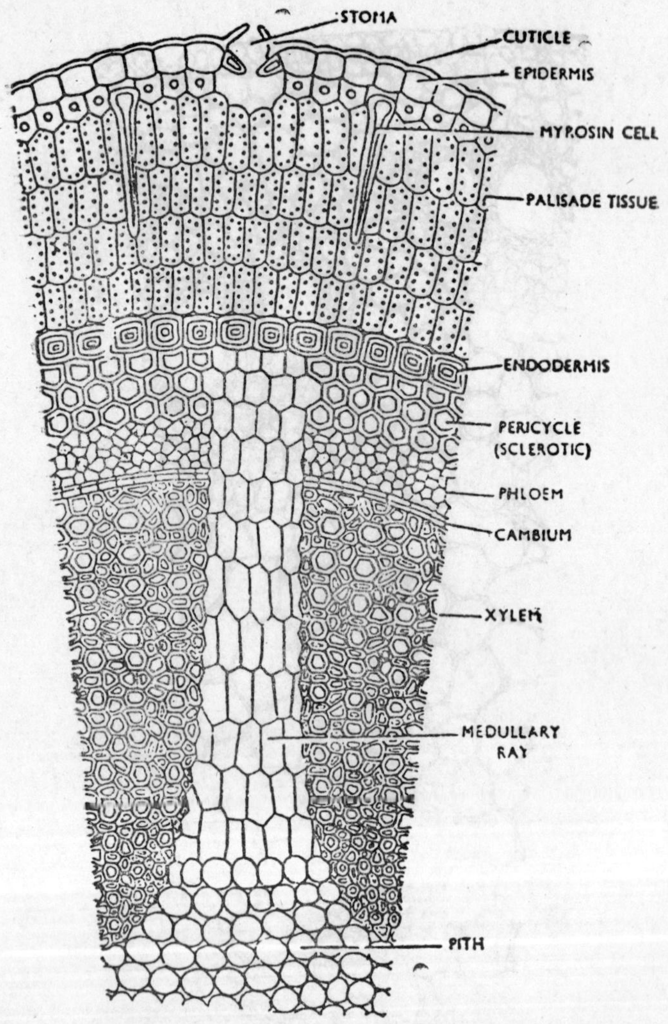

Fig. 5.4. The T S. stem of *Capparis aphylla*. A part of the same magnified.

successive stages of division, enlargement, and differentiation of cambial cells additional layers of xylem (*secondary xylem*) are laid down on the inner face of the cambium, and new layers of phloem tissue (*secondary phloem*) on its outer face. Secondary growth of woody stems is initiated during the first season of their development and continues during each growing season thereafter. Hence after a few years the great bulk of any woody stem is composed of secondary tissues.

Secondary xylem and secondary phloem also develop from the cambium in most herbaceous dicot stems, but in such species the cambium in any one stem is never active for more than one growing season.

The spring-formed xylem tissue, as viewed in cross section, is usually distinctly different in aspect from that formed later in the season. In many angiosperms the "spring" wood contains more and larger vessels, and the cell walls are generally thinner than in the subsequently formed "summer" wood. In the conifers the spring-formed tracheids are thinner-walled and of larger cross-sectional diameter than those formed later in the growing season. The transition from spring to summer wood is often a very gradual one. On the other hand, the more open xylem tissues formed each spring abut directly, upon the denser tissues which were produced during the preceding summer, thus giving rise to an abrupt line of demarcation between the zones of xylem formed in any two successive seasons. The result of this growth behaviour is that a cross section of the trunk or a branch of any tree appears as a system of concentric layers, the so-called *annual rings*, each representing an annual increment of growth. In rare cases no annual rings or more than one annual ring may be produced in a season, but usually each ring represents the xylem resulting from the activity of the cambium during one season.

In many woody species, as the xylem tissues increase in age, important changes occur in the colour, composition, and structure of the various elements, resulting in the conversion of *sapwood* into *heartwood.* As sapwood ripens into heartwood the walls of any remaining living cells of the xylem become increasingly lignified, death of these cells soon following. The water content of the tissues is generally reduced, and such compounds as oils, resins, gums, and tannins accumulate in the cells or cell walls. The darker colouration of the heartwood of most species as compared to the sapwood is caused by such accumulations.

In mature trees the heartwood becomes merely a central supporting column surrounded by a cylinder of sapwood which varies in thickness from a few to many annual layers, depending upon the species and the environmental conditions under which the tree was growing. In some species (apple, elm) the heartwood remains virtually saturated with water, while in others (ash) it becomes relatively dry. The water in the heartwood of such species as apple and elm appears to be largely static and is not directly involved in translocation.

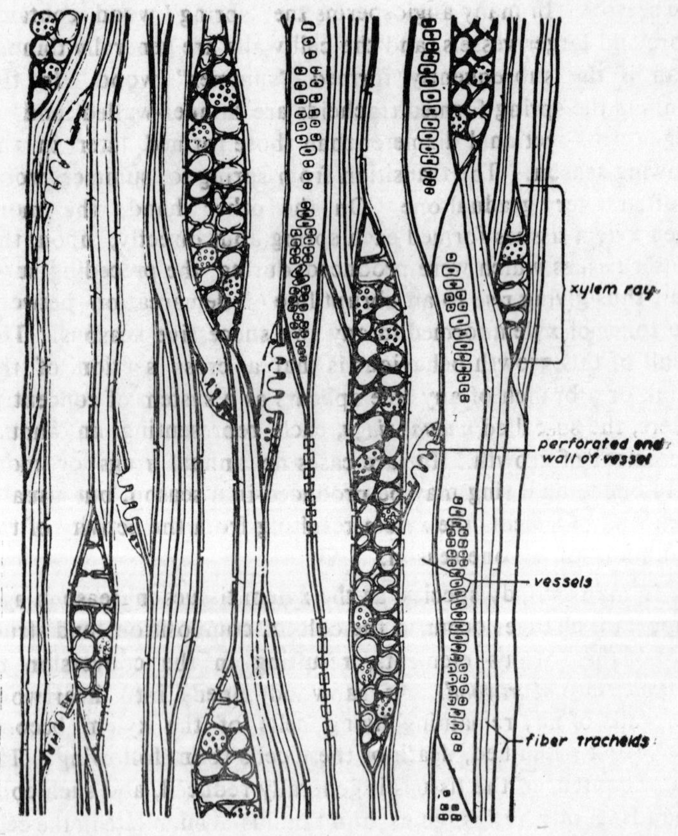

Fig. 5.5 Tangential section of a small portion of the wood of a tulip tree (*Liriodendron tulipifera*).

Coincident with the development of secondary xylem, secondary phloem tissues also develop from the cambium. Cork cambiums are also initiated in the bark which produce cork layers. Profound modifications therefore occur in the outer tissues as well as in the xylem of woody stems as they grow older.

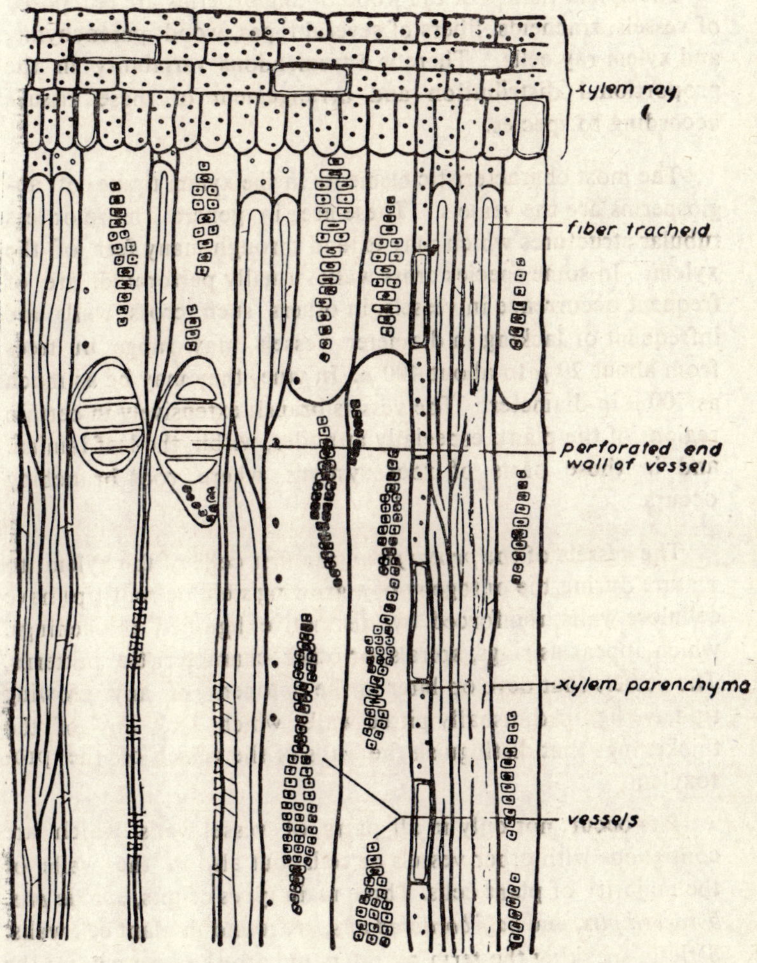

Fig. 5.6. Radial section of a small portion of the wood of a tulip tree (*Liriodendron tulipifera*).

A somewhat more detailed concept of the structure of the cells and elements of the stems of angiosperms through which the movement of water occurs is presented in Fig. 5.5 and 5.6, which represent longitudinal tangential and radial sections from the xylem of a tulip tree which may be taken as representative for this group of plants.

The xylem tissues of the wood of angiosperms are composed of vessels, tracheids, fibers of several types, wood parenchyma, and xylem ray cells. There is a tremendous variability in the proportional distribution and arrangement of these tissues according to species.

The most characteristic elements in the xylem tissue of angiosperms are the vessels. These are, in general, more or less tubular structures which may extend through many feet of the xylem. In some species cross walls, usually perforated, are of frequent occurrence in vessels, in others such cross walls are infrequent or lacking in diameter vessels may range in trees from about 20 μ to about 400 μ. In vines they may be as much as 700 μ in diameter. The vessels branch extensively in certain regions of the plant, especially at nodes, within the leaf lamina, and in those parts of root systems where root branching occurs.

The vessels of the *protoxylem* (the first cells of the xylem to mature during the ontogeny of a growing stem or root tip) have cellulose walls, reinforced by distinctive lignified thickenings, which appear as rings, spirals, or other characteristic patterns. The vessels that develop later in the ontogeny of any growing tip have lignified, usually pitted walls which lack any of the thickenings that distinguish the walls of the vessels of the protoxylem.

Pits occur, not only in all parts of vessel walls which are contiguous with other vessels or cells, but also in the walls of the majority of plant cells. Three main types of pits, *simple pits*, *bordered pits*, and *half bordered pits*, are found in plant cell walls. Strictly speaking the term pit refers only to the opening in the secondary wall of one cell, the term *pit pair* often being used to designate the two complementary pits of adjacent cells.

The stages in the development of a vessel, which is a more complex phenomenon than the formation of the other elements of the wood, are indicated in Fig. 5.7. The original cell resulting from the division of a cambium cell increases rapidly in diameter,

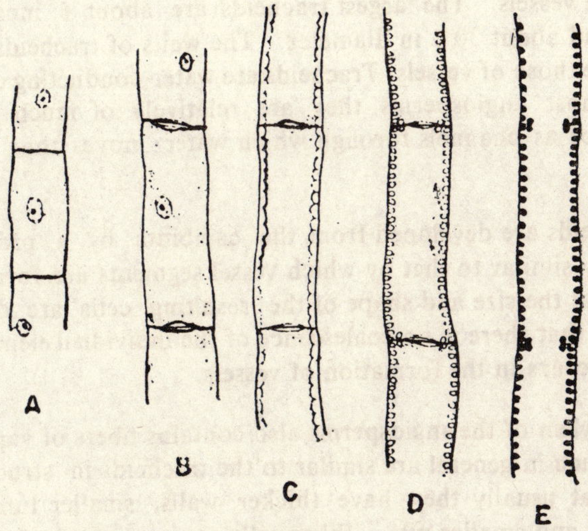

Fig. 5.7. Stages in the development of a vessel in the petiole of celery. (A) enlarged cambial cell, (B) young vessel segment showing lenticular thickenings on the end walls, (C) end walls still present in vessel segments with spiral thickenings forming on lateral walls, (D) protoplasm disintegrating in vessel segments, spiral thickening on thin portions of end walls, (E) mature vessel in which the end walls have disappeared.

simultaneously developing a rather prominent vacuole. Secondary lignified layers develop on the longitudinal walls of the *vessel segment* following which disintegration of the protoplasm and dissolution of the end walls occur. The result of this series of processes is the formation of a typical tubular, nonliving *vessel* by the coalescence of a number of vessel segments each of which has been differentiated from a single cell originating from a division of a cambium cell. In many species, especially of woody plants, vessel formation does not occur in as regular a manner as just described. In many such species the ends of the vessel segments overlap and openings develop in the side walls near the ends.

Tracheids are found in the wood of many, but not all, species of angiosperms. They are typically more or less spindle-shaped cells with thick walls which are almost always lignified. As viewed in cross section they are usually angular. Mature tracheids contain no protoplasm and hence are nonliving like the xylem vessels. The largest tracheids are about 5 mm. in length, and about 30 μ in diameter. The walls of tracheids are pitted like those of vessels. Tracheids are water-conducting cells, but in most angiosperms they are relatively of much less importance as channels through which water moves than the vessels.

Tracheids are developed from the cambium by a process essentially similar to that by which vessel segments are formed, except that the size and shape of the resulting cells are different, and that there is no coalescence of the individual elements such as occurs in the formation of vessels.

The xylem of the angiosperms also contains fibers of various types, which in general are similar to the tracheids in structure except that usually they have thicker walls, smaller lumina, and fewer and smaller pits. Fiber cells are nonliving and their walls are lignified. Because of their thick walls and small lumina the fibers do not play an important part in the movement of water through stems.

The xylem tissue of practically all angiosperms also contains wood parenchyma cells which, unlike the elements already described, remain alive for some time after their differentiation. Usually death of the wood parenchyma cells does not occur until the wood of which they are a part is converted into heartwood. Wood parenchyma cells are generally somewhat elongate, and roughly four-angled in cross section. They occur in the xylem as vertical series of cells placed end to end. The strands of wood parenchyma thus formed often extend for long distances in a vertical direction through the wood. The distribution of the wood parenchyma strands throughout the annual ring of xylem tissue varies with the species. Roshardt (1910), Overton (1911), and MacDougal, *et. al.*, (1929). All of these investigators have confirmed Strasburger's results that water would continue to ascend for some time through stems, segments of which had been exposed to one treatment of another which would kill all living cells present.

In all experiments of the type just described, it has been observed that the leaves at the top of a stem, part or all of which have been killed, sooner or later wilt and wither, although this effect is by no means an immediate one, often appearing only after several days. Proponents of the vital theories have accepted this as evidence that the living cells of a stem are essential for the passage of water through it. Dixon (1914), however, considers that this delayed lethal effect on the leaves results from either or both of two causes. Killing of the stem tissues often causes the formation of substances which plug the vessels or tracheids, thus impeding the upward movement of water. Furthermore, death of the cells in the treated regions causes the release into the conducting channels of toxic compounds which, when transported to the leaves, cause death of the leaf cells. There it still some doubt, however, whether all of the retardation in rate of translocation as a result of such experiments can be explained in these two ways. The possibility remains that the living cells of the xylem may in some way be necessary to the maintenance of the water-conducting capacity of the xylem, or even that they may contribute in some way to the upward movement of water through stems.

MECHANISM OF ASCENT OF SAP

A number of theories have been put forward from time to time to explain the mechanism of ascent of sap. The various theories can be broadly classified under three headings :

(1) Vital Theories (2) Root Pressure Theory and (3) Physical Theories.

VITAL THEORIES

All those theories which consider the living cells to be responsible for the ascent of sap are called vital theories. Despite a number of important proponents (*Godlewski*, *Bose* and *Molisch*) vital theories have not been accepted very much.

Godlewski (1884) put forward his clambering (or "relay pump") theory to explain the ascent of sap. According to him there was a rhythmic change in the O.P. of the living cells of xylem parenchyma and medullary rays which brought about a pumping action of water in an upward direction. He assumed that when O.P. of the living cell increased, water from the bordering vessel (which acted as reservoirs of water) diffused into it. The absorbed water would lower the O.P. and hence water would be forced back into the xylem element which is situated at a higher level than the living cell. Thus a sort of staircase movement of water took place in the plant. The theory of Godlewski had a considerable amount of support at one time. *Janse* (1887) supported Godlewski and showed that if the lower portion of a branch was killed, the leaves above were affected.

The theory was, however, soon discredited when *Strasburger* (1893), by means of several experiments. proved that the ascent of sap was independent of living cells and that water continued to be transported above in the plants even after the living cells had been killed by high temperature or poison treatments. Experiments conducted by *Overton* (1911) and *MacDougal* (1929) have further confirmed the results found by Strasburger. *Kurtzman* (1966) has recently shown that injection of trees with metabolic inhibitors and poisons such as picric acid and mercuric chloride fails to cause any xylem. This can be further confirmed with the help of a ringing experiment. Plants in which a girdle of bark, external to the xylem, significant change in the speed at which the sap flows. The structure of the wood also does not support this theory. Assuming xylem parenchyma to perform some kind of pumping action the living cell should be in between two xylem elements (Fig. 5.8) to give the most efficient result and not at their lateral sides as they are found in the plants.

Sir J.C. Bose (1923) strongly advocated a vital theory for the ascent of sap. He experimented to prove that the living cells of the innermost layer of the cortex, just outside the endodermis were in a state of rhythmic pulsations which caused the pumping of water from cell to cell in an upward direction.

His experimental apparatus consisted of an electric probe, a galvanometer, an electric circuit and a plant. One terminal of the galvanometer was connected with electric probe and the other at some point in the plant as shown in the Fig. 5.9. The fine needle of the probe was inserted into the stem slowly. The galvanometer showed momentary oscillations which was obviously due to local irritations. When the needle reached

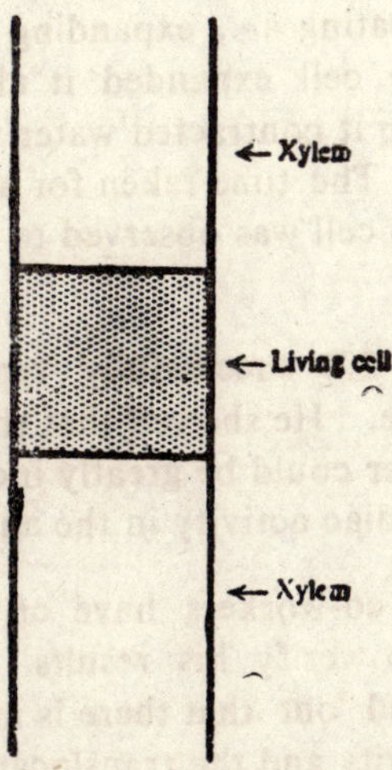

Fig. 5.8. Hypothetical arrangement of xylem element and xylem parenchyma.

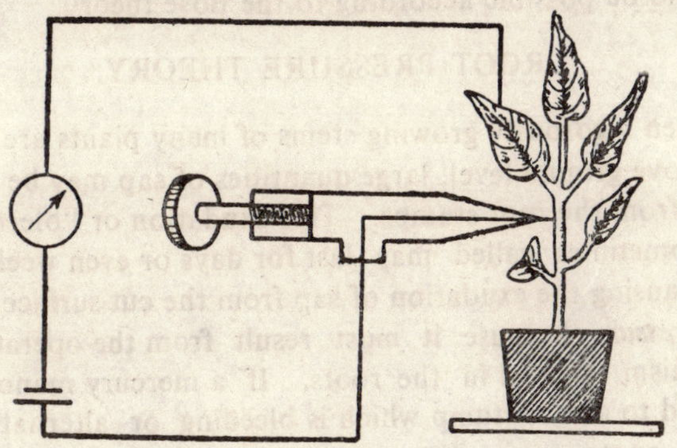

Fig. 5.9. Electric probe apparatus of J.C. Bose.

the innermost layer of the cortex, the oscillations became violent and sustained. No such movement took place on either side of this particular layer. Bose believed that the cells of this layer were pulsating *i.e.*, expanding and contracting alternately. When the cell expanded it absorbed water from the lower cell and when it contracted water was pumped into the next higher cell. The time taken for alternate expansion and contraction of each cell was observed to be from 14 second to several minutes.

Molisch (1928, 29) successfully repeated some of the experiments of Bose. He showed that the pulsatory activity of the cells of this layer could be greatly increased with drugs that cause increased cardiac activity in the animals.

Dixon and his co-workers have criticized Bose since they have been unable to verify his results. *Shull, MacDougal* and others have pointed out that there is no relationship between the pulsatory activity and the translocation of water. It has been estimated that the sap must flow through 230 to 400 pulsating cells per second to account for the normal rate of ascent of sap. This rate cannot be attained by the rate of pulsations as found by Bose. *Benedict* (1927) found that the actual rate of ascent of sap was 8,000 to 30,000 times as rapid as would be possible according to the Bose theory.

ROOT PRESSURE THEORY

When vigorously growing stems of many plants are cut off just above ground level, large quantities of sap may be seen to exude from the root stumps. This exudation or "bleeding" as it is sometimes called may last for days or even weeks. The force causing the exudation of sap from the cut surface is called *root pressure* because it must result from the operation of a mechanism located in the roots. If a mercury manometer is attached to a root stump which is bleeding or alternatively to excised roots or a plant grown in a sterile agar medium, a column of sap will rise in the manometer and its pressure can be measured. This can also be demonstrated by a simple experiment. Take a freshly cut twig of a tree and attach it to one end of a long but narrow glass tube filled with air-free water. The other end of the tube is placed in a vessel containing mercury. As water is transpired or used by the shoot, the water column will be pulled up and the mercury will rise in the tube (Fig. 5.10). Recorded root pressures seldom exceed 2 bar, although values as high as 7 bar been reported for the roots of tomato (*Lycopersicon esculentum*).

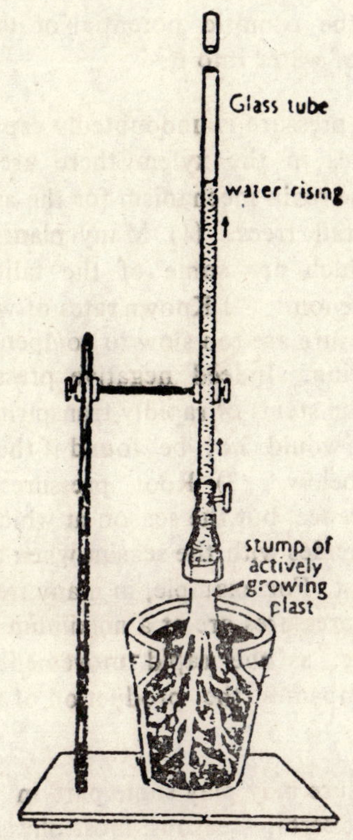

Fig. 5.10. Demonstration of root pressure.

Root pressure is though to be an osmotic phenomenon which depends on the sap in the xylem having an osmotic potential more negative than that of the soil solution, with the root tissues in between functioning as a semipermeable membrane. This interpretation is supported by the fact that root pressures do not develop if the roots are watered with a solution having an osmotic potential equal to or more negative than that of the xylem sap. However, continuous metabolic activity by the living cells of the root is necessary for the maintenance of root pressure because these are markedly influenced by factors such as oxygen supply and respiratory inhibitors, which have a more immediate effect on the respiration rather than the membrane permeability of living cells. It seems that the living cells of the root by means of energy released in respiration, actively secrete solutes into the xylem and so maintain the osmotic potential of its sap despite the continued passage of water into it.

Although root pressure is undoubtedly capable of developing positive pressures in the xylem there are several reasons why it cannot be the main mechanism for the ascent of sap in plants, especially tall trees. (1) Many plants, notably most conifers among which are some of the tallest trees, do not exhibit the phenomenon. (2) Known rates of water movement caused by root pressure are too slow to compensate for known rates of transpiration. Indeed negative pressures have often been registered in the stems of rapidly transpiring plants : such negative pressures would not be found if the sap were being pushed up from below. (3) Root pressures are strongly seasonal in occurrence, but the season at which they are most evident usually coincides with the season when the rates of sap movement are lowest. For example, in many trees which 'flush' once a year, root pressures are at a maximum just before, and not at the same time, as the rapid movement of sap up the trunk which accompanies the production of the new crop of leaves.

While root pressure may play some part in water transport in some species at certain seasons, most physiologists tend to regard the phenomenon as an indirect consequence of metabolic activity by the living cells in the root. The explanation for the ascent of sap therefore seems to demand the operation of a mechanism by which water is pulled upwards from above.

PHYSICAL FORCES THEORIES

Imbibition Theory

Unger (1868) *and Sachs* (1878) proposed that upward movement of sap and water is caused due to the capacity of imbibition possessed by materials in the cell walls of xylem and phloem tissues. It is know that, as a result of imbibition, very high imbibition-pressures (100 to 1000 atm.) can be produced which may apparantly cause the ascent of sap, but in reality this pressure is exerted equally on all sides and thus nullified. Further, rate of diffusion of water in an imbibing colloid is much less compared to the speed of sap ascent in xylem. It is also known that water actually rises through the lumen of xylem elements rather than along their walls, as was shown by experiments in which xylem ducts are blocked or plugged with air-bubbles or oil or gelatin, thus stopping the ascent of sap also. Therefore, this theory is now discarded.

Capillarity Theory

Some workers advanced the view that the ascent of sap in the lumen of xylem tracheids and vessels was due to the forces of capillarity. We know that on standing a capillary tube (of small diameter and solid walls) in such a way that its lower open end dips in a liquid, we see that this liquid often rises up in the lumen of tube and attains some level at equilibrium. This phenomenon depends upon the property of surface tension of the liquid, and also the adhesive-tension between the liquid and the internal surface of the wall of the tube. On the same basis ascent of sap in xylem tracheids and vessels has been suggested. But this view has an inherent weakness because the narrowest vessels found in plants can allow a maximum capillary rise of sap by about 150 centimeters only. Further the speed of ascent of sap is actually found to be more in broader tracheids than in narrower ones, and this is against the principles of capillarity. Therefore, the capillarity theory is also not valid.

Cohesion theory of the ascent of sap (Transpiration-pull theory)

The cohesion theory, first propounded by Dixon and Joly in 1894, supposes that the ascent of sap does not involve any expenditure of energy on the part of the plant because the motive force comes from solar heat which, by supplying the latent heat of evaporation, causes water to evaporate from the leaves. During foliar transpiration water evaporates into the intercellular air spaces of the mesophyll from the water-saturated cell walls, in which water is present in minute pores too small to be seen under the light microscope. Evaporation of water from the surface of a transpiring cell causes the menisci in these submicroscopic pores to be pulled inwards, and this has the effect of imposing a negative pressure on the water so that the water potential (ψ) of the cell decreases. As a result water diffuses into this cell from the adjacent cells, whose water potentials therefore also decrease. A gradient in water potential is thus set up, and this is transmitted from cell to cell until the cells abutting directly on the xylem elements in the vascular bundles of the leaf veins are reached. These last cells then withdraw water from the vessels or tracheids, which contain water with a low solute concentration and therefore a high water potential (*i.e.*, a ψ of nearly zero). The faster water is being transpired, the steeper will be the gradient in water potential and hence the faster will water diffuse from the xylem elements towards the evaporating surfaces. When a mesophyll cell absorbs water from an adjacent tracheid or vessel it does not empty the water-conducting unit. Since this is entirely filled with water, the only way to empty it would be to break the column of water. Such a break could only occur by the creation of a vacuum as a result of *either* the water pulling away from the cell wall *or* the water molecules pulling apart from one another.

Molecules of water, although ceaselessly in motion, are strongly attracted to one another and to any perfectly clean solid surface with which they are in contact. These properties of water are known as cohesion and adhesion respectively. They form the basis of the so-called cohesion theory of the ascent of sap. According to this theory withdrawl of water from the xylem causes a tension or negative pressure to develop in the water and this has the effect of decreasing its water potential. It is supposed that as transpiration goes on and water is withdrawn from the xylem, the water columns are stretched into taut threads but do not break because of the mutal attraction (cohesion) between the water molecules. Each column of water behaves like a tiny steel wire which is in a state of tension. Because they are saturated with water, cross-walls in the xylem conduits do not interfere with the continuity of the water columns. The tension which develops at the upper end of the water columns in the transpiring leaves is thus transmitted all the way back to their lower ends in the roots. The water columns are not pulled away from the surrounding walls because of the attraction (adhesion) between the water molecules in the columns and the structural molecules forming the fabric of the walls.

Due to transpiration pull the water in the conducting elements is thrown into a state of tension. This tension is transmitted down the columns of water in the xylem elements through the stem right upto the absorbing region just back of root tips, with the result, that the whole column of water is lifted up like a rope, just as a weight (say a bucket of water drawn from a well) tied to the lower end of a rope is pulled up by a force applied at the upper end of the rope. The factor which is responsible for the downward transfer of this pull and which maintains the unbroken continuity of the vertical column of water in the xylem ducts is the cohesive force of water. By cohesion is meant the force of mutual attraction between the water molecules. The magnitudes of this cohesive force are very large, ranging upto 350 atm. This means that a force greater than 350 atm. is required to rupture a column of water. Dixon estimated the cohesive force of the sap of *Ilex acquifoleum* at 207 atm. Due to the enormous cohesive force, the column of water, when pulled from above, holds together like a cord or wire and is hauled up as one mass.

Water has to move up a maximum height of about 400 feet. This requires a pressure of about 12 atm. which represents the maximum stress on the water column in the conducting tissues. If to this is added the resistance which the water encounters in its upward movement through the xylem and in crossing the living cells of the root and the mesophyll, the total stress on the water column would not exceed 100 atm. by even a liberal estimate. This represents the maximum cohesive force which is required for the maintenance of unbroken continuity of the water column. The estimated cohesive force of plant sap, as stated above, is far greater than this and enables the water column to hold together against all stresses.

The above mechanism of the ascent of sap was first put forward by Dixon and is the most widely accepted among plant physiologists. In essence it postulates the operation of three factors ; the primary factor is transpiration ; the second factor is the movement of water from the xylem to the mesophyll cells, replacing that lost through transpiration, thus setting up a pull on the water in the xylem ducts ; the third factor is the cohesion of water columns in the xylem ducts which transmits the pull from the mesophyll cells all the way down to the roots. Since the mainstay of this theory is the cohesive force of water, it is termed the theory of cohesion.

EVIDENCES IN SUPPORT OF DIXON'S THEORY

(1) Osmotic pressure of parenchyma cells of upper leaves in various plants has been found to be between 10–20 atmospheres. Since 1 Atm. pressure can raise water up to a height of about 32 feet, the usual tension of 10—20 Atm. developed in leaves of many plants is sufficient to explain the rise of water in the tallest known plants (about 300 ft.).

(2) Very small amount of energy is required for raising water in plants, *e.g.*, for raising 1 ml. of water to 100 metres it requires only 0.5 calories of heat energy.

(3) The force of cohesion between water molecules has been measured between 30–200 atmospheres by different workers which is sufficiently large to prevent the breaking up of the water column under the transpiration pull.

(4) Usually air bubbles are not present in any appreciable number in xylem ducts. Even if they are present they do not disrupt the water column because of considerable force of adhesion between the walls of xylem vessels and water, and the

air bubble is not able to expand much and it rises up with water column if it enters the xylem.

(5) When water is allowed to evaporate from a porous pot containing water, a tension on the water inside the pot is developed which is comparable to the tension developed in leaves on transpiration.

Experiment

The ascent of sap can be demonstrated with the help of physical instruments. Water is allowed to evaporate from a porous pot through which a long delivery tube containing water is passed. The other end of the tube is dipped in a reservoir of mercury. On evaporation from porous pot water moves up in the delivery tube followed by mercury from the reservoir. The tension developed in the delivery tube on evaporation is transmitted to the mercury below. If the porous pot is replaced by a freshly cut twig the rise in delivery tube continues (Fig. 5.11).

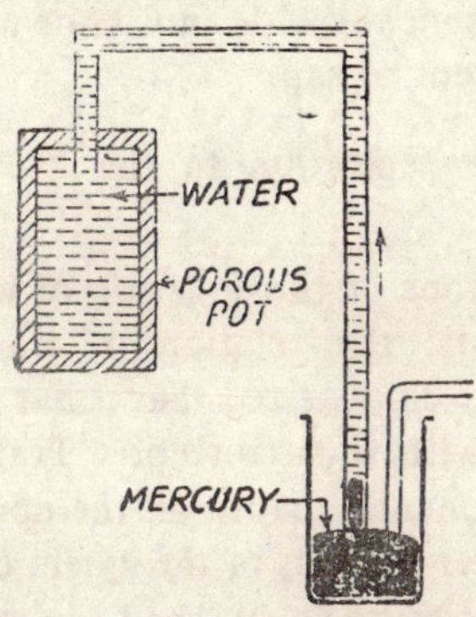

Fig. 5.11. To show the ascent of water experimentally.

OBJECTIONS TO DIXON'S HYPOTHESIS

(1) When the ascent of sap is demonstrated in laboratory the rise continues only for some time after which it stops. It is questioned how the plant is able to maintain the continuity of ascent of sap.

(2) There is a lot of variation of temperature during day and night. With increase in temperature during day there is

tendency of more of gases entering the water column from soil along with water. As such there are many chances of breaking up of the water column.

In spite of the objections plant physiologists at present agree that Dixon and Jolly's hypothesis can explain the ascent of sap in plants to a considerable extent. No doubt other forces like root pressure etc. also play some part. Perhaps the rise of water is the net result of all the forces so far accounted. An examination of the histology of stem will show that every conducting cell is intimately in contact with one or more living cells suggesting that the presence of living cells is necessary to upward translocation of water. Where the xylem consists of only fibres and vessels, the parenchyma cells ensheath the conducting vessels, they are practically absent among the non-conducting fibres. A careful study of the problem shows that although unexplainable the ascent of water ultimately depends upon the behaviour of the protoplasm. It must, however, be admitted that at present there is no known theory which can truly explain the ascent of sap.

The Relation of Transpiration to the Movement of Water Through the Plant

In most discussions of the cohesion of water theory of the rise of water in plants, the relation of transpiration to this process is so greatly emphasized that it has come to be almost indelibly associated with this concept. Transpiration is not, however, the fundamental cause of the upward movement of water in plants. Water ascends in the xylem ducts only when an adequate diffusion-pressure deficit has been created in the water of the vacuoles or the walls of cells in organs in the upper parts of the plant. Since evaporation of water from the walls of the mesophyll cells is the most frequent cause of such diffusion-pressure deficits, the process of transpiration has been generally linked in discussions with the mechanism of the ascent of sap. It is only because of its effect in increasing the diffusion-pressure deficit of the water in the mesophyll cells that transpiration sets in motion the entire train of water through the plant. Any other process which results in an increase in the diffusion-pressure deficit in the cells at the terminus of any plant axis may also induce the translocation of water. Upward movement of water often continues during the night hours after transpiration has virtually ceased. This lag results from the residual high diffusion-pressure deficit of the leaf cells at the end of the daylight period. Water will continue to enter these cells until they reattain the maximum turgidity of which they are capable under the conditions prevailing. Similarly movement of water will occur into any rapidly growing item tip or fruit because the binding up of water in certain phases of the growth process creates a diffusion-pressure deficit in the cells of such organs, thus inducing the migration of water toward such centers of growth activity.

At the present time most plant physiologists are agreed that the cohesion theory is a correct representation of the principal mechanism by which water is transported through plants, be they the tallest of trees or herbs only a few feet in height. It is also generally agreed that root pressure plays a minor role in this process at least under certain conditions and in some species. There is no doubt that living cells are involved in the phenomena generally described under the term root pressure, and there is at least a possibility that living cells of the xylem are in some way essential to the maintenance or operation of the cohesion mechanism.

Rates of Movements of Water Through Plants

The rate at which water ascends through the xylem ducts may vary from a movement so slow as to be almost imperceptible to speeds of at least 75 cm. per minute and probably higher. In general, daily variations in the rate of ascent of water approximately parallel daily variations in transpiration rate ; the maximum velocity of the transpiration stream is often not attained, however, until several hours after the peak of transpiration.

Lateral Movement of water

A cell-to-cell lateral movement of water in a radial direction undoubtedly occurs along the vascular rays in the stems of most species of plants. In woody stems there is probably also a lateral movement of water around the stem in a tangential direction. Except in trees in which the "grain" of the wood is twisted, the conducting vessels on one side of the tree generally connect with branches on that side of the tree at their upper extremity, and with roots on the same side of the tree at their lower extremity. If no lateral movement of water occurs in woody stems, it would be expected that removal of the roots from one side of a tree would result in a dearth of water, or perhaps even death of the leaves or branches on that side of the tree.

Experiments have been performed on apple, peach, oak, and other woody species in which the roots on one side of the plant were removed in an attempt to ascertain whether or not lateral movement of the water occurs. Although the water content and growth of the plants treated in this manner diminished, there was no difference in the moisture content of the leaves on the two sides of the tree. Neither did the leaves on the side from which the roots had been removed show any greater tendency to wilt on clear, warm days than those on the other side. These results indicate very strongly that lateral movement of water occurs in woody stems, and that the water conductive system of plants acts as a unit system.

Downward Movement of water

There are a number of experiments on record which indicate that downward movement of water can occur in stems. For example, Dixon (1924) found that, if the tip of the leaf of a potato plant be cut off under an eosin solution, the liquid will descend through the xylem tissue of the leaf, petiole, and stem, and eventually reach the underground organs. In general, however, such an effect is to be expected only when an internal water deficit exists in a plant.

The cohesion theory of the movement of water will account equally well for the conduction of water in either an upward or a downward direction through the plant. Whenever conditions are such that the diffusion-pressure deficit of the cells of an organ, which is basally situated relative to another, is greater than that of the cells of a more nearly apical organ, a reversal in the direction of the movement of the water will occur.

Sap Flow in the Maple

Sap flow from cuts or bore holes in maple trees occurs only during the late winter and early spring months when cold nights alternate with days upon which the temperature rises above freezing. The flow is largely or entirely confined to the daylight hours. A single sugar maple tree will generally yield from 25 to 75 liters of sap per season, the sucrose content of which is usually between 2 and 3 per cent. In other woody species that exhibit a prefoliation sap flow, such as birch and grape, exudation usually will not occur until later in the spring at a period when freezing temperatures have become uncommon. Xylem sap flow in such species is a root pressure phenomenon and continues during both day and night hours.

Sap flow in the red maple is not a result of root pressure because sap exuded as freely from bourings into the xylem of cut trees with their bases immersed in water as from similar borings into rooted trees during the late winter and early spring whenever the necessary alternation of cold nights and warm days occurred. These investigators suggest that the crystallization of water within the trunk and branches, especially in the outer layers of the xylem, indirectly causes upward movement of water through the tree. As a result of the conversion of some of the water present into ice, the osmotic pressure of the cells increases, which in turn induces upward movement of water through the xylem. It must also be assumed that at least some of the roots are in soil layers warm enough to permit the absorption of water. When, under the influence of daytime rising air temperatures or direct incolation, the temperature of the trunk and branch tissues reaches or exceeds the freezing point, the ice melts and the tissues become flooded with sap. This sap exudes through any sort of an incision made into the outer xylem of the tree.

6

Loss of Water in Plants

Water in very large amounts is absorbed from the soil by the roots. Only a small fraction of this enters into the composition of the plant or the manufacture of carbohydrates. Most of it is lost by the aerial parts or the plant in the form of water vapour. This evaporation of water from the surface of the aerial parts of the plant is known as *transpiration*. Transpiration may be easily demonstrated by placing a potted plant under a bell jar (Fig. 6.1). The soil in the pot is well watered and the pot is covered with rubber sheeting or the soil and the walls of the pot are coated with paraffin to prevent evaporation

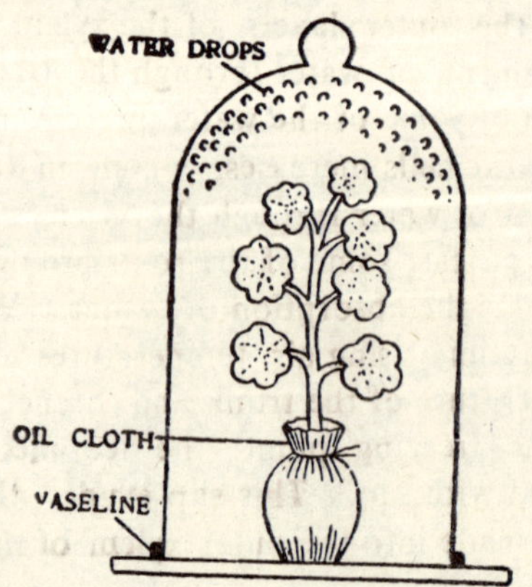

Fig. 6.1. Demonstration of Transpiration.

from them. In a short time, drops of water will appear on the inner surface of the bell jar. This water has come from the aerial parts of the plant.

The amount of water lost by transpiration is surprisingly large. It exceeds the final dry weight of the plant by several hundred times. A corn plant during its life of a few weeks may transpire enough water to fill a barrel and a medium sized elm tree may similarly lose more than a ton of water every day. Ball has estimated that the amount of water transpired from an Egyptian cotton crop is about 50 tons per acre per day, or 3 pints per plant. It has also been estimated that for every ounce of dry matter in the aerial parts of a crop plant, as much as 20-50 pounds of water are transpired.

The rate of transpiration. Numerous estimates have been made of the quantity of water lost by transpiration. The rate of water loss in a number of flowering plants, during the middle hours of the day, averages 1.25 grams of water per 100 square centimeters of leaf surface. The average daily loss of a red maple 53 feet high has been estimated at more than 6 gallons, and in the 150 days of the growing season this amounts to more than 900 gallons. A single corn plant may transpire more than 4 quarts of water in a day. An acre of corn gives off more than 300,000 gallons of water during the growing season. The rate at which water is lost from a plant varies greatly during the day and season. It is governed by leaf structure, the nature of the protoplasm of the leaf cells, and several environmental factors.

The mechanism of transpiration. Transpiration, except for certain differences resulting from the structure of the leaf, is essentially identical with evaporation from a free water surface, and the physical processes involved are the same. The wet walls of the mesophyll cells exposed to the intercellular spaces within the leaf provide the evaporating surface. When the stomates are closed, as at night or under low light intensities during the day, the water vapour within the intercellular spaces of the leaf will tend to reach equilibrium with the water in the wet cell walls of the leaf. When the stomates are open, the concentration of water vapour in the intercellular passec is greater than in the outside air (unless the outside air is also saturated). A greater number of water molecules then leave the leaf through the stomates than enter in any given period, and water is lost from the plant.

The mechanism of foliar transpiration

The mechanics of foliar transpiration can be adequately discussed only in reference to the anatomy of the leaves from which it occurs (Fig. 6.2). It should be recalled that the vacuoles of all of the living cells of a leaf are filled with water, which also

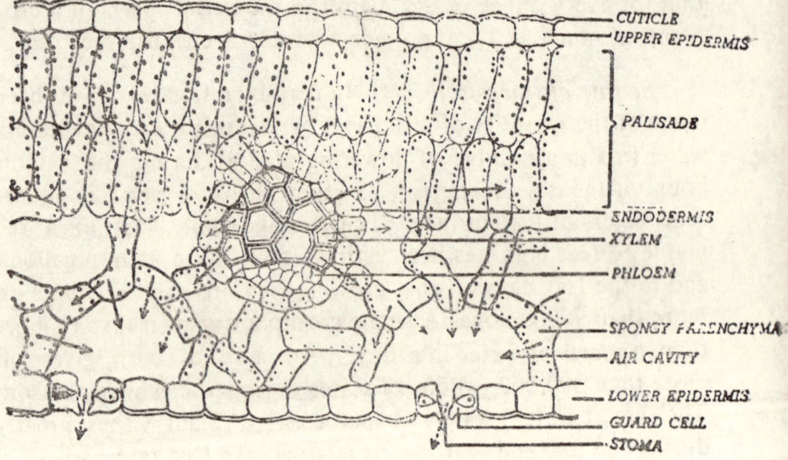

Fig. 6.2. Cross-section through a typical leaf showing the movement of water during transpiration. Solid arrows indicate the movement of liquid water and dotted arrows of water vapour.

saturates the protoplasm. The wet cell walls, this water being supplied to the leaf cells through the water conducting tissues of the vascular bundles. Hence evaporation of water will occur from these wet cell walls into the internal atmosphere of the intercellular spaces just as it will occur from any wet surface into the surrounding air. The intercellular spaces constitute a connected system, raminding throughout the leaf. Under certain unusual conditions the intercellular spaces can become injected with liquid water but under all normal conditions they are occupied by air.

If the stomates are closed the only effect of evaporation from the mesophyll cell walls will be the saturation of the entire volume of the intercellular spaces with water-vapour. When the stomates are open, however, diffusion of water-vapour may occur through them into the outside atmosphere. Such outward diffusion will always take place unless the atmosphere has a vapour pressure equal to, or greater than, that of the intercellular spaces, a condition which does not commonly exist during the daylight hours. The rate of such diffusion will depend principally upon the excess of the vapour pressure in the leaf over that of the atmosphere, although the "diffusive capacity" of the stomates is also an important factor. The process of stomatal transpiration therefore involves *evaporation* from the cell wall surfaces bounding the intercellular spaces and the *diffusion* of this water-vapour from the intercellular spaces into the atmosphere through the stomates.

One side of every epidermal cell on a leaf is also exposed to the atmosphere. Evaporation of water occurs into the atmosphere directly from these cell surfaces. The surfaces of practically all aerial leaves are covered with a layer of wax-like substance known as *cutin* (Fig. 6.3). This is not readily permeable to water and hence reduces transpiration directly through the walls of the epidermal cells to a magnitude much

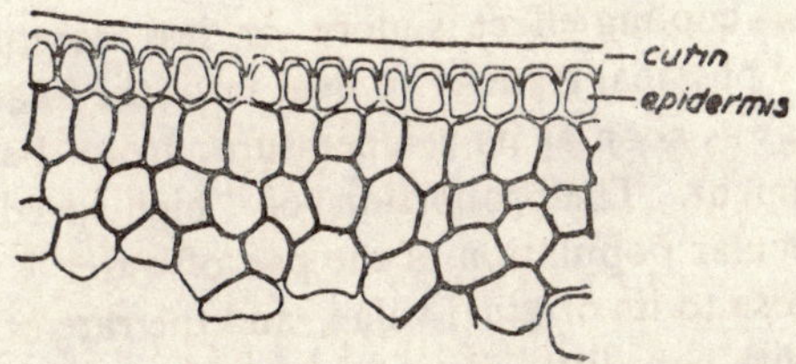

Fig. 6.3. Cutin layer on the upper epidermis of a leaf of *Clivia nobilis*.

less than it would be, were there no such layer present. The thickness of the cutin layer varies with the species of the plant and the environmental conditions under which the leaves have developed. The layer of cutin is usually thicker on leaves which have developed in bright sunlight, for example, than on leaves of the same species which have developed in shade. Even in leaves which are heavily coated with cutin, some cuticular transpiration occurs, possibly largely through tiny rifts in the cutin layer. In most species of plants of the temperate zone less than 10 per cent of the foliar transpiration occurs through the cuticle, the remainder being stomatal transpiration.

Evaporation. The fact that transpiration may be regarded as essentially a modified form of the process of evaporation makes desirable a fuller consideration of the dynamics of this process. When an open pan of water is exposed to the atmosphere, the level of the water in the pan slowly drops. Evidently water molecules are being slowly lost into the atmosphere. All of the molecules in a mass of water are not traveling with the same velocity, although for any given temperature the *average* speed of all of the molecules in the liquid mass is a constant. Some of the moleoules in the liquid water attain sufficient momentum to overcome entirely the attractive forces holding them in the liquid, and escape into the surrounding atmosphere in the form of vapour. The swiftest molecules present are most likely to be able to overcome the attraction of the other water molecules, therefore they are the most likely to be lost from the liquid. During evaporation any body of water is thus slowly being depleted of its more rapidly moving molecules. The residual water becomes progressively richer in relatively sluggish molecules ; in other words, it becomes cooler. This cooling effect is more or less completely offset, however, by physical transfer of heat into the water from its surroundings as soon as its temperature drops below that of the environment. The proportion of "high-speed" molecules in the molecular population of the pan of water is thus maintained at close to its original value, and the rate of evaporation continues with very little diminution.

In the preceding paragraph we have focused our attention only on the water molecules which escape from the liquid. Water-vapour molecules are also returning to the liquid from the atmosphere during the process of evaporation. If a water-vapour molecule in the atmosphere above the evaporating, surface, particularly one of the more sluggish ones, strikes the surface of the water at the proper angle and with not too great a velocity, it will be held there by the attractive forces exerted by the liquid water molecules, and again become part of the liquid water. "High-speed" molecules, on the other hand, are much less likely to be captured by the attractive forces exerted by the molecules at the surface of the body of liquid. They impinge upon the bounding film of the liquid with such velocity that, unless they hit that surface at right angles or nearly so, they usually glance off along a new pathway.

This picture of the kinetics of the evaporation process holds for any evaporating surface whether it be the exposed surface of a body of water, a moist piece of cloth, paper, or porous clay, or the mesophyll cell walls of a leaf. Ordinarily the term evaporation is used to refer only to conditions in which the rate of escape of molecules exceeds their rate of return, and its use will be restricted to this sense in this discussion.

If the air above the water surface is confined, as, for example, when a dish of water is covered by a bell jar, the number of water-vapour molecules in the confined space will gradually increase as a result of evaporation from the surface of the water. Since their movement is a random one the water-vapour molecules will be continually colliding with walls of the container, each other, the molecules of other gases present in the air, and surface of the liquid water. Some of those which strike the surface of the liquid will be held there by inter-molecular attractive forces. As the concentration of water-vapour molecules in the air increases, the number which plunge back into the water in any unit interval of time increases. Eventually a dynamic equilibrium will be attained at which the number of molecules leaving the surface and the number returning to it in a unit time will be equal. At this point the air will be *saturated* with water-vapour and evaporation will no longer be occurring.

The water-vapour molecules exert a definite pressure against the walls of the container and the surface of the water. This is known as *vapour pressure*. In the illustration given in the preceding paragraph, the vapour pressure of the water increases progressively until the saturation point is reached. The vapour pressure of water under the conditions of such a dynamic equilibrium may conveniently be termed the *saturation vapour pressure*. The vapour pressure of a liquid is usually expressed in terms of millimeters of mercury. The saturation vapour pressure of water at 20°C., for example, is 17.54 mm. Hg. This means that at this temperature the water-vapour exerts a pressure equal to the pressure exerted by a column of mercury 17.54 mm. high. The saturation vapour pressure of any liquid is independent of the area of the evaporating surface, but increases with increase in temperature. In physics and chemistry the saturation vapour pressure is usually referred to simply as "vapour pressure," but in biological work it is necessary to distinguish between two different senses in which this latter term is used.

Botanists are often interested in measuring the rate of evaporation under a given set of conditions in connection with studies of transpiration and of plant distribution. Many measurements of evaporation rates have been made with instruments called *atmometers*. These consist essentially of surfaces of porous clay moulded in the form of hollow cylinders or spheres. Water evaporates from such surfaces in essentially the same way that it evaporates from a free surface of water. Atmometers are attached to a reservoir of water, and are usually provided with mercury valves which prevent absorption of rain. The loss of water from such instruments can be determined either by measuring the decrease in volume of water or the loss of weight of the instrument. Details of atmometric technique are discussed by Livingston (1935).

Evaporation from Moist Surfaces

As a physical process, transpiration by plants follows the rules governing the *evaporation* of water from moist surfaces.

An exposed water surface gives off more water vapour per unit time and area, the steeper the *gradient of vapour pressure* between surface and air. A vapour-pressure gradient arises when the water-vapour content of the air at the evaporating surface is greater than that at some distance from this surface. This is always the case when the evaporating surface is adequately supplied with water and is warmer than the air. Strong irradiation *warms the surface* and thus leads to a sharper vapour pressure gradient and to more rapid evaporation. In this way transpiration (and thus movement of water through the plant) occurs even under conditions of high humidity—in fact, even when the air is saturated with water. This is an important factor in the transport of minerals, particularly in humid tropical regions.

Evaporation under conditions of unlimited water supply is called *potential evaporation* (E_p). Potential evaporation is determined by measuring the weight of water evaporated under standard conditions, or is computed from known values for energy balance and mass air exchange. Formulas for applications in agrarian meteorology have been developed by C.W. Thornthwaite and H.L. Penman. In the area of plant ecology standardized evaporating devices (evaporimeters, or more correctly *atmometers*) are used to estimate potential evaporation. Atmometer measurements are useful for relative comparisons, but are not suitable for analysis of the physical process of evaporation. In arid subtropical regions potential evaporation reaches 10—15 mm . d^{-1} (that is, 10—15 l of water per m^2 surface area per day), in regions with Mediterranean climate during the dry season 5—6 mm . d^{-1}, and in the equatorial zone 3—4 mm . d^{-1}. In the temperate zone, potential evaporation of fine summer days can amount to as much as 4 mm . d^{-1}, whereas the average over the growing season is around 2 mm . d^{-1} and in winter, only 0.1—0.2 mm . d^{-1}.

The *actual evaporation* from moist surfaces (soil, the cell walls of plants) is usually less than the potential evaporation, because the water is almost never replenished as rapidly as it is lost.

The Site of Transpiration in the Plant

Water evaporates from the entire outer surface of a plant, and from all interior surfaces that come into contact with air. In thallophytes the outer surfaces of the thallus are involved, whereas in vascular plants external transpiration occurs by way of the cutinized epidermis (*cuticular transpiration*) and suberized surfaces (*peridermal transpiration*). Within the organs of the plant water evaporates from the surfaces of cells bordering on intercellular air space. In this case the water first is converted from the fluid phase to the vapour phase, and then the water vapour escapes through the stomata (*stomatal transpiration*). From the surface of the plant the vapour diffuses into the adjacent layer of air and thence into the open air. The movement of water vapour from the evaporating surface to the open air is brought about by diffusion.

Transpiration as a Diffusion Process

Transpiration can be considered as a process of diffusion. Thus *transpiration intensity* is proportional to the difference between the *concentration of water vapour* at the evaporating surfaces and the water-vapour content of the atmosphere ($gH_2O \cdot m^{-3}$), and is inversely proportional to the sum of the *diffusion resistances* Σr :

$$Tr = \frac{\Delta C}{\Sigma r}.$$

Transpiration intensity is expressed in terms of water loss (mg H_2O) per unit time (usually minutes or hours) and unit surface area (usually dm_2^2, the total leaf area—*i.e.*, both upper and lower sides). It can be determined directly—for example, gravimetrically by measuring the water lost during a certain time from plants or parts of plants – but it can also be computed by equation above if the humidity gradient ΔC (between the transpiring surface and the surrounding air) and the leaf diffusion resistance Σr are known. In *leaves* the diffusion resistance to water vapor is given by the boundary-layer resistance and the cuticular and stomatal diffusion resistances. At *surfaces lacking stomata* the diffusion of water vapour is limited by the boundary-layer resistance and by cuticular or peridermal diffusion resistances.

The *boundary-layer resistance* r_a to water-vapour transport, like that to CO_2 exchange, is very much dependent on the size, shape and surface properties of the leaves, as well as on wind velocity. Small leaves (*e.g.*, conifer needles) rarely have r_a greater than 1 s . cm^{-1}, even in still air ; in large leaves such as those of banana plants the boundary-layer resistance in still air can be 3 s . cm^{-1} or more. As air movement increases these differences are progressively reduced (Fig. 6.4). At wind velocities of about 2m . s^{-1} or more the boundary-layer resistance is in general less than 0.5 s . cm^{-1} and is no longer an appreciable factor as compared with stomatal diffusion resistance.

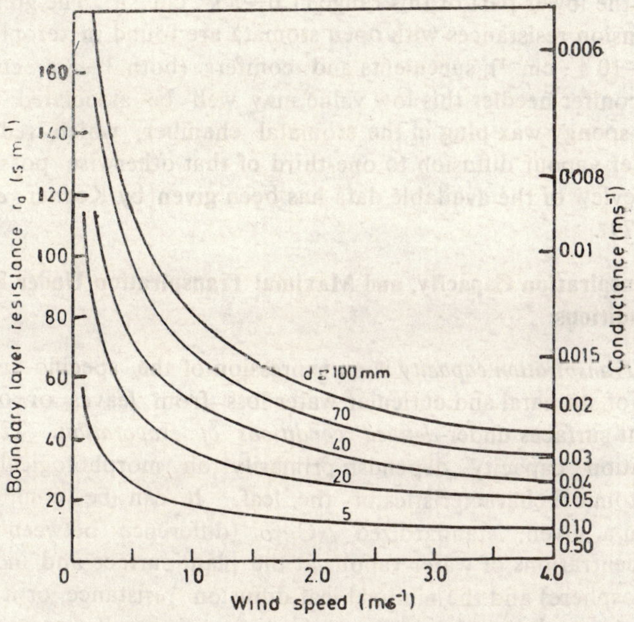

Fig. 6.4. Boundary-layer resistance over leaves of different sizes, as a function of wind velocity. *d* indicates the leaf dimension parallel to the direction of air movement. From Grace (1977). See also the classical study of the effect of wind on transpiration by Gaumann and Jaag (1939).

The *stomatal diffusion resistance* r_s depends on the degree of opening of the stomata. The *minimal stomatal diffusion resistance* is determined by anatomical features such as the size, structure, arrangement and density of the stomata of the species concerned ; it can thus be used to characterize the maximal transpiration capacity of the species. In the great majority of herbaceous plants the *minimal leaf diffusion resistance* (referred to the total surface area, both upper and lower sides of the leaf) lies between 0.7 and 5 s . cm^{-1} ; in herbaceous crops, including cultivated grasses, it ranges from 0.7 to 4.5 s . cm^{-1}. The corresponding values for wild grasses are 1.2—6.6 s . cm^{-1}, for herbs of sunny habitats 0.8—3.3 s . cm^{-1}, and for shade herbs 3—8 s . cm^{-1}. Broad-leaved woody plants, both deciduous and evergreen, have minimal leaf diffusion resistances in the range 1.6—8 s . cm^{-1} ; fruit trees account for the lower part of this range (1.6—3 s . cm^{-1}). The greatest diffusion resistances with open stomata are found in xerophytes (2—10 s . cm^{-1}), succulents and conifers (both 3—8 s . cm^{-1}). In conifer needles this low value may well be associated with the spongy wax plug in the stomatal chamber, which reduces water-vapour diffusion to one-third of that otherwise possible. A review of the available data has been given by Korner *el al.* (1979).

Transpiration Capacity, and Maximal Transpiration Under Field Conditions

Transpiration capacity is an expression of the specific intensity of stomatal and cuticular water loss from leaves or other plant surfaces under *defined conditions of evaporation.* Transpiration capacity depends primarily on morphological or anatomical characteristics of the leaf. It can be computed from a given, standardized ΔC_{H_2O} (difference between the concentrations of water vapour at the plant surface and in the atmosphere) and the minimal leaf diffusion resistance, or it can be measured directly under standard conditions. Transpiration capacity can also be expressed by the specific *relative transpiration* E/E_p, as suggested by B. E. Livingstone ; here E is the amount of water transpired and E_p is the potential evaporation at the same time and place. The *maximal transpiration* is the average peak transpiration rate measured *in the natural habitat*. In order to evaluate specific water consumption it is important to know the transpiration capacity ; data on maximal transpiration, like that given in Table 6.1, are useful in estimating the water consumption typical of a particular habitat.

All these measures express water loss per unit area of the transpiring leaves. If the total surface area or the mass of the foliage of a plant is known, they can be used to estimate the water consumption of the entire plant (Table 6.2). In the case of trees, extrapolation from the transpiration of a twig to that of the whole crown provides only a rough estimate, because the movement of water vapour away from the transpiring surfaces is impeded by the high degree of overlapping within the crown. To avoid such errors, it is preferable to use procedures for more direct measurement of overall transpiration—for example, by weighing pot plants or by calculations based on measurements of water flow through the xylem.

The *specific transpiration capacity* of leaves is greatest in swamp and floating plants. Herbaceous species in sunny habitats transpire most strongly, shade herbs only half as strongly, and trees and sclerophyll plants more weakly still. Plant organs also vary in the rate at which water is lost from the stomata ; the stems give off considerably less water than the leaves, as do the flowers (which have very few stomata). The *maximal rates* of transpiration of plants in the field follow fairly closely the specific transpiration capacities. The only conspicuous departures from this rule are the trees of tropical forests and desert plants. Because of the high air humidity in their habitats, the plants of tropical rain forests and particularly elfin forests transpire at moderate intensities. By contrast, desert plants exposed to intense radiation and dry air evaporate off a great deal of water as long as their stomata are open, even if their transpiration capacity is low.

Cuticular Transpiration

Cuticular transpiration can be regarded as diffusion through a hydrophobic medium, since the water molecules must pass

Table 6.1. Maximal transpiration and average cuticular transpiration from the leaves of morphologically and ecologically different plant types and from the shoot surface of leafless succulents under the evaporation demand in the habitat. These are typical values, drawn from the data in original works of many authors. Transpiration intensity is referred to the total surface area (in the case of leaves, upper and lower surface) and given in mg H_2O . dm_2^{-2} . h^{-1}

Plant type	*Transpiration with stomata open*	*Cuticular transpiration*	
		after stomatal closure	*as % of total*
Hydrophytes	1800—4000		60—70
Herbaceous dicotyledons			
Sunny habitats	1700—2500	(100) 200—300	(6) 10—20
Shady habitats	500—1000	50—250	10—25 (30)
Mountain habitats	800—1600		
Grasses			
Tundra	70—120		
Reeds and meadows	1000—1500		
Dry habitats	(600) 1500 (3000)	250—300	(5) 15—25 (40)
Halophytes	400—800 (1500)		
Dwarf shrubs			
Tundra	50—150		
Heath	600—1000	50—60	5—10
Evergreen conifers	450—550	12—15	ca. 3

Winter-deciduous forest trees, temperate zone			
Light-adapted species	(500) 800—1200	90—110	10—20
Shade-adapted species	(250) 400—700	(30) 60—100	12—18
Temperate fruit trees			
Pomoidea	700—1000	120—160	10—20
Prunoidea	400—700	80—110	15—20
Grapevines	400—500	80—90	17—24
Tropical forest trees			
Sunny sites	100—600		
Shady sites	60—120		
Mangrove trees	200—600		
Palms	400—600 (900)		
Sclerophylls	(200) 500—1000 (1400)	(10) 50—100	(4) 7—15
Shrubs of subtropical deserts	900—3400	30—100	3—10
Succulents			
Leafy succulents	300—600	5—12	1—2
Cacti	200—600	8—10	1—2

Table 6.2. Water consumption by trees of various size in the field; for transpiration of tropical rain forest trees in relation to stem size.

Plant species	*Height (m)*	*Age (yrs)*	*Leaf area (m^2, both sides)*	*Foliage mass (kg dry matter)*	*Leaf number*	*Transpired water (l per tree)*		*Region*
						Maximum per day	*Average per year or season*	
Betula pendula	12	28	115	4.6		140		Denmark
Populus canadensis	13	18	135	5.4		128		Denmark
Quercus petraea	12	25	90	3.5		73		Denmark
Fraxinus excelsior	15	27	60	2.4		47		Denmark
Alnus incana	11	19	116	3.7		47		Denmark
Fagus sylvatica	12	27	100	2.8		32		Denmark
Picea abies	13	33		10.9		27		Denmark
Betula pendula	15	21		3.0			5307	Southwest Germany
Fagus sylvatica	14	21		2.3			3801	Southwest Germany
Pinus sylvestris	15	21		4.4			3037	Southwest Germany
Pseudotsuga menziesii	15	21		6.1			2368	Southwest Germany
Pseudotsuga menziesii	28	45	400			65	23,760	USA (Washington)
Salix alba	3.5		10		25,000	9.1	800	Southwest Germany
Alnus glutinosa	2.5		8		2500	3.7	350	Southwest Germany
Fraxinus excelsior	3		7		500	3.6	340	Southwest Germany
Acer platanoides	2.5		6		900		160	Southwest Germany
Acer pseudoplatanus	2.5		3		300		130	Southwest Germany
Elaeis guineensis	5	11	4		32	30	23,300	Zaire

through the cutinized layers of the outer wall of the epidermis and through the cuticle itself. The ***cuticular diffusion resistance*** is usually very high ; it varies in different species with arrangement, density, and number of the cutin and wax lamellae embedded in the outer wall of the epidermis, and with the thickness of the cuticle. The conditions under which the individual plant has grown greatly affect the development of the leaf surface ; plants adapted to dry air and soil have leaves with thicker layers of cuticle and a thicker wax coating than those that developed in higher humidity. In hygromorphic leaves the cuticular diffusion resistance is 40—100s . cm^{-1}, but in leaves and needles with massive protection against transpiration it can reach values around 400s . cm^{-1}. When the outer epidermis dries out and shrinks the hydrophobic layers are brought closer together, and as a result the cuticular diffusion resistance can double. At low temperatures, too, the cuticular diffusion resistance increases. Cuticular protection against transpiration is very effective. The specific *effectiveness of stomatal closing* can be expressed by the *modulation amplitude of transpiration*—that is, the ratio of unrestricted total transpiration to cuticular transpiration. Even in plants of shady and humid habitats, cuticular transpiration amounts to little more than a third of the total ; in sclerophylls, evergreen conifer needles and desert shrubs it is 3—10%, and in succulents only 1—2%. This means that in those plants best protected against transpiration the cuticular transpiration capacity is reduced to 0.1—0.05% of the potential evaporation from a moist surface. These proportions have been derived from measurements made under the conditions of evaporation in the habitat. When particular plant species, varieties and states of adaptation are to be characterized, the effectiveness of stomatal closing is better expressed by a *comparison of the diffusion resistance* when the stomata are maximally opened with that when they are completely closed. Such a comparison shows that the maximal leaf diffusion resistance (stomata closed) in broad-leaved trees is 30—40 times the minimal diffusion resistance (stomata fully open). In herbaceous dicotyledons leaf diffusion resistance is increased by a factor of 10—50, and in grasses by a factor of 5—15, when the stomata close.

Peridermal Transpiration

The loss of water through the surfaces of *suberized* shoot axes is comparable in magnitude to cuticular transpiration; the amount of such loss, in any case very small, depends upon the species-specific structure of the periderm, the permeability of the lenticels, and the presence or absence of cracks in the bark. For this reason the trunks of oaks, maples, and pines give off more water than those of spruce, beech, and birch trees, with their smoother and denser bark. In trees of the temperate zone peridermal transpiration in summer is of the order of magnitude of 10—100 mg. dm^{-2}. h^{-1}, about 1% of the potential evaporation. No one has yet made a comparative study of peridermal transpiration in smooth-barked trees of the wet tropics and thick-barked woody plants of dry regions.

Variability in Rate of Transpiration

Dependence of Transpiration on External Factors. External factors affect transpiration to the extent that they alter the *steepness of the water-potential gradient* between plant surface and surrounding air. The rate of transpiration rises with decreasing *humidity* and increasing *temperature* (Fig. 6.5), and is greater when the *air is moving*. Within closed stands of plants, dense crowns of trees, and other closely packed vegetation

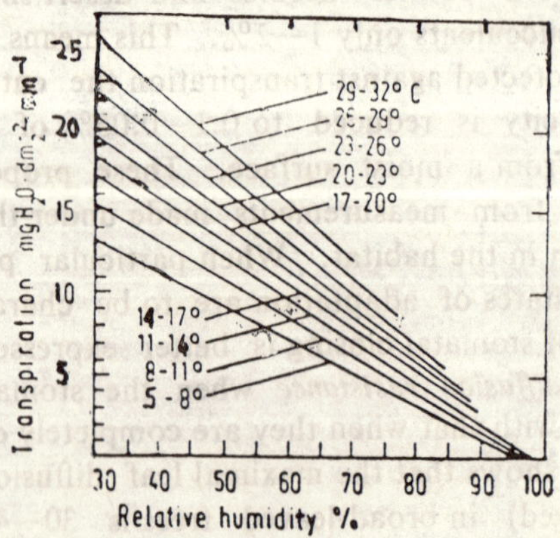

Fig. 6.5. Leaves of reeds in relation to the relative humidity and the air temperature.

humidity is higher than outside and convection is slower, so that the rate of transpiration is reduced.

Physiological Control of Transpiration

Transpiration is strictly dependent on the physical conditions affecting evaporation only as long as the degree of stomatal opening does not change—that is, as long as the stomata remain either open to a *fixed* degree or firmly closed. Only then is the amount of water lost proportional to the evaporative power of the air. The ability of a plant to regulate stomatal opening enables it to adjust the rate of transpiration to the requirements of its water balance. Changes in stomatal opening can be determined *porometrically*. By diffusion porometry one can obtain a quantitative measure of the opening. It is useful to describe stomatal regulatory processes directly in terms of a change of the stomatal diffusion resistance (or conductivity) rather than as a change in rate of transpiration.

The *response threshold*, the *rapidity*, and the *effectiveness* of stomatal regulation vary among species and with the degree of adaptation to the habitat. Trees and herbaceous sciophytes narrow their stomatal openings even when there is only a slight water deficiency, and briskly complete the process of closing. Herbs of sunny habitats restrict their stomatal transpiration only under much drier conditions, and even then closing is slow. There are differences in the way stomata behave in different plant species and even in different individuals of the same species. In fact, even within a single plant, leaves vary appreciably in this respect, depending on their form and their position on the shoot.

Transpiration and CO_2 Exchange

Transpiration and CO_2 acquisition by a plant are linked by the stomata, through which both water vapour and CO_2 diffuse. In order to take up CO_2 the plant must give off water, and when water loss must be reduced the influx of CO_2 is reduced as well.

The chief problem associated with gas exchange, as O. Stocker has put it, is that of "tacking between thirst and starvation". For the management of plantations in agriculture and forestry, where the target is the greatest possible productivity, it is important to know the relationship between water consumption and productivity. As an expression of this relationship Hellriegel (1883) introduced the term *Transpirationskoeffizient* (transpiration ratio), the amount of water in liters consumed by a plant or a stand during the growing season per kg dry matter produced (Table 6.3). One can also compute the number of grams of dry matter synthesized per liter of evaporated water. The water required per unit of dry matter formed varies among different species and varieties and is very much dependent on the state of development, the density of the stand and the environmental conditions—the water supply and the rate of evaporation, in particular. Knowing the specific transpiration ratios of crop plants, the grower can select species and varieties appropriate to the situation and in dry areas, can accurately adjust the amounts of water used for irrigation.

Table 6.3. Average transpiration radius for productivity ($1/\omega_P$, in liters of transpired water per kg dry matter produced).

Herbaceous C_3 plants	
Grain	500—650
Legumes	700—800
Potatoes and beets	400—650
Sunflowers (young)	280
Sunflowers (flowering)	670
Woody plants	
Tropical foliage trees (crop plants)	600—900
Temperate-zone foliage trees	200—350
Conifers	200—300
Oil palms	ca 300
C_4 plants	220—350
Maize species in field experiments :	
INRA 260 (early variety)	320
Pioneer 3567 (late variety)	266
Maize in growth chamber :	
INRA F7 × F2	136
CAM plants	50—100

A measure that has been employed since its introduction by Polster (1950) to analyze the influence of environmental factors and the structural and functional characteristics of certain plants upon the optimization of dry-matter production and water consumption is the *photosynthesis/transpiration* (*F/E*) *ratio*. Holmgren *et al.* (1965) have proposed that the specific diffusion behaviour of a plant be described by the ratio of the sums of all diffusion resistances to water vapour and to carbon dioxide, $\Sigma r_{H_2O}/\Sigma r'_{CO_2}$ Recently the expression *water-use efficiency* ω has been used in place of all these ratios. But it must not be overlooked that the relationships are quite different in the different expressions : the transpiration ratio in the sense of Hellriegel and the "efficiency of transpiration" are measures of the dry-matter production and water consumption of plants or stands over periods that are ordinarily quite long (weeks or entire seasons), whereas the *E/F* or *F/E* ratios refer only to gas exchange in assimilating organs over short periods of time (hours or days). The ratio of diffusion resistances, in contrast to the other two ratios, does not take into account the influence of variations in the concentration gradients of H_2O and CO_2 that depend on environmental factors. Here we propose, for the sake of clarity, to use the terms "transpiration ratio" and "water-use efficiency" in the broadest sense but to make them more precise by stating the frame of reference. In general it holds that the transpiration ratio and the water-use efficiency are inversely proportional (transpiration ratio = $1/\omega$).

The *water-use efficiency of productivity* ω_P is defined as

$$\omega_P = \frac{\text{Dry-matter production}}{\text{Water consumption}} \; [g\ \text{DM} \ . \ 1^{-1}\ H_2O].$$

Dry-matter production and water consumption can be expressed with reference to a single plant or to a stand ; and the value for water consumption is the overall evapotranspiration.

The *water-use efficiency of photosynthesis* ω_F is defined as

$$\omega_F = \frac{F}{E} \; [\text{mg } CO_2 \ . \ g^{-1}\ H_2O].$$

$$\omega_r = \frac{\Sigma r_{H_2O}}{\Sigma r'_{CO_2}}.$$

With the stomata maximally opened, ω_r in most C_3 plants is 0.1—0.4, in some trees 0.5, and in C_4 plants as great as 0.8.

Conversion of ω_F to ω_P, using the conversion factors for CO_2 in the dry matter can be useful as an estimator but is not quite accurate, because the dry-matter production of a plant does not depend only on the intensity of CO_2 exchange ; it is affected much more by the CO_2-exchange balance and the carbon allocation pattern typical of the species.

ω_F and ω_r are interconvertible if the concentration gradients for CO_2 and H_2O are known. The rates of diffusion of the two substances, however, are not identical. The CO_2 concentration difference between the outside air and the chloroplasts is much less than the water-vapour pressure difference between the interior of the leaf and the atmosphere, as long as the outside air is not saturated with water. At a temperature of 20°C and a relative humidity of 50%, the water-vapour gradient is about 20 times as sharp as the CO_2 gradient. For this reason alone, the evaporation of water proceeds much more rapidly than the uptake of CO_2. Moreover, the water molecules, being smaller, diffuse 1.5 times as fast as the larger CO_2 molecules, given identical gradients. There is also a fundamental difference with respect to the diffusion pathways. The route is longer for CO_2, which must enter the chloroplasts, and there is an additional obstacle in that CO_2 movement in solution occurs exceedingly slowly. The *ratio of transpiration to photosynthesis* (*E*/*F*) is therefore always changed whenever any of the factors affecting diffusion is altered.

When the *stomata are open,* CO_2 uptake is limited more severely than transpiration by the diffusion resistances inside the leaf (above all by the carboxylation resistance). When the *stomata are closed,* CO_2 uptake is blocked, but water continues to escape through the cuticle so that the *E*/*F* ratio, for practical purposes, approaches infinity. The most favourable compromise between water consumption and CO_2 uptake is achieved when the stomata are partially open. This is evident not only in the light-dependence relation but also in the behaviour during progressive desiccation ; as the plant begins to dry out, the *E/F* ratio is lowest when both exchange processes are somewhat restricted (Fig. 6.6 and 6.7). In the *natural*

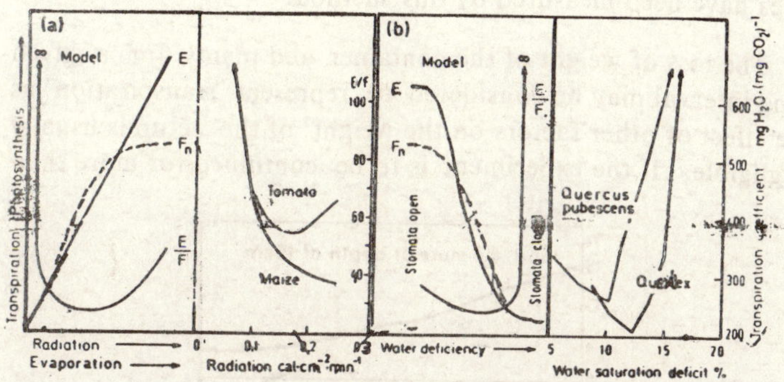

Fig. 6.6. Transpiration, photosynthesis and *E/F* ratio as a function of insolation, evaporative power of the air and water supply.

habitat, the expenditure of water necessary to obtain the CO_2 for photosynthesis is least in the early hours of the morning, since photosynthesis gets off to a more rapid start than does transpiration. As warning by the sun progresses and humidity decreases, the rate of water loss rises more rapidly than that of CO_2 uptake, and the *E/F* ratio increases.

Measurement of Transpiration. Four general methods are in use for measuring transpiration, but only the first two of these as listed can be used for quantitative determinations of the rate of loss of water-vapour from plants which are rooted in the soil.

1. *Method of Weighing Potted Plants.* This method can be employed only with plants which are rooted in pots or other suitable containers. For laboratory experiments potted plants are often used, the pot generally being enclosed in a metal shell, and the soil surface sealed off so that evaporational water loss can occur only through the plant. For field or large-scale experiments, metal receptacles have been found convenient, in which case it is only necessary to seal off or cover the soil surface in such a way as to prevent evaporation. The

method is limited in practical application to plants which can be grown in readily portable containers. The transpiration rates of plants as large as mature maize plants and 5-foot coffee trees have been measured by this method.

The loss of weight of the container and plants for a given time interval may be considered to represent transpiration as the effect of other factors on the weight of the setup is usually negligible. If the experiment is to be continued for more than

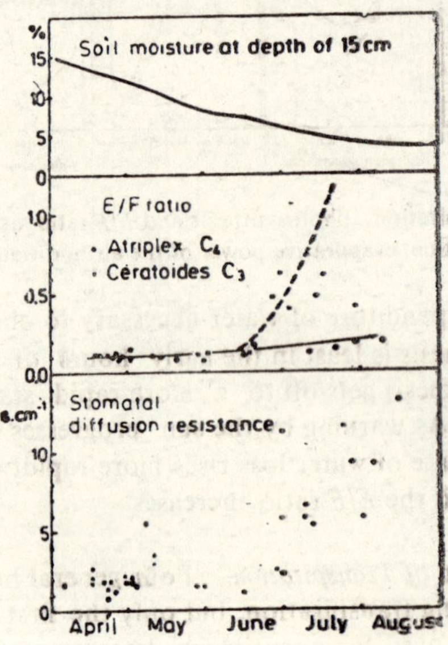

Fig. 6.7. Change in the E/F ratio and in the stomatal diffusion resistance, measured in the morning, of leaves of the C_3 plant *Ceratoides lanata* and the C_4 plant *Atriplex confertifolia*. during progressive drying of the soil in a cool-temperate saline desert (Utah). At first, when there is adequate water, the *E/F* ratio is similar in the two plants; during the summer drought stomatal diffusion resistance increases steadily in both species. Because of its greater carboxylation capacity the C_4 plant exhibits a smaller effect on photosynthesis than does the C_3 plant, in which CO_2 uptake is markedly reduced beginning in June. Accordingly, the *E/F* ratio increases much more rapidly in *Ceratoides* than in *Atriplex*. Modified from Caldwell *et al.* (1977).

a relatively short period it is necessary to provide the container with a watering tube through which known volumes of water can be introduced into and distributed throughout the enclosed soil at appropriate intervals.

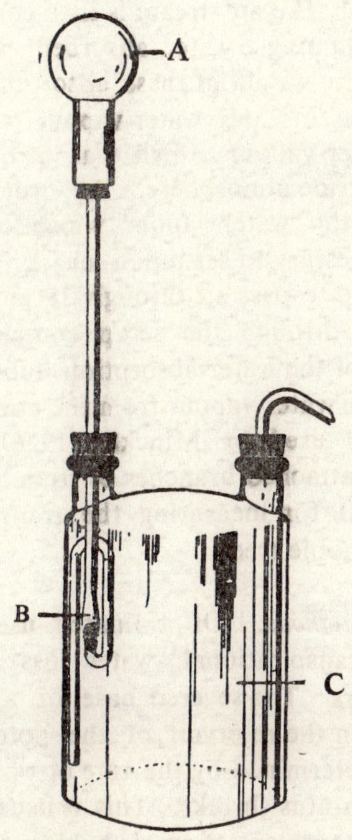

Fig. 6.8. Arrangement for the measurement of evaporation rates by means of an atmometer. (*A*) Porous clay hollow sphere, (*B*) mercury valve, (*C*) reservoir.

In employing this method the receptacle in which the plants are growing may be weighed at selected intervals by manual manipulation or may be placed on a balance which is arranged so that each loss of a definite increment of weight (for example, 1 g.) is automatically registered on a recording device.

2. *Method of Collecting and Weighing Transpired Water-Vapour.* This method requires a rather elaborate experimental

setup, but is the only way in which the rate of transpiration can be determined quantitatively for plants rooted in out-of-door habitats. A stream of atmospheric air is passed through a glass or cellophane chamber containing the plant or portion of a plant, or through "leaf cups", attached to the stomate-bearing side of a leaf. The air stream is then conducted through tubes or vessels containing a water absorbent such as calcium chloride. The gain in weight of these absorption tubes during the experiment represents the water-vapour transpired by the plant plus the water-vapour which was introduced into the system from the outside atmosphere. In order to determine the proportion of the water-vapour which comes from the atmosphere it is necessary to set up a check apparatus, containing no plant, and to pass air through it at the same rate that it is circulated through the setup containing the plant. The gain in weight of the water-absorption tubes in the second apparatus represents water-vapour from the atmosphere. Such a method has been used by Minckler (1936) for measuring transpiration from attached branches of trees and by Heinicke and Childers (1936) for measuring the transpiration rate of attached leaves on apple trees.

3. *Potometer Methods.* Of a limited usefulness for the measurement of transpirational water loss are instruments known as potometers. The severed base of a leafy shoot is immersed in water in the reservoir of the potometer and the rate of water loss determined by the rate at which the volume of water in the apparatus shrinks. This is usually followed by noting the rate of movement of an air bubble in the water in the capillary sidearm of the instrument. Some potometers are constructed so that the entire root system of plants which have been specially grown for the purpose in solution cultures can be immersed in the reservoir of the instrument (Fig. 6.9). A potometer actually measures the rate of absorption of water rather than the rate of transpiration. While under many conditions the rates of these two processes are virtually equal, this is not always true, particularly if an internal water deficit exists in the plant. The rate of transpiration as measured for a cut shoot in a potometer does not necessarily bear any relation to its

rate of transpiration while it was still attached to a plant. The reasons for this will become clear in the discussion of the internal water relations of plants. The principal utility of potometers in the laboratary demonstrations of the effect of various environmental factors upon the rate of transpiration.

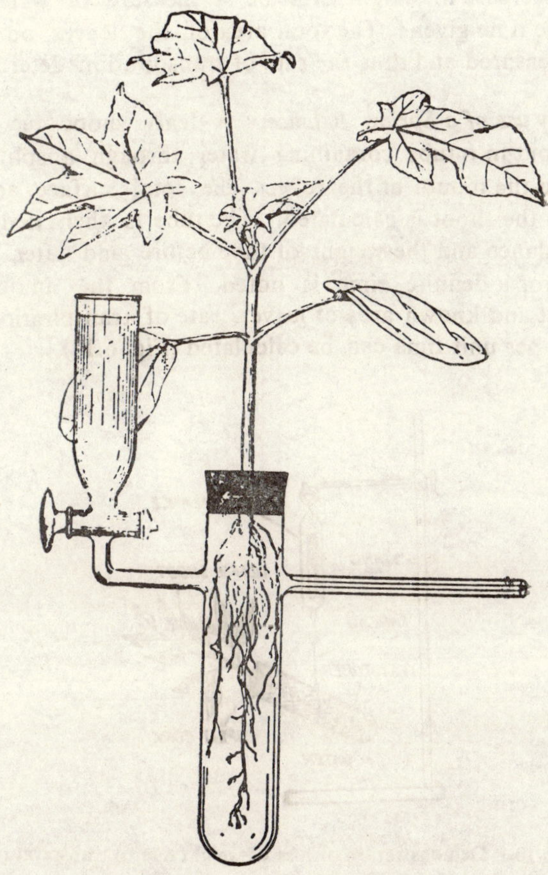

Fig. 6.9. Potometer as set up for measurement of rate of absorption of water by the root system of a seedling plant. The full length of the capillary tube is not shown.

Some of the gravimetric methods commonly used are described below :

(1) *By use of an Ordinary Balance.* Soil of a pot with a plant growing in it is covered with wax or oil cloth after wetting the soil and weight of the pot is taken. The plant is allowed to transpire for a definite length of time after which the pot is weighed again.

The decrease in weight will be a measure of water lost during the time given. The total area of the leaves on plant can be measured and thus the rate of transpiration determined.

(2) *By use of a Spring Balance.* A leafy shoot cut under water is put in a tube containing water through a split cork fitted into the mouth of the tube. The total surface area of leaves on the shoot is calculated. The tube is then tied to a spring balance and the weight of tube before and after transpiration for a definite time is noted. From the amount of water lost and known area of leaves, rate of transpiration per unit area per unit time can be calculated (Fig. 6.10).

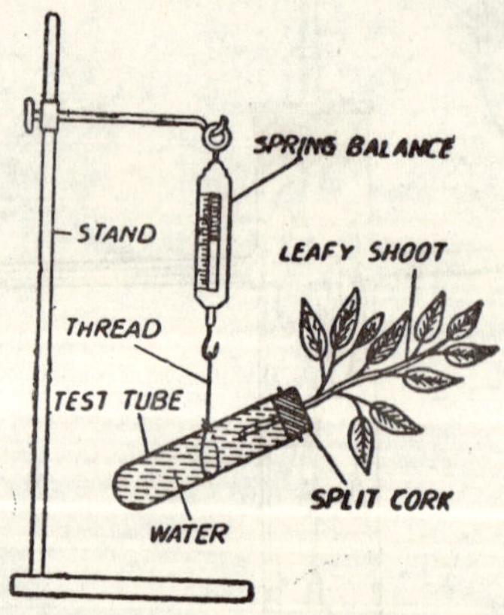

Fig. 6.10. Demonstration of loss of weight due to transpiration.

Thus in the gravimetric methods a cut twig suitably fixed in water or a potted plant whose rate of transpiration is to be determined, is weighed at intervals and the loss in weight due to transpiration is recorded.

(3) *By Measuring the Water Vapours.* A plant is suitably enclosed in a chamber and vapours given off by it are collected

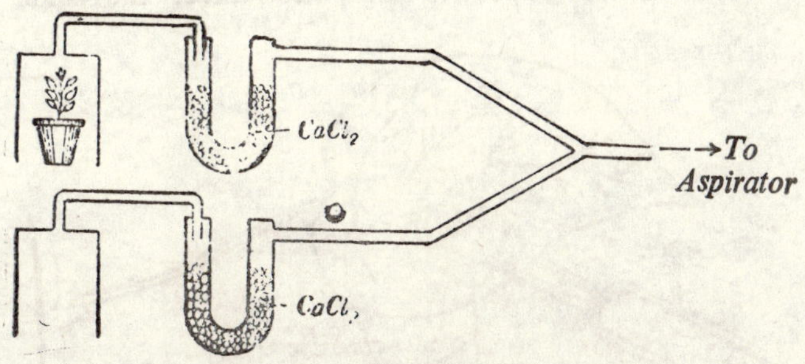

Fig. 6.11. Measuring the loss of water vapours.

by $CaCl_2$ in tube '*a*' (Fig. 6.11). Another similar chamber is attached to the same aspirator and vapours present in the air are collected in tube '*b*' containing $CaCl_2$. The increase in the weight of tube '*a*' minus the increase in the weight of tube '*b*' gives the net weight of vapours lost due to transpiration. Dividing it by time and the total area of the leaves will give the rate of transpiration.

(4) *Cobalt Chloride Method.* This method makes use of the difference in the colour of cobalt chloride in dry and moist conditions. Anhydrous cobalt chloride is blue in colour. The colour changes to pale pink on moistening. In practice, filter papers soaked in a 3% solution of cobalt chloride and dried in an oven are used. Small pieces of these cobalt chloride papers covered over by glass plates are clipped on to the transpiring leaf surface and the change of colour of the cobalt chloride paper from blue to pink shade is noted (Fig. 6.12).

As originally used by Stahl (1894) the cobalt chloride method was used to demonstrate transpiration. The original method has been greatly improved by Livingston and Shreve (1916). Slightly modified by Rubel (1922) the method, as now used, consists in determining, by means of a stop watch, the time required for dry cobalt chloride paper when placed on a leaf to assume a pink colour of a definite shade. The filter paper must be thin and uniform ; the cobalt chloride solution of a definite concentration (usually 3% solution is used) ; and the paper must be thoroughly dried before use. Along side of the cobalt chloride paper are also placed under the same clip

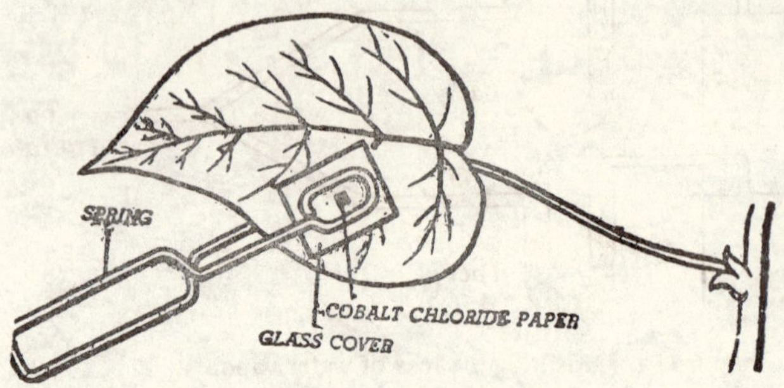

Fig. 6.12. Cobalt chloride experiment for the demonstration of transpiration.

two ordinary papers of standard colours, one representing the initial standard tone and the other the final standard tone. The stop watch is started when the cobalt chloride paper has lost the deep blue colour and has acquired a shade of the initial standard tone and it is stopped when the shade of the final standard tone is reached. The rapidity of the change is an indication of the rate of transpiration. Thus, if one leaf induces the change in 20 seconds and another in 40 seconds, the rate of transpiration of the former is twice that of the latter. Livingston has compared this time interval with the time interval taken by the cobalt chloride paper for a similar change over a moist blotting paper. This ratio has been termed by him the *index of transpiration* and has been considered important for the ecological classification of plants. For instance, if the time interval for cobalt chloride paper in contact with a leaf is 50 seconds and for a similar change over a moist blotting paper 10 seconds, then the index of transpiration of that leaf is 5. Another leaf which brings about the change in cobalt chloride paper from the initial to the final standard tone in 40 seconds will have an index of transpiration equal to 4. The rates of transpiration of the two leaves will then be in the ratio of 4 to 5.

The cobalt chloride method is convenient and has been lately widely used in comparing the rates of transpiration of different plants and also from the two surfaces of dorsiventral leaves under field conditions. But it has its drawbacks. It does not measure the absolute rate of transpiration. Moreover, the leaf under examination is placed under unnatural conditions of reduced light and is isolated from wind action.

Potometer Methods

Potometer methods are based on the principle that the rate

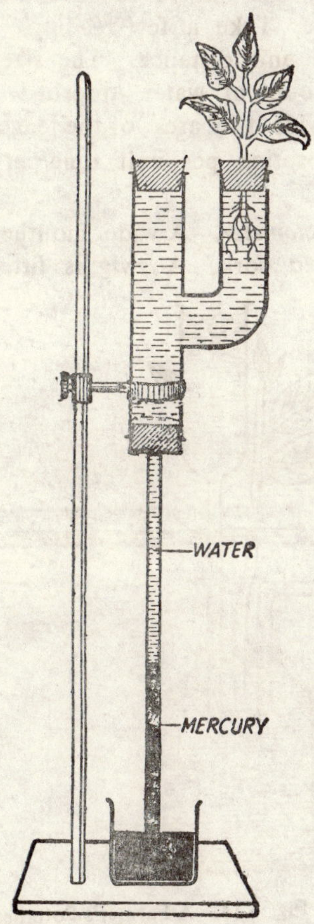

Fig. 6.13. A simple potometer to demonstrate suction due to transpiration.

of absorption of water is always equal to the rate of transpiration. Various types of potometers have been developed. Some of them are discussed here.

(i) Simple Potometer. As shown in Fig. 6.13, the apparatus consists of 'ꟼ', shaped tube, the long arm of which is fitted with a delivery tube.

The apparatus is filled with water. A freshly cut twig which is cut under water is fitted. The connections are made airtight. A bubble of air is introduced in the capillary tube. As the twig transpires, the bubble moves upwards. Note the time taken by the bubble to travel a certain distance in the capillary tube. Take a few readings and find out the average time for a unit distance. The size of the capillary is known. The amount of water absorbed per unit time can be calculated. Divide it by area of the leaves. That will give the rate of absorption per unit time/per unit area by the twig.

(ii) Farmer's potometer. A wide mouthed bottle is fitted with a three holed cork. A twig is fitted in the central

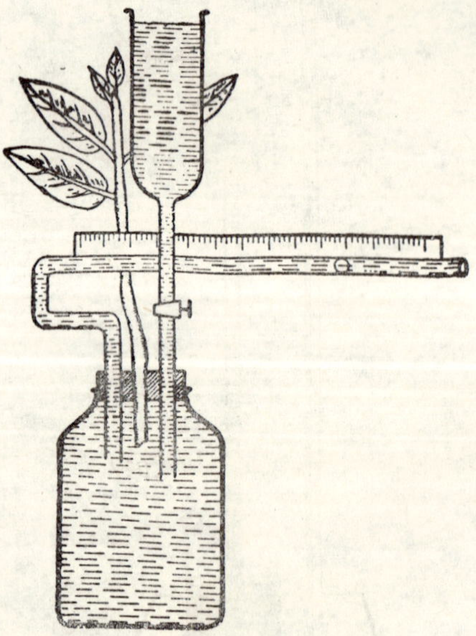

Fig. 6.14. A Farmer's Potometer.

hole. The other two holes are fitted with a thistle funnel or reservoir and a bent tube. A graduated scale is attached at the back of the bent tube (Fig. 6.14). The whole apparatus is filled with water and is then kept is sun light. After sometime it is seen that air enters into the bent tube from the open end. The rate of movement of air in the bent tube is calculated by the distance covered by it in a minute. This is equal to the rate of absorption of water and hence to the rate of transpiration of the plant. By opening the stopper of the reservoir the tube is again filled with water and the air is pushed back. Several readings can, thus, be taken to measure the rate of transpiration.

(*iii*) *Ganong*'s *Potometer*. The apparatus consists of a fine graduated capillary tube with a wide vertical end on one side, and a bent end on the other side. A reservoir is fixed to it near the wide end to store water. The apparatus is fitted as shown in the figure 6.15. The whole apparatus is filled with water. A twig is cut under water and is immediately fixed to the wide end of the potometer through a hole in the cork. The bent end of the potometer has an opening which is dipping in coloured water kept in a small

Fig. 6.15. A Ganong's Potometer.

tube. The experiment is set up in the sun light. To start with, the bent end of the potometer is taken out from the coloured water. After few minutes it is seen that air enters into the tube through the opening. After this the bent tube is again dipped in water. Water transpired by the plant will be compensated by the water of the capillary tube. Absorption of water by the twig will cause movement of the water from the other end. This will be indicated by the movement of the closed air bubble. The distance covered by the air bubble in a minute gives the rate of transpiration. By opening the reservoir the air bubble can be brought back at the initial point and another reading can be taken. The mean of all such readings is then taken.

Comparision between the rate of Transpiration of the two Surfaces of a Leaf by Calcium Chloride Method.

Two narrow bell jars are fixed on the two sides of a dorsiventral leaf by means of a clamp stand. The contact between the leaf and the bell jars is made airtight by vaseline. Within each jar is kept a small tube of calcium chloride (Fig. 6.16). The

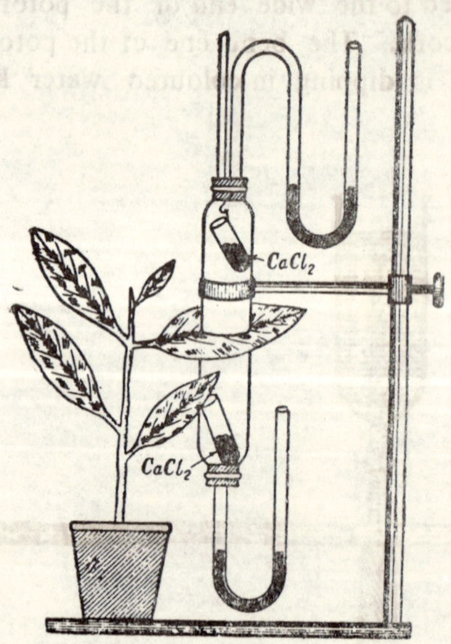

Fig. 6.16. Apparatus to compare stomatal and cuticular transpiration.

weights of the two tubes are taken and noted. U-tubes fitted at the ends of the two jars are fitted with mercury or glycerine to disconnect the inner air from the outer atmosphere. After few hours the two tubes are weighed again and the differences in initial and final weights are noted in either cases. The gain in weight of the tube is equal to the amount of water absorbed by calcium chloride. In a dorsiventral leaf the gain in weight is more in the lower tube than that of the upper tube. This proves that the stomatal transpiration is more than cuticular transpiration.

FACTORS WHICH AFFECT THE RATE OF TRANSPIRATION

The rate of transpiration changes from time to time because it is conditioned by a number of internal or plant and external or environmental factors.

Internal or Plant Factors

In the early part of this century it was considered that xerophytes possess structural characteristics like thick, cutinized leaves, etc., which presumably reduced water loss. Maximov (1929) reported that this generalization is not true and in many cases xerophytic characters prevent only wilting or death of leaves. Structure of leaf, however, does have important effect on the rate of water loss from the plant. Some of the important plant factors are discussed below :

1. *Leaf Area.* As most of the water loss takes place from leaves, plants with large leaf area lose more water than plants with a small leaf area. But the rate is usually not proportional to the area of the leaves ; smaller plants tend to transpire more rapidly per unit of leaf area than larger plants of the same species. Decreasing the leaf surface by pruning usually increases the rate of transpiration. The leaves of some grasses roll when they begin to wilt thus decreasing the transpiring surface.

2. *Leaf Structure.* If a number of detached leaves of different species are allowed to dry in air, it will be found that some of them dry up in a few hours while others take a few

days or even a few weeks. This difference can be usually correlated with (*i*) thickness of leaves ; (*ii*) efficiency of cuticle and number of epidermal hairs ; and (*iii*) internal surface exposed to intercellular spaces in the leaf.

Plants with a thick cuticle survive much longer than those with a thin cuticle. Turrell (1944) showed that in xerophytic thick leaves the total cell wall area exposed to internal atmosphere of the leaf is more than what it is in mesophytic leaves. This larger internal exposed cell wall surface provides larger evaporating surface and increases the rate of transpiration when the stomata are open and plant is supplied with adequate amount of water.

The presence of hairs on leaves was supposed to reduce transpiration, but Sayer (1926) showed that removal of hairs from leaf of *Verbascum* had no effect on transpiration. Indeed if the hairs are saturated with water they may even increase transpiration.

3. *Stomata*. As stomata are the openings through which transpiration takes place, their distribution, structure, size of the stomatal aperture and periodicity of opening and closing greatly influence the rate of transpiration. More the number of stomata per unit area and bigger the size of their pore, higher is the rate of transpiration.

4. *Orientation of Leaves*. Higher the temperature of a leaf, higher is the rate of evaporation from its surface. In most of the plants, leaves are oriented perpendicular to the incident light hence they are affected to the maximum extent by the heating effect of the sun. In plants like *Lactuca*, *Quercus*, etc., the leaves are arranged parallel to the average incident light. Such an orientation perhaps reduces leaf temperature but it is doubtful if this has much significance in loss of

5. *Root Shoot Ratio*. The loss of water takes place from the plant surface only if it is efficiently absorbed. The ratio of water absorbing surface to transpiring surface is, therefore, more important than the total leaf area. Parker (1949) found

that transpiration per unit area of leaf increased as the ratio between root surface and leaf surface increased.

6. *Water Contents of Leaves.* Whether reduction in the water contents of leaves reduces transpiration directly or indirectly (by bringing about stomatal closure) is yet a controversial point. Livingston and Brown (1912) suggested that high transpiration decreases the water contents of cells surrounding the intercellular spaces in the leaf and thereby the rate of transpiration is decreased. Gregory (1950) found that this incipient loss must be very great before it can be directly effective in reducing transpiration. It is clear that when dehydration of cells is severe enough to appreciably reduce the vapour pressure of water in them, the rate of loss of water into the intercellular spaces is reduced and consequently the rate of transpiration also falls.

7. *Diseases.* Wheat plants infected by loose smut transpired more than healthy plants. In general fungal infection increases transpiration.

External or Environmental Factors which affect Transpiration

The rate of transpiration is basically dependent upon all those factors which affect the steepness of the water vapour pressure gradient from the evaporating surface of plant to the outer atmosphere. Humidity of air, temperature of plant and air, wind velocity are such few factors. In addition light, available soil moisture, etc., are other external environmental conditions affecting transpiration. Let us discuss these factors in some detail.

Solar Radiation. This term refers to the visible light and other forms of radiant energy (infrared and ultraviolet radiations) reaching the earth from the sun. The principal effects of solar radiation upon transpiration result from the influence of light upon the opening and closing of the stomates. In most of the species of plants which have been studied the stomates are usually closed in the absence of light, thus causing a virtually complete cessation of stomatal transpiration during the hours of darkness. Since none of the other environmental factors, can have any influence upon stomatal transpiration except when the stomates are open, light occupies a position of prime importance among the environmental conditions influencing transpiration.

A second important effect of solar radiation upon transpiration operates through its influence on leaf temperatures. This effect will be analyzed later in this chapter.

Humidity. Several units are used for designating the humidity conditions of an atmosphere. One of these is the *actual* vapour pressure of the atmosphere. This should not be confused with the *saturation* vapour pressure. Since the rates of diffusion and evaporation are influenced directly by the vapour pressure of the atmosphere, this is usually the most satisfactory unit in which to express humidity values for physiological purposes.

A more familiar humidity unit is the *relative humidity*, which is the percentage saturation of an atmosphere. Since, at a given temperature, the vapour pressure of an atmosphere is proportional to the concentration of water-vapour molecules present, the relative humidity is equal to the ratio of the actual vapour pressure of the atmosphere to its saturation vapour pressure at the same temperature. For example, a saturated atmosphere at 30°C. has a vapour pressure of 31.82 mm. Hg ; its relative humidity is 100 per cent. If only half the amount of vapour is present that would be present at saturation at this temperature, *i.e.*, if the vapour pressure is 15.91 mm. Hg., then the relative humidity of the atmosphere is 50 per cent. Change in either the vapour pressure or the temperature of an atmosphere will result in a change in its relative humidity. Change in vapour pressure at constant temperature results in a proportionate change in relative humidity. Increase in temperature, unaccompanied by a change in the amount of water-vapour present, results in a decrease in the relative humidity of an atmosphere, because of the increase in saturation vapour pressure. Contrariwise, decrease in temperature of an atmosphere, without any accompanying change in water-vapour content, results in an increase in relative humidity *vice versa*.

Table 6.4. The relation between relative humidity and vapour pressure at different temperatures.

Temperature		*Actual vapour pressure (mm. Hg) at indicated relative humidity*										
°C.	°F.	0	10%	20%	30%	40%	50%	60%	70%	80%	90%	100%
0	32	0	0.46	0 92	1.37	1.83	2.29	2.75	3.21	3.66	4.12	4.58
5	41	0	0.65	1.31	1.96	2.62	3.27	3.92	4.58	5.23	5.89	6.54
10	50	0	0.92	1.84	2.76	3.68	4.60	5.53	6.45	7.37	8.29	9.21
15	59	0	1.28	2.56	3.84	5.12	6.40	7.67	8.95	10.23	11.51	12.79
20	68	0	1.75	3.51	5.26	7.02	8.77	10.52	12.28	14.03	15.79	17.54
25	77	0	2.38	4.75	7 13	9.50	11.88	14.26	16.63	19.01	21.38	23.76
30	86	0	3.18	6.36	9.55	12.73	15.91	19.09	22.27	25.46	28.64	31.82
35	95	0	4 22	8.44	12.65	16.87	21.09	25.31	29.53	33.74	37.96	42.18
40	104	0	5.53	11.56	16.60	22.13	27.66	33.19	38.72	44.25	49.79	55.32
45	113	0	7.19	14.38	21.56	28.75	35.94	43.13	50.32	57.50	64.69	71.88
50	122	0	9.25	18.50	27.75	37.00	46.26	55.51	64.76	74.01	83.26	92.51

Expression of humidity values in terms of relative humidity, although a common practice, is unsatisfactory for physiological purposes because the same relative humidity, 50 per cent for example, may refer to widely different vapour pressures (Table 6.4). For a relative humidity of 50 per cent at 10°C., the vapour pressure is only 4.60 mm. Hg. whereas that for a saturated atmosphere at the same temperature is 9.21 mm. Hg. The difference between these two values—4.61 mm. Hg.—is an index of the steepness of the vapour pressure gradient between an evaporating surface and the atmosphere. For a relative humidity of 50 per cent at 50°C., however, the difference between the saturation vapour pressure at an evaporating surface and the vapour pressure of the air above would be 46.25 (92.51-46.26). Hence at a relative humidity of 50 per cent, evaporation from a moist surface would take place many times more rapidly at 50°C, than at 10°C. Only when all relative humidity values are recorded at the same temperature are they an expression of relative differences in vapour pressures.

In general, the greater the vapour pressure of an atmosphere, other factors remaining unchanged, the slower the rate of transpiration. Whenever the stomates are open the rate of diffusion of water-vapour out of a leaf depends upon the difference between the vapour pressure in the intercellular spaces and the vapour pressure of the outside atmosphere, since the vapour pressure is a measure of the diffusion pressure of the water-vapour. Let us suppose that the vapour pressure of the intercellular spaces is 31.82 mm. Hg. which is the value for a saturated atmosphere at 30°C. Such a vapour pressure is frequently attained in the internal air spaces of leaves. Let us further assume that at the same time the vapour pressure of the atmosphere is only half as great (15.91 mm. Hg). Such a value would be a representative one for a warm summer's day in the eastern United States. In very quiet air the vapour pressure in the neighbourhood of transpiring leaf surfaces may be greater than that in the atmosphere in general but in this discussion it is assumed that there is sufficient air movement to prevent any appreciable accumulation of water-vapour in the vicinity of the leaves.

Under the conditions as stated diffusion of water-vapour would occur through open stomates at a relatively rapid rate. If the vapour pressure of the atmosphere were lowered below this value, the rate of diffusion of water-vapour out of the leaf would be increased ; conversely, increase in the vapour pressure of the atmosphere would result in a decrease in the diffusion rate of water-vapour out of the leaf. Similarly, an increase in the vapour pressure of the intercellular spaces would result in an increase in the rate of transpiration, whereas a decrease in the vapour pressure of the intercellular spaces relative to that of the atmosphere would have the opposite effect. On the rare occasions when the vapour pressures of the atmosphere and of the intercellular spaces are equal, no transpiration will occur, even if the stomates are open.

Martin (1943) found the transpiration race of sunflower (*Helianthus annuus*) and ragweed (*Ambrosia trifida*) in the dark, but with open stomates, to show a linear relation to the relative humidity of the atmosphere at 27° and 38°C. The rate at a relative humidity of 50 per cent, for example, was about twice the rate at a relative humidity of 75 per cent. These results are in agreement with the foregoing theoretical discussion, since, assuming a saturation or near saturation vapour pressure in the intercellular spaces, which the conditions of the experiment favoured, the vapour-pressure gradient from the intercellular spaces to the outside atmosphere would be about twice as great at 80 per cent relative humidity as at 75 per cent relative humidity.

Temperature Effects on Transpiration

1. *Thermal Relations of Leaves.* While leaf temperatures often do not deviate greatly from surrounding atmospheric temperatures, the discrepancy between the two is often sufficiently great to make it necessary to take it into account in careful experimental work.

Theoretically the temperature of a leaf may be regarded as conditioned by four different influences : (1) thermal absorption, (2) thermal emission, (3) internal endothermic (energy-storing processes, such as photosynthesis and transpiration, and (4) internal exothermic (energy-releasing) processes such as respiration. The influence of this last factor upon leaf temperatures is practically always negligible and will be disregarded. Similarly the quantity of energy transformed in photosynthesis is relatively so small that it need not be considered in evaluating the factors determining the temperature of leaves.

Thermal emission refers to the loss of heat from a leaf by the processes of conduction, convection, and radiation. Simple and depends in part upon the other prevailing environmental conditions. Usually, however, increase in wind velocity, within limits, results in an increase in the rate of transpiration. This is usually explained by assuming that water-vapour often accumulates in the vicinity of transpiring leaves in a quiet atmosphere, especially if they are not exposed to direct sunlight. The result of such an accumulation of water-vapour is a decrease in the steepness of the vapour-pressure gradient through the stomates and hence a decrease in the rate of transpiration. If, however, the leaves are exposed to a wind, any accumulation of water-vapour molecules in the immediate

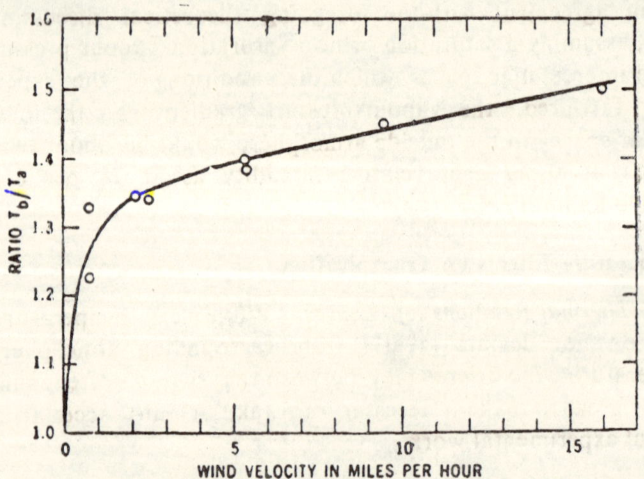

Fig. 6.17. Relation between wind velocity and rate of transpiration of sunflower expressed as ratio between plants in wind (T_b) and plants in quiet air (T_a).

vicinity of the leaf surfaces will be dispersed. The effective result will be an increase in the steepness of the vapour-pressure gradient through the stomates, and a consequent increase in the rate of loss of water-vapour.

Whenever a temperature differential exists between a leaf and the surrounding atmosphere, convection currents are set up in the gases in the vicinity of a leaf which may largely or entirely prevent any accumulation of water-vapour in the immediate vicinity of the leaf. The influence of wind in raising the transpiration rate of leaves is probably more effective, therefore, when they are subjected to such conditions that their temperature does not depart appreciably from that of the surrounding atmosphere. The swaying of branches and shoots, and the bending, twisting and fluttering of leaf blades in the wind also contribute to higher rates of transpiration in moving than in quiet air. It has been shown experimentally that immobile leaves usually transpire less than similar leaves allowed to bend and move freely when both are exposed to wind of equal velocity. Such bending and contortion of leaves may increase the rate of water-vapour loss in part by compressing the intercellular spaces, thus forcing water-vapour and other gases out through the stomates.

A gentle breeze is relatively much more effective in increasing the transpiration rate than winds of greater velocity (Fig. 6.17). Winds of very high velocity have been observed to have a retarding effect upon transpiration. This probably results from closure of stomates under such conditions. Wind may also exert indirect effects on the rate of transpiration through its influence on the temperature of the leaves, as previously described.

Soil Conditions Influencing the Availability of Water. Although transpiration can continue for short periods at rates considerably in excess of the rate of absorption of water, in general, if soil conditions are such that absorption of water is appreciably retarded, the rate of transpiration will soon show a corresponding retardation. The availability of soil water to the plant is therefore an important and, in fact, often the limiting factor in transpiration. The principal soil factors which affect the rate of absorption of water by plants are : (1) available soil water, (2) soil temperature, (3) aeration of the soil, and (4) concentration of solutes in the soil solution. All of these factors indirectly influence the rate of transpiration.

Atmospheric Pressure. It has been demonstrated experimentally that a reduction in atmospheric pressure results in an increase in the rate of transpiration. This result would be predicted theoretically since reduction in the density of the atmosphere would be expected to permit diffusion of water-vapour to occur into it more rapidly. In any given locality variations in atmospheric pressure are too slight to have any significant effect upon the rate of transpiration. Plants growing at high altitudes are subjected to distinctly lower atmospheric pressures than the plants of lowlands, and in comparative evaluations of the transpiration rates of species growing in these two types of habitats the influence of this factor must be considered.

Effects of Structural Features of Plants on the Rate of Transpiration. Different kinds of plants, even when growing side by side under virtually identical environmental conditions, may have very different rates of transpiration. In part, such differences in transpiration rate from one kind of a plant to another result from species differences in internal processes and conditions such as osmotic pressures of the leaf cells, imbibitional capacities of the protoplasm and cell walls, and behaviour of the stomates. Structural differences, particularly in the leaves, also account in part for unlike rates of water-vapour loss from plants of different species growing in the same environment. It has proved impossible, however, to draw correct inferences regarding the relative rate of transpiration of a plant on the basis of its observed anatomical peculiarities. Many species which, on the basis of their distinctive structural features, have been judged to have a low rate of transpiration, proved upon experimentation to transpire very rapidly when environmental conditions were favourable.

Certain structural features of plants influence the rates of both stomatal and cuticular transpiration. The total leaf surface of the plant is one such factor. The distribution and gross morphology of the root system is another. In a habitat where deep-rooted and shallow-rooted plants grow side by side, the former may transpire more rapidly than the latter during drought periods because their roots penetrate to a soil horizon which still contains available water, while those of the shallow-rooted plants are in a moisture deficient soil. The completeness with which the soil mass is interpenetrated by roots may also influence the rates of water absorption and hence of transpiration. Sorghum, for example, has almost twice as many fibrous roots as maize. Rates of water absorption and transpiration can be better maintained, especially in relatively dry soils, by plants with the sorghum type of root system than by those with the maize type. For a given kind of plant, a low shoot-root ratio is more favourable to the maintenance of relatively high rates of transpiration than a high shoot-root ratio.

The rates of cuticular transpiration are undoubtedly held to a low value in many species by the cutin layer itself as well as by leaf coatings of waxy or resinous substances which are present on many kinds of plants. The presence of dead epidermal leaf hairs also contributes to maintenance of low rates of cuticular transpiration, but living epidermal hairs have the contrary effect. The presence of dead epidermal hairs on the leaf may also have a retarding effect on the stomatal transpiration of some species, but Sayre (1920) could find little evidence for such an effect in the leaves of mullein (*Verbasaum thapsus*).

Rates of stomatal transpiration are influenced by the size, spacing, distribution, and structural peculiarities of the stomates. In some species the stomates are sunken below the general level of the epidermis. In general, this should result in a slower rate of diffusion than through stomates of the non-sunken type because of the greater length of the diffusion gradient of water-vapour through sunken stomates. The area of the internal evaporating surface of the mesophyll cell walls in proportion to the surface area of the leaf probably has an influence on the rate of stomatal transpiration, at least under some conditions. Values ranging from 6.8 to 31.3 have been obtained for the ratio of internal exposed surface area to external surface area in leaves of a number of species. The ratio is, in general, higher for xeromorphic than for mesomorphic leaves, and for leaves of a given species which developed under xeric conditions as contrasted with leaves of the same species grown under mesic conditions. Turrell (1944) has shown that xeromorphic leaves of oleander, (*Nerium oleander*), with relatively high internal-external surface ratios, transpire more rapidly than the mesomorphic leaves of periwinkle (*Vinca rosea*), with lower internal-external surface ratios. It was also clear from the results, however, that the higher internal-external surface area of the oleander leaves was not the only factor responsible for their higher transpiration rates.

The Daily Periodicity of Transpiration. All plants exhibit a daily periodicity of transpiration rate which varies somewhat with the species and is greatly influenced by the environmental conditions to which the plant is exposed. We shall first consider the transpiration periodicity upon a standard day as defined in the preceding chapter, since most experiments upon daily variations in transpiration rates have been conducted upon plants growing under approximately such conditions.

During the hours of darkness the transpiration rate (Fig. 6.18) is generally low and in most species water loss during this period may be regarded as almost entirely cuticular. It is not justifiable to assume that absolutely no stomatal transpiration occurs at

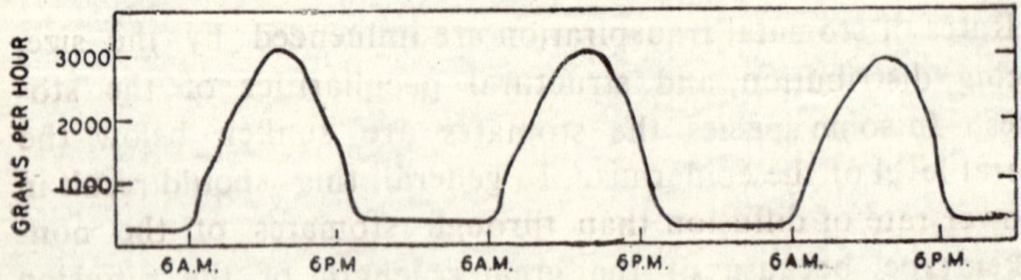

Fig. 6.18. Daily periodicity of transpiration of alfalfa on 3 successive days under approximately standard-day conditions. Transpiration expressed as grams per hour per 6-foot square plot of alfalfa (average of two plots).

night in any species unless this is definitely known to be true. In some species of plants, as we have already seen, night stomatal opening is a common occurrence, while in others it occurs under certain environmental conditions. Even in those species in which the stomates are normally closed during the hours of darkness, a few may remain open. The transpiration rate shows a steady, and usually consistent, rise during the morning hours which culminates rather than the exception, transpiration periodicity curves of this type (Fig. 6.19) are for many species of plants, much of the time more nearly representative than the curves shown in Fig. 6.18.

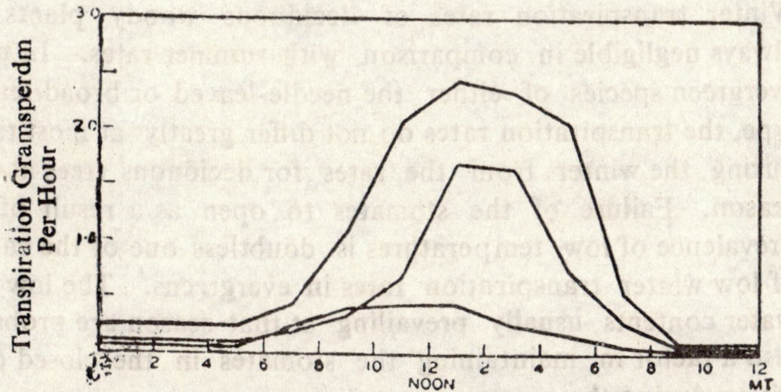

Fig. 6.19. Daily periodicity of transpiration of bean (*Phaseolus vulgaris*) for 3 successive days during a period when the soil was gradually becoming drier.

Internal factors may also be responsible for transpiration periodicities of a different trend from those which have already been discussed. In some species of plants, as has already been described, the stomates remain open to a greater or lesser extent during the hours of darkness. Such plants have higher transpiration rates at night than those in which the stomates are closed. In some species of cacti there is a complete inversion of the usual transpiration periodicity curve, transpiration rates being regularly greater at night than in the daytime. This appears to be a result of a complete or nearly complete stomatal closure during the hours of daylight, while the stomates are, as a rule, open at night.

Seasonal Variations in Transpiration Rates. In temperate regions transpiration occurs predominantly during the warmer months of the year, and especially during those periods of the warm season when the soil water supply is abundant. Among the deciduous group of woody perennials, the branches are defoliated during most of the autumn, all of the winter, and the earlier part of the spring. Although the twigs and branches which remain exposed to the atmosphere are completely encased in corky layers of bark and the buds are enclosed within cutinized bud scales, some water loss occurs from such species even during the winter months. Young twigs lose water-vapour under such conditions faster than older ones. Winter transpiration rates of deciduous woody plants are always negligible in comparison with summer rates. In most evergreen species of either the needle-leaved or broad-leaved type, the transpiration rates do not differ greatly at most times during the winter from the rates for deciduous trees at that season. Failure of the stomates to open as a result of the prevalence of low temperatures is doubtless one of the causes of low winter transpiration rates in evergreens. The low leaf water contents usually prevailing at that season are probably also a factor in maintaining the stomates in the closed condition during the winter.

Mild periods of any considerable duration often have a detrimental effect upon evergreen species in regions where a cold climate usually prevails during the winter months. The warm air temperatures induce stomatal opening, and a relatively high rate of transpiration ensues. The water in the surface layers of soil may remain frozen, and thus unavailable, during a period of warm air temperatures. Even if none of the soil water is frozen, soil temperatures will be low, and this greatly retards the rate of absorption of water. The combined effect of a relatively high transpiration rate and relatively low absorption rate results in a gradual desiccation of the leaves and branches of the plant. This diminution in the water content of the aerial organs during warm periods in winter is more severe in windy weather and is more likely to occur if a sequence of mild days follows immediately after a cold spell.

If severe enough, this desiccation results in the death of some of the branches, or in extreme cases of the entire tree or shrub. This is one cause of the phenomenon known as *winter-killing*. Winter-killing resulting from desiccation of the tissues should not be confused with cold injury resulting from low temperatures, which is a different phenomenon.

Structural and physiological modifications to check excess of transpiration :

Some plants have certain structural and physiological characteristics which enable them to check excess of transpiration and to withstand drought. Such structural modifications are summarized.

1. Reduction in leaf surface area

In some plants foliage leaves are either altogether absent and their function taken up by the stem itself (*e.g.*, cacti and other desert plants) or their size very much reduced (*e.g.*, Pines and Firs etc.) Reduction in leaf area brings about reduction in transpiration.

2. Periodic reduction of leaf area

The deciduous trees and shrubs drop their leaves in autumn and many perennial plants shed their annul shoots at the end of the growing season. This adaptation enables these plants to come over the coming drought caused by low soil temperature.

3. Temporary reduction in leaf area

Some plants like corn and certain grasses roll up or turn the edges of their leaves when exposed to bright sun and fast breeze. This causes reduction in the rate of transpiration.

4. Structural modifications

Plants growing in such areas where drought occurs quite frequently or at definite seasons develop certain structural modifications which enable them to conserve water by reducing the rate of transpiration. These structural modifications are the development of thick cuticle, multilayered epidermis, compact arrangement of mesophyll cells and the presence of sunken stomata covered with hairs. Some plants such as cacti and agave have water storage tissue in their body. Cells of the water storage tissue are highly mucilagenous which retain water avidly. Such succulent plants are of common occurrence in deserts where they are exposed to scorching sun interrupted only occasionally by heavy rains.

5. Drought resistant protoplasm

Protoplasm in some plants (*e.g., Pelargonium*) is very much drought resistant and can withstand considerable loss of water without being killed.

Distinction between transpiration and evaporation

Although there is a great similarity between transpiration and evaporation, as water is lost in the form of vapours in both the cases, but there are certain important differences between the two. These differences are summarised below :

Transpiration	*Evaporation*
1. It is a physiological phenomenon that takes place only in living plants.	1. It is purely a physical phenomenon that may take place from any exposed surface.
2. It takes place mostly through certain specialized pores (*i.e.*, stomata) whose activity is controlled by guard cells.	2. This may take place through and opening or pore.
3. It is affected by several factors such as light, water contents of soil and plant, temperature and minerals which have great influence on the opening and closing of stomata and therefore also affecting the rate of transpiration.	3. It is affected by temperature and water contents of the plant.
4. Surface area from which transpiration takes place is controlled and regulated by guard cells.	4. Surface area from which evaporation takes place is constant.

The Magnitude of Transpiration. Transpiration may be computed per unit of leaf surface or leaf area, per unit of fresh or dry weight, per plant, or per unit area of field or forest. Rates may be calculated for hourly, daily, seasonal, or yearly periods. Transpiration rates show an enormous variation from one kind of a plant to another, and for the same kind of a plant under different environmental conditions.

Transpiration rates in broad-leafed plants of temperate regions range up to about 5 g. per $dm.^2$ of leaf area per hour. Usual rates, under conditions favourable for stomatal transpiration, fall within a range of 0.5 to 2.5 g. per $dm.^2$ per hour. At night, or during periods when a dry soil, a low temperature, or other conditions unfavourable to stomatal transpiration prevail, the rate may fall to 0.1 g. per $dm.^2$ per hour, or even less. Under favourable conditions many herbaceous plants transpire several times their own volume of water in a single day.

The transpiration rates of large plants such as mature trees obviously cannot be measured directly, but can only be estimated from data on the leaf population of the tree and the known rates of transpiration of some of the leaves. To arrive at an accurate estimate of the leaf population of a large tree is in itself a laborious procedure. Among the few such estimates which have been made are those of Turrell (1934), who found a 33-foot open-grown catalpa (*Catalpa speciosa*) to bear 26,000 leaves having an aggregate leaf area of 19,500 $dm.^2$, and of Cummings (1941), who found a 47-foot open-grown silver maple (*Acer saccharinum*) to bear 177,000 leaves having an aggregate leaf area of 68,000 $dm.^2$

Accurate measurements of the transpiration rates of even a small part of a tree can be made only by the air-flow method described previously. Results obtained by the potometer or hygrometric paper methods on small branches or individual leaves may be grossly in error for reasons which have already been discussed. Even if accurate measurements have been made of the transpiration rate of certain branches, a further complication in attempts to estimate the transpiration for the entire tree is introduced by the fact that the rate of the process is known to differ greatly from one leaf or branch to another. All calculations of the transpiration rates of large trees can, therefore, only be approximations, although such estimates are not only of scientific interest but also of practical value.

Likewise, calculations of transpirational water loss per acre of cropped land or of natural vegetation are necessarily only approximations. Results of some such calculations indicate that sufficient water may transpire from maize plants during the course of a growing season to cover the field to a depth of 15 inches and from apple trees to cover the orchard to a depth of 9 inches. Hoover (1945) estimates the transpiration of a deciduous forest, largely oak, in the southern Appalachian mountains, to be equivalent to 17-22 inches of rainfall per year. This estimate was made on the basis of the effect of removing vegetation from a watershed on the runoff.

Significance of Transpiration. Opinions regarding the significance of transpiration have ranged all the way from those which would put it practically on a par with such processes as photosynthesis and respiration, to those which would relegate it to the category of a "necessary evil". The principal roles which have been ascribed to the transpiration of plants can be summarized under the following three headings :

1. *Supposed Role in the Movement of Water.* It is often claimed or assumed that the movement of water through the plant requires the occurrence of transpiration from the leaves. That this concept is entirely erroneous will be clear from the discussion of the mechanism of the conduction of water through plants. Under conditions of high transpiration the movement of water through plants is, in general, more rapid than under conditions of low transpiration. The mechanism causing ascent of water through a plant operates in such a way that any increase in the diffusion-pressure deficit of the mesophyll cells favours a more rapid movement of water toward those cells. Hence a rapid transpiration rate which invariably

results in a considerable increase in the diffusion-pressure deficit of the mesophyll cells, usually speeds up the rate at which water ascends through the plant. However, translocation of water to the extend that it is used in restoring the turgor of leaf cells, in photosynthesis or in growth, continues even during periods when the transpiration rate is negligible.

2. *Supposed Role in the Absorption and Translocation of Mineral Salts.* It has often been assumed that the more rapid the rate of transpiration, the greater the rate of absorption of mineral salts. This view implies that the dissolved mineral salts are swept into the plant along with the absorbed water, a postulation which ignores much evidence that the mechanisms operating in the absorption of water are very different from the mechanisms operating in the absorption of mineral salts. The results of certain experiments do indicate that a somewhat larger quantity of mineral salts accumulates in plants under conditions favouring high transpiration rates than in similar plants growing under conditions favouring low transpiration. The work of Broyer and Hoagland (1943), however, indicates that such results are only obtained under certain metabolic conditions, and that there is no consistent correlation of the rate of transpiration with the rate of absorption of mineral salts.

Upward translocation of the mineral salts absorbed by the roots occurs principally in the xylem. The dissolved mineral salts are passively carried upward in the ascending water columns. Since the rates of translocation of water show a fairly close correlation with the rates of transpiration, a similar correlation also exists between rates of upward movement of mineral salts through the plant and rates of transpiration. Although high transpiration rates undoubtedly often result in more rapid upward transport of mineral salts through the plant than low rates, there is no reason to attach any critical significance to this fact. There is no evidence that inadequacies in the distribution of absorbed mineral salts throughout plants ever result from low transpiration rates.

3. *Supposed Role in the Dissipation of Radiant Energy.* Leaves exposed to direct sunlight absorb large quantities of

radiant energy which, unless dissipated in some other way, will be converted into heat energy and raise the temperature of the leaf. The possibilities of such an effect are indicated by the following approximate calculations : In direct noon-day summer sunlight, the rate of receipt of solar energy is often as much as 1.3 g.-cal. per square centimeter of leaf area per minute and not uncommonly even greater. The proportion of this actually absorbed varies with the kind of leaf, but will be assumed to be only 50 per cent, which is probably a representative value. The incident radiant energy which is not absorbed by the leaf is all transmitted or reflected. Such a small proportion of the absorbed energy is used in photosynthesis that in can be neglected in a rough calculation. Hence about 0.65 g.-cal. of energy is absorbed per square centimeter of leaf area per minute. If the mass of a square centimeter of a leaf is taken as 0.020 g. and its specific heat as 0.879 g.-cal.[2], the rise in temperature per minute would be $\frac{0.65}{0.020 \times 0.879}$ or about 37°C. Since the thermal death point of most plant protoplasm lies between 50° and 60°C, it should not require more than a few minutes to heat the leaves of most plants to a lethal temperature. Actual measurements of leaf temperatures, however, show that they seldom even approach their thermal death points. Leaf temperatures usually do not exceed atmospheric temperatures by more than a few degrees centigrade. Evidently some efficient energy-dissipating mechanism is at work which prevents the accumulation of heat energy in leaves.

Since transpiration is an energy-consuming process it has often been assumed that it is in the evaporation of water from the leaves that most of the energy absorbed by them is dissipated. We might well inquire, therefore, regarding the possible efficiency of transpiration as an energy-dissipating process. The evaporation of a gram of water at 20°C. requires 584 g.-cal. For the dissipation of 0.65 g.-cal. of heat, therefore, the evaporation of 0.0011 g. (0.65/584) of water per square centimeter of leaf area per minute is required. This is equivalent to 6.6 g. ($0.0011 \times 100 \times 60$) of water per square decimeter of leaf area per hour, a rate of transpiration which is seldom attained by plants for any sustained period of time under natural conditions. Evidently transpiration, even when occurring at its maximum rate, can seldom account for dissipation of more than part of the radiant energy which is absorbed by leaves in intense sunlight.

The fact that transpiration is often inadequate in direct sunlight to account for the dissipation of all the absorbed radiant energy leads naturally to the question of whether it is at all essential for this process. Observations have shown that leaves in which the transpiration rate is greatly reduced—as, for example, those in which the stomates are plugged with vaseline, leaves in a wilted condition, or the leaves of plants in xeric habitats during dry seasons, in which the occurrence of any appreciable amount of transpiration is precluded by the lack of soil water—seldom have temperatures which are anywhere near the thermal death point, even when exposed to direct sunlight. Although transpiration often accounts for the dissipation of some or even most of the absorbed radiation, in so doing it apparently plays no essential role since absorbed radiation can be dissipated by purely physical means. As soon as the temperature of a leaf exceeds that of the surrounding atmosphere, it will lose heat to the atmosphere in the same way that any other object heated above its environmental temperature does, *i.e.*, by one or more of the purely physical processes of conduction, radiation, and convection. The term *thermal emission* is frequently applied to this physical loss of heat energy by leaves and other objects.

Actually, instead of benefiting plants, transpiration may often be detrimental. Under conditions of deficient soil water or of high transpiration rates even when the soil water supply is adequate, this process results in a diminution in the water content of a plant and in the turgidity of its cells. Prolonged drought conditions ultimately result in a severe desiccation with the consequent death of all except the most drought resistant species. When the diminution in water content is less severe, a train of other effects such as a decrease in the turgidity of the cells, stomatal closure, and reduction in rate or cessation of photosynthesis are induced, all of which have the end result of checking the growth of the plant. It is probably true that lack of water is more often the limiting factor in plant growth than any other single factor. Furthermore, deficiency of water is probably responsible for the death of more plants under natural or even cultural conditions than any other single cause.

The fundamental effects of transpiration upon the plant are not to be sought in any hypothetical "advantages" of the process to the plant, but in its very real and readily ascertainable influences upon the water relations of plant cells and tissues, and through these its effects upon other plant processes. In spite of the fact that transpiration may be regarded in a sense as an incidental phenomenon, its indirect influence upon the metabolic processes of plants is a profound one. It is this fact, more than any other, which justifies intensive and critical studies of this process.

The Loss of water from Plants in Liquid form. If a pot of young oat plants is copiously watered and then enclosed in a bell jar, in a relatively short time a slow exudation of water begins at the tip of each leaf. The drops which form at the leaf apexes gradually enlarge and eventually may run down the side of the leaf or fall off. This process of the escape of liquid water from uninjured plants is called *guttation* (Fig. 6.20). It is of very general occurrence, having been recorded in plants of more than 300 genera, although there are many species in which it has not been observed. Guttation occurs most frequently and abundantly under conditions which favour rapid absorption of water by the roots, but which result in a reduced rate of transpiration. In most temperate regions such conditions occur most frequently during the late spring when there is often an alternation between relatively cool nights and relatively warm days. Guttation is frequently observed at that season, usually taking place at night or during the early hours of the morning. The drops of guttation water which form at the tips of grass blades and the tips or edges of the leaves of other plants are often erroneously considered to be dew.

The process of guttation occurs from pore-like structures known as *hydathodes* (Fig. 6.21). These structures are also sometimes referred to as *water stomates* or *water pores*. As is indicated, a typical hydathode consists of an enlarged stomate-like opening below which is a rather large chamber, bordered by a mass of thin-walled, loosely arranged cells called the *epithem*. The xylem elements of a vascular bundle terminate just below the epithem.

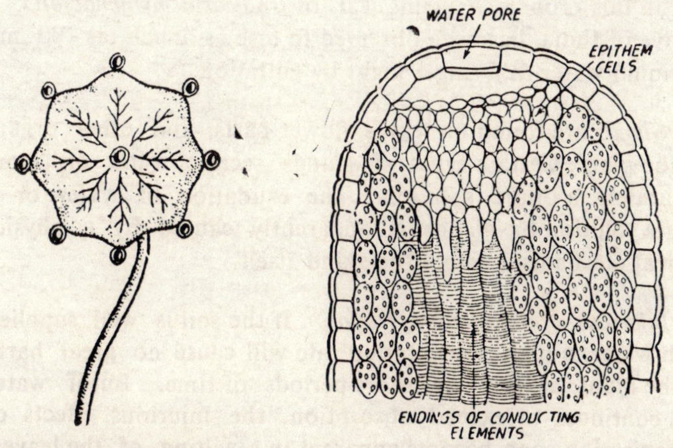

Fig. 6.20. Exudation of drops of water from *Nasturtium* leaf (Guttation).

Fig. 6.21. V. S. near a vein ending to show Guttation through hydathodes.

The exudation of water through hydathodes is considered to result from a pressure which develops in the sap of the xylem elements, and not to any locally developed pressure in the hydathode itself. This pressure is generally believed to be identical with the so-called "root pressure". It is supposed that the water is forced from the vessels through the intercellular spaces of the epithem layer and out of the plant through the pore of the hydathode. This water is not pure but contains at least traces of solutes, including sugars, amino acids, and mineral salts. Guttation water sometimes evaporates so rapidly as to leave deposits of salts on the margins or tips of leaves. Reabsorption of drops of guttation water also sometimes occurs, such reabsorbed water often containing a greater concentration of salts than the original exudate because of solution in it of previously deposited encrustations of salts. Either evaporation, or reabsorption of guttated water, or both, may result in injurious effects on plants and are a probable cause of tip burn in leaves of some species. While the volume of water exuded by most plants of temperate regions in guttation is usually small, some tropical species lose large quantities in this process. A young leaf of *Colocasia nymphaefolia*, a native of India, has been observed to lose as much as 100 ml. of liquid water in a single night by guttation.

Glands are found on leaves, flower parts, and other organs of the plant. Certain types of glands secrete water or, more accurately, a dilute solution. The exudation of water or a dilute solution by glands is apparently caused by a physiological mechanism within the gland itself.

Injurious effects of transpiration. If the soil is well supplied with water, a high transpiration rate will cause no great harm to the plant, at least over short periods of time. But if water loss continues to exceed absorption, the injurious effects of transpiration soon become apparent in a wilting of the leaves, a result of the loss of turgor. The degree of wilting and the water loss necessary to produce wilting vary with different plants. Thin leaves composed chiefly of thin-walled parenchyma cells become visibly wilted in a short time. Leaves of the jewelweed (*Impatiens*) wilt in less than an hour after picking. Mountain laurel (*Kalmia*), on the other hand, shows no sign of wilting after a much longer period, as the leathery leaves contain a considerable amount of supporting tissue.

Wilting is classified as *temporary* and *permanent*. Temporary wilting occurs when there is still water available in the soil, although there is a temporary excess of transpiration over absorption. Plants usually recover from temporary wilting as soon as the rate of transpiration decreases. Leaves which have wilted during the day will recover at night and appear fully expanded in the early morning. Leaves may also regain their turgor during the day if the temperature drops or if a light rain falls, even though the water does not penetrate to the roots.

Permanent wilting, on the other hand, results from a deficiency of water in the soil. The roots are unable to absorb water, and the plant dies unless the water supply of the soil is renewed. In exceptional circumstances, however, as when a dry, hot wind is blowing, plants may wilt and die from the drying out of their tissues, even though soil water is still available to the roots.

Repeated temporary wilting will have an adverse effect on the metabolism of the plant, and plants wilted repeatedly become stunted. An important reason for this is that water deficiency reduces the growth rate of young tissues, affecting particularly the processes of cell division and cell enlargement. Another factor is that retardation of the growth rate results in a decrease in the amount of food used by growing tissues, and carbohydrate accumulation almost invariably accompanies drought.

It is through such processes as the above that drought affects crops. There are many areas in the world, including some in our own country, where rainfall is barely sufficient or is insufficient for the growth of wheat, corn and other cereals, and where drought is a constant menace. The stunting of growth that results from excessive transpiration affects all parts of the plant. Growth in height and diameter is retarded, and older leaves wilt and die. The yield of fruit or seed is markedly reduced, and plants may die before they can bear at all. It is no wonder that drought and famine have gone hand in hand through the centuries.

Water Balance : A Shifting Equilibrium

The water balance of a plant is given by the *difference* between the rates of water intake and water loss :

Water balance=water absorption—transpiration.

Here transpiration is viewed as a measure of the expenditure of water and not as a physical process ; therefore it is expressed in terms of the amount of water lost per unit mass (usually fresh weight) and not per unit area. It would be better, however, to compare absorption and transpiration with the water content of the plant, expressing the *water turnover* as mg water evaporated per g water content. The *water turnover rate* gives the percentage of the water present in the plant (in the leaf) which is lost in a given period of time (minute, hour, day) and must be replaced if the water balance is to be maintained.

A satisfactory *water balance* can be maintained only if the rates of uptake, conduction and loss of water are suitably adjusted. The balance becomes *negative* as soon as the supply of water no longer meets the transpiration requirements. If the stomata narrow as a result of this deficit, so that transpiration is decreased while uptake continues as before, the *balance is restored* after a transient overshoot to positive values. Thus the water balance of a plant oscillates continually between positive and negative deviations. It is useful to distinguish the short-term oscillations from the long-term disturbances of this equilibrium. *Short-term fluctuations* reflect the interplay of the various water-regulatory mechanisms, particularly the changes in stomatal aperture (Fig. 6.22). There are more marked departures from equilibrium in the *course of a day*, particularly in the alternation between day and night. In the daytime, in a natural habitat, the water balance is almost always negative ; the water content of the plant is not restored until evening or during the night (and then only if there are sufficient water reserves in the soil). During dry periods the water content is not entirely restored overnight, so that a deficit accumulates from day to day, until the next rainfall ; similarly, there are *seasonal fluctuations* in water balance.

Determination of water Balance

Water balance can be *computed directly* from quantitative determinations of water up-take and transpiration. One can also make an *indirect estimate* of the water balance through its effect upon water content or water potential of the plant. A negative balance always eventually produces a decrease in turgidity and water potential of the tissues. These changes appear first in the leaves, which are the site of most intensive evaporation and moreover are the furthest removed from the roots.

Change in Water Content as an Indicator of Water Balance. A water deficit can be demonstrated by repeated measurements of the water content of leaves and other parts of the shoot. The actual water content at any given time (W_{act}) must be given with respect to a standard measure—for example, the water content of the leaves under conditions of saturation (W_s). The water *content* at any particular time of observation can be expressed as the "*relative water content*", a percentage of the water content at saturation :

$$RWC = \frac{W_{act}}{W_s} . 100 \, [\%].$$

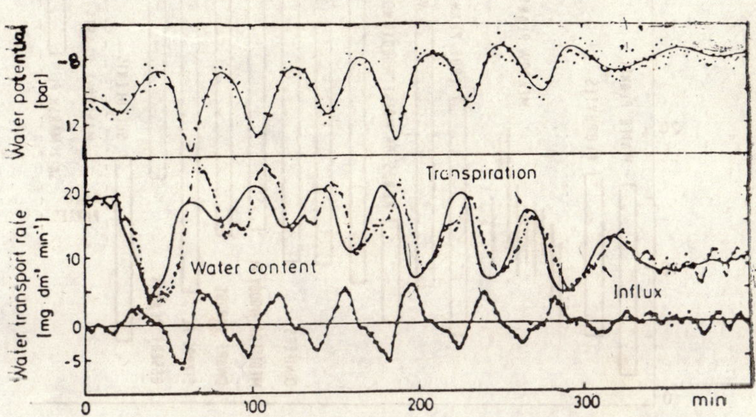

Fig. 6.22. Short-term fluctuations in water turnover, water balance and water potential of cotton leaves. During the phase of rapid transpiration the water content of the leaf falls and its water potential is reduced. The amount of water passing through the petiole (influx) follows a curve 180° out of phase with that for water potential. The short-term fluctuations in transpiration are brought about by oscillation of the stomatal aperture.

A measure of water *deficiency* is the *water saturation deficit*. The water saturation deficit indicates how much water a tissue lack s as compared with complete saturation.

$$WSD=\frac{W_s-W_{act}}{W_s}.100.$$

The water saturation deficit increases as a result of maintained negative water balance, and the relative water content under the same conditions decreases.

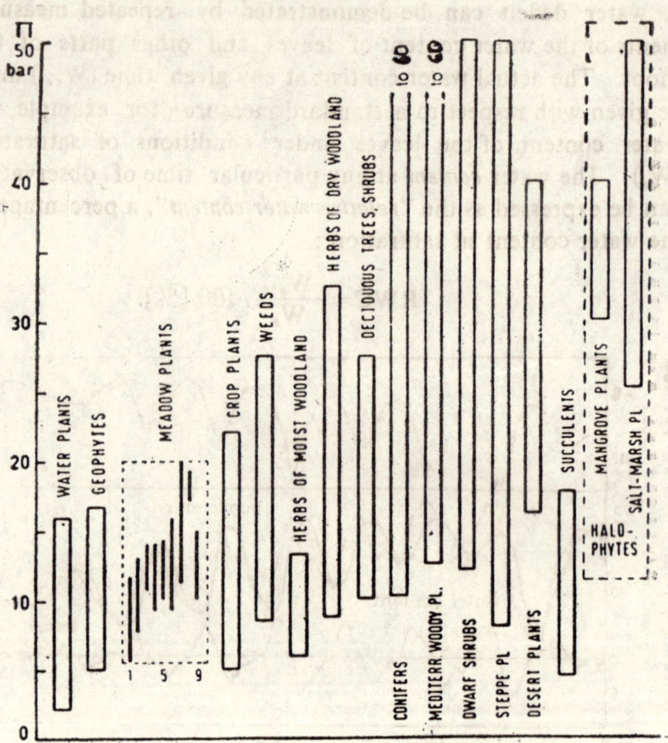

Fig. 6.23. Ranges for the values of osmotic pressure of leaves of ecologically different types of plants (the osmotic spectrum). The sub-ranges (*black bars*) shown for the meadow plants illustrate how to interpret the osmotic range given for each plant group ; that is, it is derived from the difference between the lowest and the highest osmotic pressures found among all plants studied in the ecological group. *1, Polygonum bistorta* ; *2, Taraxacum officinale* ; *3, Galium mollugo* and *Campanula rotundifolia* ; *4, Achillea millefolium* ; *5, Tragopogon pratensis* ; *6, Poa pratensis* ; *7, Melandrium album* ; *8, Cynodon dactylon* and *Lilium perenne* ; *9, Arrhenatherum elatius*.

Changes in Water Potential. Fluctuations in water content necessarily affect the concentration of the cell sap and the water potential of the cells. The *osmotic pressure*, as a component of the water potential of the cell, provides an indication of changes in the water balance. The osmotic pressure rises as long as water balance is negative. The absolute values of π differ according to plant species, form, state of development and the tissue concerned. The range within which osmotic pressure fluctuates also varies in these respects. There are plants which can support great osmotic stress without damage (*euryhydric species*). In contrast, *stenohydric* species suffer impairment of their vital functions with even slight increases in osmotic pressure. The optimum value (π_{opt}) is considered to be the osmotic pressure in a state of optimum water content; the osmotic maximum (π_{max}) is the value found in nature under conditions of extreme

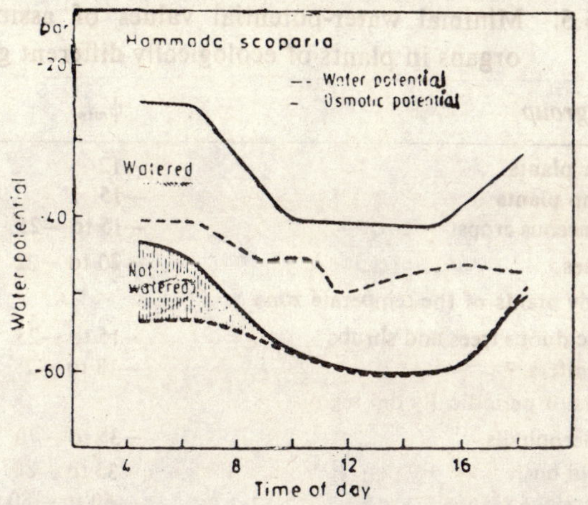

Fig. 6.24. Changes in water-potential and osmotic potential in the leaves of *Hammada scoparia* during the course of a day at the end of the dry period in the Negev Desert, in one case under the natural drought conditions and in the other with irrigation of the shrubs. After Kappen *et al.* (1976). The relationships in salt-filtering and nonfiltering mangrove plants are discussed by Miller *et al.* (1975). For variations during the life cycle of soybeans see Sionit and Kramer (1976); of sorghum and maize, Fereres *et al.* (1978).

drought. To obtain a picture of the water balance in plants of different climatic regions and habitats, one can summarize the full range of variation of π values in the graphic form of "osmotic spectra". In such a diagram (Fig. 6.23) one can see that the plant species fall into ecologically distinct groups.

A measure of changes in water balance that is more sensitive than the osmotic pressure is the leaf water potential. The immediate result of water deficit is loss of turgor accompanied by a distinct decrease of water potential (Table 6.5). It is especially in the range of slight water dificits that $\triangle \psi$ undergoes greater change than π, so that in this range water potential (usually determined by measuring the xylem water potential) is the most reliable indicator of balance. As a result the range of variation of water potential, the immediate response of the shoot to an inadequate water supply, is always greater than that of the osmotic potential (Fig. 6.24).

Table 6.5. Minimal water-potential values of assimilative organs in plants of ecologically different groups.

Plant group	ψ_{min}
Water plants	—12
Swamp plants	—15
Herbaceous crops	—15 to —25
Grasses	—20 to —22 (—45)
Woody plants of the temperate zone	
Deciduous trees and shrubs	—15 to —25
Conifers	—18 to —25 (—60)
Plants of periodically dry regions	
Sclerophylls	—35 to —70
Arid bush	—35 to —80
Garrigue plants	—40 to —80
Desert plants	
Shrubs	—55 to —90 (—160)
Succulents	—18 to —20
Mangroves	—50 to —60
Halophytes	—30 to —55 (—90)

The numbers in parentheses are extreme values.

Water Balance in Different Plant Types

In every climatic zone and habitat, plants with quite diverse water relationships grow side by side. They can be classified in two fundamental categories with respect to water balance, the hydrostable (isohydric) and hydrolabile (anisohydric) types (Fig. 6.25).

Hydrostable species can, to a great extent, maintain a favourable water content throughout the day, the water balance remaining near zero ; their stomata respond with great sensitivity to lack of water, and their root systems as a rule are extensive and efficient. A further factor stabilizing protoplasmic water content is the presence of water reserves in storage

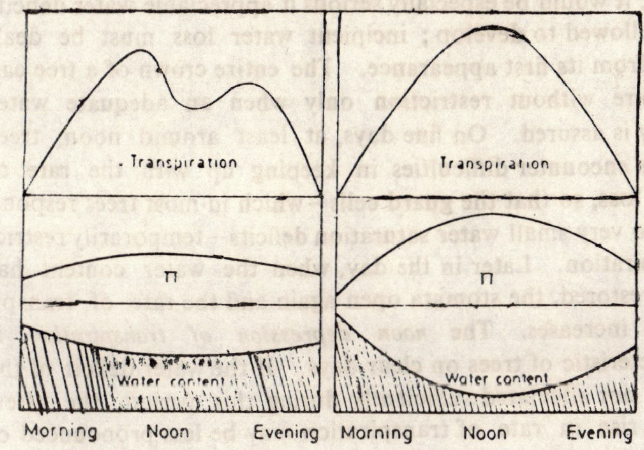

Fig. 6.25. Schematic diagram of the two basic categories of water-balance mechanism as proposed by Berger-Landefeldt (1936). The isohydric type (*left*) avoids pronounced fluctuations in water content and osmotic pressure during the day by stomatal regulation of the rate of water loss. The anisohydric type does not restrict transpiration until it has become very dry ; the latter mechanism allows the water balance to stay negative for extended periods, as is reflected in the wide fluctuations osmotic pressure and water content in the course of a day. For characterization of water balance by the amplitude of water-potential fluctuation see Ritchie and Hinckley (1975).

organs, roots, the wood and bark of the stems and the leaves. Hydrostable plants include trees, some grasses, shade plants and succulents.

Hydrolabile species are in danger of larger losses of water and greater increases in cell sap concentration. Such vascular plants are always *euryhydric* ; their protoplasm must be able to tolerate these rapid and extensive fluctuations in water potential without injury. Many herbs of sunny habitats are hydrolabile, as are the poikilohydric plants.

Trees. In mature trees, with their extensive evaporating surfaces and the long distances water must travel from roots to leaves, it would be especially serious if appreciable water deficits were allowed to develop ; incipient water loss must be dealt with from its first appearance. The entire crown of a tree can transpire without restriction only when an adequate water supply is assured. On fine days, at least around noon, trees always encounter difficulties in keeping up with the rate of water loss, so that the guard cells—which in most trees respond even to very small water saturation deficits—temporarily restrict transpiration. Later in the day, when the water content has been restored, the stomata open again and the rate of transpiration increases. The *noon depression of transpiration* is characteristic of trees on clear days. If the water deficit of the tree is not alleviated sufficiently during this period, the afternoon rise in rate of transpiration may be less pronounced or completely absent. Water-conservation measures are not taken in all parts of the crown at once, but occur in a stepwise manner and in an orderly sequence ; the first and most prominent *reduction of transpiration* occurs in the shady parts of the crown, where the threshold for stomatal closure is lowest, then in the base of the crown, and finally even the leaves of the tree top limit their evaporation. Correspondingly, the rise in water potential in the sunny tops of trees is sharper than in the shaded parts, and these topmost leaves are therefore preferentially replenished with water. The range of *daily fluctuation of osmotic pressure* is very small in trees, rarely exceeding an amplitude of 3 bar ; this means that in spite of the increase in the forces drawing water upward during the course of a day the protoplasm itself is not subjected to any great fluctuations in water potential. Not all trees are hydrostable to the same extent. Trees characterized by good maintenance of water content are conifers, sciophytic species and some heliophytic species such as oaks. There are also trees—for example, species of ash—that do not behave so conservatively; during dry spells the leaves wither prematurely.

Herbaceous Dicotyledons. Among the herbaceous plants one finds all intermediate forms between hydrostability and extreme hydrolability. This *diversity of types* is particularly evident in sunny habitats with a tendency to dryness of the soil. When there is a fairly adequate water supply, some of the dicots transpire without restriction throughout the day, at a rate determined by the evaporative power of the air. Others behave more like trees, depressing the rate of transpiration at midday or even limiting the period of maximal stomatal opening to the early morning and late evening. The extravagant transpiration displayed in the first group can be afforded only by plant species which either can draw upon abundant water reserves through an extensive root system or are able to endure a high degree of desiccation. In the latter case, the osmotic pressure of the cell sap can rise by 3—6 bar during the course of a day. Such species are found frequently among the herbaceous heliophytes.

Grasses and Sedges. These, too, comprise both hydrostable and hydrolabile species. The stomata of hydrostable grasses respond with extreme sensitivity to the first sign of negative water balance. The stomata therefore *close gradually* during the morning so that often no sharp depression is observed around midday. As the leaves age—at the beginning of the dry period in savanna grasses—the stomata lose their mobility, and the grasses gradually lose control of the water balance. Even though the soil is dry, the plants continue to transpire without restraint until the leaves wither.

Water Balance During Drought

Drought as a Stress-Inducing Factor. The term "drought" denotes a *period without precipitation,* during which the water content of the soil is reduced to such an extent that the plants suffer from lack of water.

In dry regions drought is of such regular and prolonged occurrence that annual evaporation exceeds total annual precipitation. The climate in this situation is called *arid*, as opposed to the *humid* climate in regions with surplus precipitation. About ⅓ of the earth's land area has a rain deficit, and half of this (about 12% of the land area) is so dry that annual precipitation is less than 250 mm, not even a quarter of the potential evaporation. Extensive dry regions are found mainly between 15° and 30° north and south latitude, and on the lee sides of high mountain chains intercepting rain-bearing winds (Fig. 6.26). In lands far from an ocean there is a gradual transition from a humid climate, through a semi-arid inter-

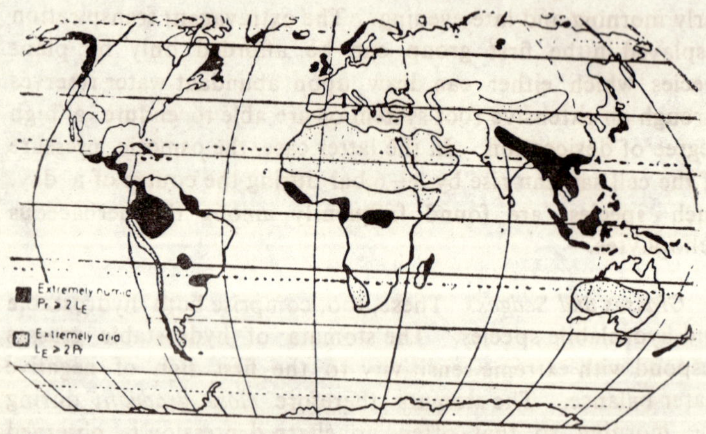

Fig. 6.26. The distribution of extremely humid and extremely arid regions on the earth. Extremely humid : annual precipitation at least twice the amount of water evaporated annually. Extremely arid : annual evaporation at least twice as great as annual precipitation. The demarcation of extremely humid and arid regions is based on the maps of Geiger (1965), giving amount of precipitation and actual evapotranspiration.

mediate region having occasional or periodic dry periods, to an arid region characterized by permanent drought and increased salinity of the soil.

A method used by M.I. Budyko and A.A. Grigorjev to determine the *degree of aridity* is based on the notion that over large areas, and on average over the year, the radiation input (positive radiation balance Q_R) is used up to evaporate water. The energy consumed for evaporation can be computed by using the specific heat of evaporation of water λ. It follows that Q_R is a measure of the potential evaporation in a region. The "Radiational Aridity Index" calculated from the radiation balance relates the annual precipitation Pr, multiplied by λ in order to obtain units of energy, to the radiation balance Q_R as follows :

$$\text{Radiational Aridity Index} = \frac{\lambda Pr}{Q_R} \ [\text{MJ} \ . \ m^{-2} \ . \ yr^{-1}].$$

Arid regions are characterized by an index greater than 1. For steppes and savannas the aridity index lies between 1 and 2, for semi-deserts between 2 and 3.5, and for deserts over 3.5. Values below 1 indicate a humid climate. In the temperate zone the indices range from 0.3 to 1 ; in tropical rain forests they are about 0.5 and in the subarctic, about 0.3.

The relationship between annual precipitation and annual evaporation gives only a rough indication of the humid or arid character of an area. As far as the plants growing there are concerned, the important thing is that there should be an assured water supply at the time of greatest need—during the growing season. To provide a picture of *relatively humid and relatively arid seasons,* one can construct climatic diagrams like those suggested by H. Gaussen and Walter (Fig. 6.27). In these, the monthly totals for precipitation and the monthly average air temperature are plotted for the year, the scale being such that *1°C corresponds to 2 mm precipitation.* In such a plot, the temperature curve serves as an indicator of the progressive change in *evaporative power of the air* during the year. That part of the year during which the precipitation curve lies below the temperature curve is a time of drought for the majority of plants that are not irrigated and cannot utilize the groundwater. This procedure offers the advantage that it can be used at stations for which no data on evaporation or radiation are available. Limited precipitation does not in itself bring about

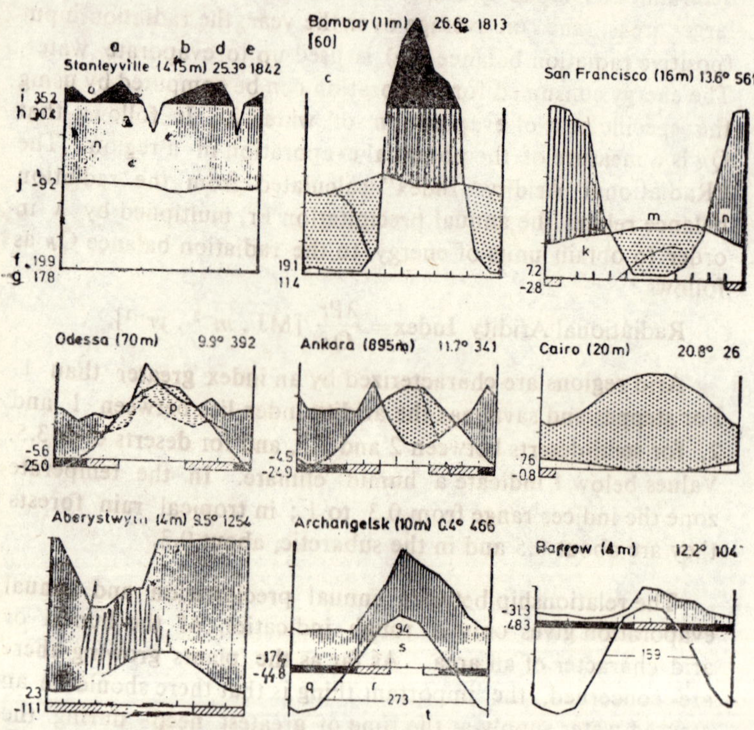

Fig. 6.27. Climate diagrams for *Stanleyville* (Congo, permanently wet equatorial climate); *Bombay* (India, tropical summer-rain climate); *San Francisco* (California, winter-rain region with summer drought); *Odessa* (Black Sea coast, semi-arid steppe climate); *Ankara* (Turkey, Mediterranean climate type with equinoctial rain); *Cairo* (Egypt, subtropical desert climate); *Aberystwyth* (Wales, maritime-temperate climate); *Archangelsk* (taiga zone on the White Sea, cold-temperate climate); *Barrow* (Alaska, arctic tundra climate).

aridity; the cold polar regions have little precipitation but are not arid, since there the evaporative power of the air is low.

Restriction of Water Consumption During Drought

Plants in humid regions, when no rain falls for days or weeks at a time, so that the water reserves in the soil are depleted and their water balance becomes increasingly unfavourable, progressively reduce their water consumption by *opening their stomata less and for shorter periods.* At first transpiration is reduced during the hottest hours of the day, then the afternoon resumption of transpiration is omitted, and finally the stomata open only in the morning. Eventually, while their water content is still adequate, the plants transpire only through the cuticle.

Plants of dry regions, as a rule, have roots that grow deep into the soil or have water-storing tissues, so that they are not forced to a drastic restriction of transpiration (and simultaneously of CO_2 uptake) so soon. This property reflects the adaptation of such plants to arid habitats. *Desert shrubs* in the dry season consume 20—40% of the amount of water given off in the wet season, and in *subshrubs and herbaceous species* this value can be as much as 60—80% (Table 6.6). Some ephemeroid species never restrict evaporation, continuing to transpire at a high rate until they dry up. By contrast, evergreen *sclerophyll trees and shrubs* in regions with periodic summer drought can very effectively limit water consumption, reducing it to 10—20% of that when water is readily available (Table 6.6). The most effective restriction of transpiration is found in *CAM plants*, the stomata of which are completely closed all day during dry periods : if they open at all, it is at night.

Precipitation in solid form cannot be taken up immediately by the plants. Winter, for plants projecting above the *snow cover*, is thus not only a cold but also a dry season. Wherever the snow is blown away in winter the young trees perish, and where the rooted soil strata are frozen too long, or permafrost rises too high, no trees can grow. The forest becomes sparse and gives way, in the north, to tundra, and in the mountains to scrub, dwarf shrubs, or alpine herbaceous communities.

Table 6.6. Total daily transpiration of plants in periodically dry regions, comparing the rate under conditions of adequate water supply with that during prolonged drought, under the conditions of evaporation prevailing in the habitat in each case.

Plant	*Total daily transpiration* ($g\ H_2O \cdot dm_s^{-2} \cdot d^{-1}$)		*Level to which transpiration is reduced* (%)
	Rainy season	*Dry season*	
Mediterranean maquis			
Olea europaea	10	2.8	28
Laurus nobilis	3.3	0.5	15
Phillyrea media	10.4	1.5	14
Quercus ilex	9.5	1.0	11
Australian sclerophyll woodland			
Acacia acuminata	12.4*	2.9*	23
Acacia craspedocarpa	18.6*	3.9*	21
Banksia menziesii	7.4	0.6	8
Australian bush			
Hakea preisii	19.8	13.4	68
Eremophila miniata	33.3	17.9	54

Halophyte desert in Utah			
Atriplex confertifolia (C_4)	14.2	2.7	19
Ceratoides lanata (C_3)	14.0	0.2	1
North African desert plants			
Nitraria retusa	19.1	15.0	78
Zilla spinosa	22	14	64
Zilla spinosa	38	15	39
Zygophyllum coccineum	15	7	47
Pennisetum dichotomum	15	6	40
Haloxylon persicum	26	10	38
Hammada scoparia	4†	1.5†	38
Anabasis articulata	3.1†	1.0†	32
Retama retam	25	7	28
Artemisia herba-alba	6†	1.6†	27
Noea mucronata	5.5†	1.0†	18
Cactus desert in California			
Ferocactus acanthodes	1.54	0.32	21

*g H_2O . g^{-1} dry weight . d^{-1}. †g H_2O . g^{-1} fresh weight . d^{-1}.

Water deficiencies can thus occur *in winter* and can even cause injury if replenishment of the water supply is prevented by prolonged frost. In regions with long cold winters (mountains, the arctic and subarctic), the soil remains frozen near the surface for months, and the water ducts in shoots and roots become filled with ice (the water in the tracheary elements is frozen at temperatures below—2°C). The twigs not protected by snow, however, continue to lose water. The water balance

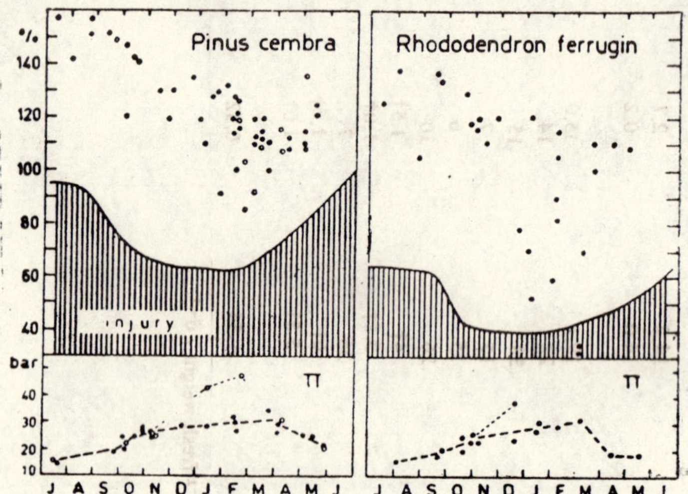

Fig. 6.28. Water content (as % of dry weight), osmotic pressure and drought resistance of pine needles and the leaves of *Rhododendron ferrugineum* at the alpine treeline. The *circles* in the pine diagram denote small trees (up to 2 m high), and the *dots* refer to mature trees. Pine needles and *Rhododendron* leaves acquire an enhanced protoplasmic resistance to desiccation at the beginning of their winter dormant period, and do not lose it until the beginning of the growing season in the spring. In late winter, when the ground is still frozen but the sun warms more strongly the branches not protected by the snow cover, the water content of the *Rhododendron* and of small pine trees falls, approaching and sometimes passing the limit for safety of the plant. Because of the water reserves in their trunks, larger trees are not as greatly endangered as are young trees and shrubs.

becomes negative, and the water content deteriorates progressively. The problem is greatest in late winter, when the soil has not yet thawed but the sun increasingly warms the branches and stimulates transpiration (Fig. 6.28).

Drought resistance is the *capacity of a plant to withstand periods of dryness.* This capacity is a complex characteristic. The prospects for survival of a plant under extreme stress due to drought are better, the longer a dangerous decrease in the water potential of the protoplasm can be delayed (the *avoidance* of desiccation) and the more the protoplasm can dry out without becoming damaged (the capacity to *tolerate* desiccation). In the terminology of Levitt (1958), drought resistance is the result of desiccation avoidance and desiccation tolerance. Among vascular plants desiccation tolerance is very slight, so that differences in the drought resistance of different species are chiefly due to differences in desiccation avoidance. In regions with regular dry periods, and in deserts, there are plants that evade drought by timing their growth and reproduction to occur in the brief period during which sufficient water is available.

Drought-Evading Xerophytes

Drought-evading plants as a rule are not drought-resistant. Survival of dry periods requires only the appropriately timed production of desiccation-resistant seeds or organs specially protected from drying out. *Pluviotherophytes* are short-lived vascular plants that germinate after fairly heavy rainfall and rapidly complete their cycles of development. They pass the dry season as seeds, unharmed by desiccation. *Geophytes* have underground organs full of water (rhizomes, tubers, bulbs) that can survive. When the rainy season begins they sprout immediately, utilizing stored carbohydrates, and soon flower and bear fruit.

Desiccation Avoidance

Desiccation is delayed by all those mechanisms that enable the plant to maintain a favourable tissue water content as long as possible despite dryness of air and soil. This is achieved by improved *uptake* of water from the soil, by reduced *loss* of water (early and effective increase in diffusion resistance and reduction of the transpiring surface), by a high capacity for water *conduction* and by the *storage* of water. These behavioural measures for desiccation avoidance are also reflected in the morphology of the plant.

Water uptake by an extensive root system with a large active surface area is improved by rapid growth into deeper soil strata. There the roots reach horizons containing water on which the plants can draw for some time. The seedlings of woody plants in dry regions have taproots ten times as long as the shoot, while the grasses in such places elaborate a dense root system and also send their threadlike roots to depths measured in meters. The ratio between the masses of shoot and root is shifted further in favour of the roots, the more the exposure to drought. The situation becomes critical when there is not enough room for expansion of a root system. Plants with extensive roots (woody plants in particular) growing on shallow soils are especially endangered by drought. Analogous problems arise when one attempts to establish plants on and over man-made structures (flat roofs, tunnels and other underground construction). Even in humid regions only drought-resistant plants are suitable for such purposes, for they can manage with the small amounts of water stored in the limited volume of soil.

Increased diffusion resistance of the shoot surface is achieved by early closure of the stomata and effective cuticular protection against transpiration. As a *modulative* adaptation, this takes the form of a rise in the abscisic acid level in the plant, which enhances the readiness of the stomata to close. A *modificative* change is the development, on leaves that grow when water is scarce, of smaller stomata more densely arrayed. Such leaves reduce transpiration more rapidly following the onset of stomatal regulation. Moreover, the epidermis of such leaves is made more impermeable by thicker cuticular layers and, in many cases, a dense covering of hairs. Restriction of transpiration by these natural means can be supplemented artificially by *antitranspirants,* which form surface films or trigger closing of the stomata. Substances that cause *stomatal closure* are abscisic acid, inhibitors of energy-supplying electron transfer (*e.g.*, phenyl mercury acetate, acetyl salicylic acid, thiocarbamate, and other urea derivatives) and chemicals that affect permeability (partially oxidized fatty acids, sulphonates). Water-vapour release through the stomata, and to an even greater extent that through the cuticle, is diminished by spraying to form a *coating* of emulsions of polyethylene, acrylic, silicon and polyterpene compounds. Because of the undesirable side effects on CO_2 exchange and temperature regulation, the use of antitranspirants is practicable only as an emergency measure.

Reduction of the transpiring surface is brought about rapidly and reversibly by folding and *rolling up of the leaves.* This mechanism is widespread among grasses and sedges; in *Stipa tenacissima* it cuts transpiration to 40% of that in outspread leaves. A particularly effective reduction of the transpiring surface of a plant is the partial or complete *abscission of the leaves.* A number of woody plants in dry regions shed their leaves regularly during the drought season. Through peridermal transpiration the trunks and branches of large trees lose only 1/300 to 1/3000 of the amount of water evaporated from the leaves when the water supply is good. Species of *Ephedra, Sarothamnus, Cytisus, Spartium, Retama,* and desert shrubs like *Zilla macroptera* and *Zollikoferia arborescens* shed their leaves as required, reducing the surface area to $^1/_3$—$^1/_5$ of its usual value. The relationship between transpiring surface and water stored in the tissues is expressed by the *specific leaf area* (*SLA*) of the leaves or shoots :

$$SLA = \frac{\text{Surface area } (dm^2{}_1)}{\text{Fresh weight of leaf (g)}}$$

Alternatively, one may compare leaf area with the dry weight of the entire plant ; this is the leaf area ratio (*LAR*).

Since the rate of water loss increases with the transpiring area, a limited leaf area conserves the water in the leaf. Leaves developed under conditions of poor water supply are, as a rule, correspondingly smaller and more deeply laciniated, and have a smaller specific leaf area (*xeromorphosis*).

The *capacity for water conduction* is increased by enlargement of the area of the water-conducting system (more vessels, dense leaf venation) and reducing the transport distance (shorter internodes). When the transpiring surface is reduced the *relative* area of the conducting system is increased even though the *absolute* area (conducting cross section) is unchanged.

With *water storage*, mechanisms for desiccation avoidance reach their peak of perfection—especially when they are coupled with surface reduction and high transpiration resistance of the epidermis. A measure of storage capacity is given by the degree of succulence :

$$\text{Degree of succulence} = \frac{\text{Water content at saturation (g)}}{\text{Surface area (dm}^2\text{)}}$$

Under certain circumstances succulent plants can survive years of drought, and after the last rain sufficient water is stored that months can pass before the stomata are permanently closed (Fig. 6.29).

Desiccation Tolerance

Desiccation tolerance refers to the species-specific and adaptable *capacity of protoplasm to endure severe loss of water*. A water deficiency causes progressive loss of protoplasmic turgor and an increase in solute concentration. The initial result of these two effects is a disturbance of cell function ; then functional deficits appear, and finally protoplasmic structures (biomembranes in particular) are damaged. The resistance to desiccation is measured by equilibrating parts of plants or pieces of tissue (sections or cut-out disks), unprotected against transpiration, with air of known humidity. The lowest relative humidity (or the corresponding water-potential values) at which the cells are just capable of survival (the "critical limit") or are damaged by a certain amount (*e.g.*, 50% injury is the *desiccation lethality*, DL_{50}) is used as the *measure of tolerance*.

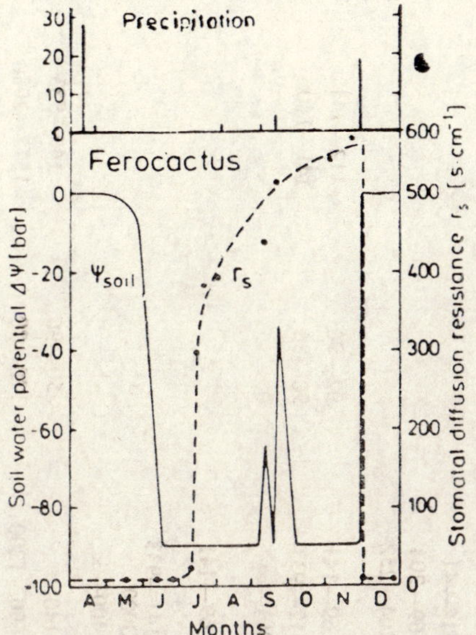

Fig. 6.29. Soil drying (ψ_{soil} at depth of 10 cm) and slowly rising stomatal diffusion resistance (r_s, daily minimum) of *Ferocactus acanthodes* in the Colorado desert during the months of drought between the relatively heavy rainfall in April and November.

In the literature one also finds tolerance given in terms of water content—for example, critical and sublethal (5—10% injury) water content or water deficit. Such data, if carefully interpreted, are informative with respect to the detection of adaptive and seasonal changes in desiccation tolerance, but they should not be used for comparisons of different species since their absolute values are influenced by the anatomical peculiarities of the sample investigated. For example, the fraction of the measured weight associated with sclerenchymatous structures can alter. The desiccation tolerance of plant protoplasm varies over a wide rang (Table 6.7).

Desiccation Tolerant Species. All orders of *thallophytes* contain species capable of tolerating complete desiccation. Many representatives of the bacteria, cyanophytes and lichens can withstand a state of dryness (in equilibrium with the surrounding air) for months and even years, resuming metabolic activity

Table 6.7. Desiccation tolerance of plant cells after 12—48 h in vapour chambers with different relative humidities.

Plant		*Tolerated* % *RH*	*without injury* *bar*	*Moderate injury* % *RH*	*bar*
Marine algae					
Sublittoral algae		99—97	14—41		
Algae of the ebb line		95—86	69—204		
Intertidal algae		86—83	204—252		
Liverworts					
Hygrophytes	usually	95—90	69—141	92—90	112—141
Mesophytes	usually	92—50	112—933	90—36	141—1400
Xerophytes	usually	(36)—0	(1400)—∞	0	∞
Mosses					
Water mosses and hygrophytes		95—90	69—141		
Mesophytes	usually	90—50	141—933		
	extreme	10	3000		
Xerophytes	usually	5—0	400—∞		
Fern gametophytes					
Forest ferns		>90	>140	50—90	140—933
Rock ferns		40—60	690—1200	20—30	1575—2080
Tracheophytes (tissue sections)					
Leaf epidermis				96—92	55—112
Mesophyll		96	55	95—90	69—141
Root cortex				97—95	41—69

as soon as they are moistened again. Some of them can survive weeks in absolutely dry air over concentrated sulphuric acid or phosphorus pentoxide. Other completely desiccation-tolerant plants include some fungus mycelia, various mosses, and some pteridophytes. Among the *flowering plants* there are relatively few desiccation-tolerant species—*e.g.*, *Ramonda serbica* and some species of *Haberlea* (both Gesneriaceae) in the Balkan region, and species of Scrophulariaceae, Myrothamnaceae, Velloziaceae, Cyperaceae, and Poaceae in South Africa.

Desiccation-Sensitive Species. The protoplasm of most plants is extraordinarily sensitive to loss of water and dies after even a slight reduction in water content. The cells in the leaves of homoiohydrous vascular plants perish if they are exposed without protection for a few hours to a relative humidity between 92 and 96% (which in a closed system would be in equilibrium with a solution having $\pi = 55-110$ bar); roots are more sensitive, while buds are more resistant. Plants may become "hardened" to dry conditions, but this shifts the tolerance limit only slightly (3—4% RH). During growth periods the cells are especially sensitive to desiccation; during dormant periods they are somewhat more resistant. Among the *thallophytes* the most sensitive are the planktonic algae and those seaweeds attached so far below the surface that they are normally always covered by water; mosses are also sensitive and are found only where the prevailing humidity is continuously high. Most mosses are adapted to a narrow range of water potential, so that the various species (there are differences even between species of a single genus) are of value as *indicators of the humidity of a habitat*. It should be said, of course, that "hardening" can considerably increase the resistance of mosses to desiccation.

Specific Survival Time

Drought can be so extreme that plants are no longer capable of extracting any water at all from the soil. Drought resistance after complete cessation of water replacement is termed *specific survival time* (*Uberdauerungsvermogen*). This is a measure of the characteristic degree to which a plant species can conserve the water stored in its shoots. According to A. Pisek, specific survival time is computed from the cuticular transpiration (E_c) and the available water (W_{av}) in the plant—that consumed between the time of stomatal closure and the appearance of the first signs of desiccation injury.

$$\text{Specific survival time} = \frac{W_{av}}{E_c}.$$

Survival time is measured in hours or days and indicates how long after stomatal closure the leaves of a plant species can remain undamaged without a supply of water, for a given evaporative power of the air. The amount of available water is greater if the stomata close as soon as there is a slight water deficit, if the tissues can store an abundance of water, and if desiccation injury is incurred only with large water deficits. Survival time of the *leaves* thus depends both on the effectiveness of *desiccation avoidance* and of *desiccation tolerance*. It can be seen that of the factors prolonging life, the *cuticular surface protection* varies most among different species. Only the succulents have a remarkable specific survival time, attributable primarily to their *ability to store water*.

During a prolonged dry spell the water *stored in massive organs* (the trunks and larger branches of trees, and the underground storage organs of herbaceous plants) becomes very significant. Movement of water from the wood and cortex of 10—20-year-old winter-deciduous trees can prolong the survival of the *leaves* by factor of 5—10, depending on the amount of foliage, thickness of the cortex, and anatomical features of the wood ; survival of young trees 0.5—1 m tall can be prolonged by a factor of 2—3. Once trees are fully grown, interspecific differences in ability to protect the leaves from drying by providing water from axial organs tend to disappear ; at this stage the decisive factor is the *ratio of foliage volume to axial volume*. Redistribution of water, from axis and roots to the leaves, is also a useful mechanism in that it is fully effective even *after* the stomata have closed, when the leaves are consuming water most economically.

7

Biochemical Aspects of Phloem Transport

The conducting systems which ars used by cormophytes for the long distance transport of assimilates are composed of living cells. They differ from normal parenchymatous cells in their cytological structure, but they still have a selectively permeable ectoplasm with a plasmalemma. As specialised living cells the sieve elements are supposed to have a metabolism which differs from that of the other cells of the phloem. Sinçe the transport substances in the sieve elements are quantitatively and qualitatively different from the compounds in the donor and acceptor tissues, the loading and unloading of the conducting cells must be accompanied by chemical transformation of the substances. Finally, the maintenance of the plasmatic structures which are involved in the transport processes, and the transfer of the assimilates over long distances with relatively high velocity demand a considerable amount of energy. It is, therefore, clear that various biochemical and bioenergetic processes are involved in phloem transport.

BIOCHEMICAL REACTIONS INVOLVED IN LOADING OF THE SIEVE ELEMENTS

In most plant species sugars or sugar alcohols form the bulk of the transport substances in the phloem. (An exception is

Cacurbita, where nitrogenous substances are quantitatively the most important fraction.) Up to date about 400 species have been analysed for the sugar composition of the sieve tube sap, belonging to 43 orders of the Magnoliophyta (angiosperms) (Fig. 7.1). 145 species from 70 genera and 36 families were studied by Helga Kluge (1967), while most of the other species were examined by Zimmermann.

On the basis of the composition of the sugars in the sieve tube sap three main types of species can be distinguished.

1. *Species with sucrose as the only or at least as the predominant sugar*. In small amounts oligosaccharides of the raflinose type may also be present. To this group belong most of the analysed species, *e.g.* all Fabaceae (which show only sucrose), the only fern species studied up to date and all the species of Pinophyta (gymnosperms) and Liliopsida (monocotyledons) so far examined.

2. *Species which have considerable amounts of oligosaccharides as well as sucrose*. These oligosaccharides belong to the raffinose family (Fig. 7.2) which is characterised by the attachment of one or more galactose residues to the sucrose. These raflinose type sugars were first demonstrated in sieve tube sap by Zimmermann (1957). According to this author they occur in species belonging to Bignoniaceae, Celastraceae, Combretaceae, Myrtaceae, Oleaceae, Onagraceae, and Verbenaceae. In addition H. Kluge (1967) found these oligosaccharides also in members of Anacardiaceae, Buxaceae, Clethraceae, Corylaceae, Rutaceae, Tiliaceae, and Ulmaceae.

3. *Species which contain considerable quantities of sugar alcohols in addition to the sugars mentioned above*. To this group belong species of the Oleaceae (*e.g. Fraxinus, Syringa*), which contain mannitol in the sieve tube sap. In *Prunus serotina* and *Malus domestica* sorbitol was found beside sucrose. A detailed study of a large number of Rosaceae species revealed that all analysed species of the subfamilies Spiroideae, Pomoideae, and Prunoideae contained sorbitol in the sieve tube sap, but not the species of the Rosoideae. Dulutol was found in the sieve tube sap of some Celastraceae.

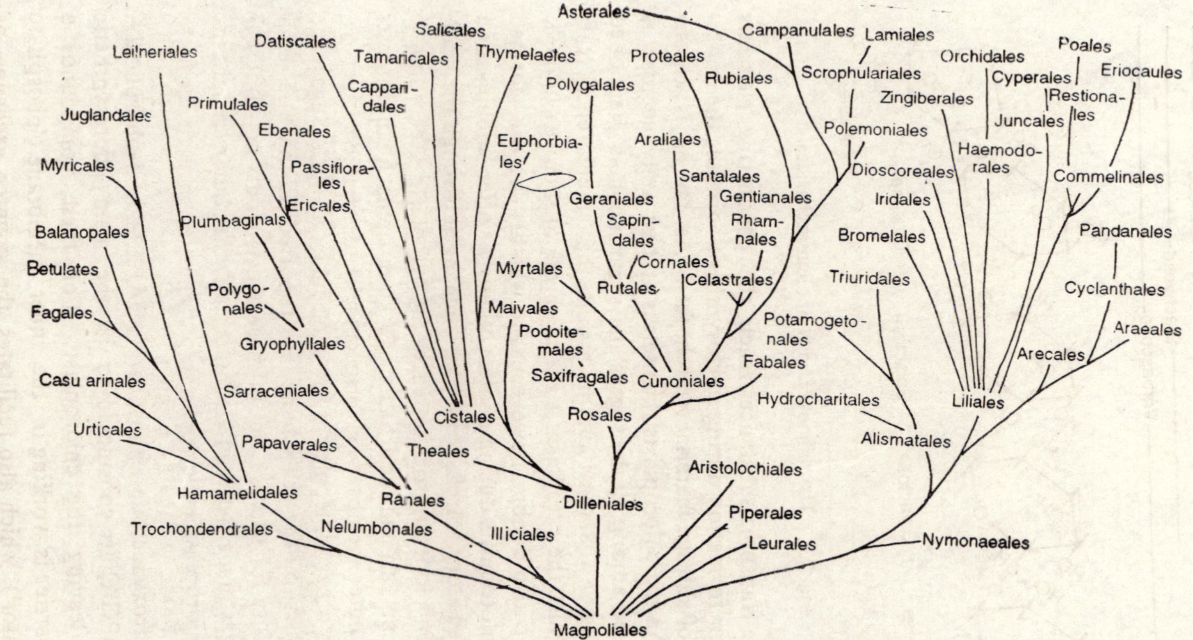

Fig. 7.1. System of the Magnoliophyta (Angiosperms) according to Takhtajan (1959). **Species from the orders in bold type have been analysed for the sugar composition of their sieve tube saps.**

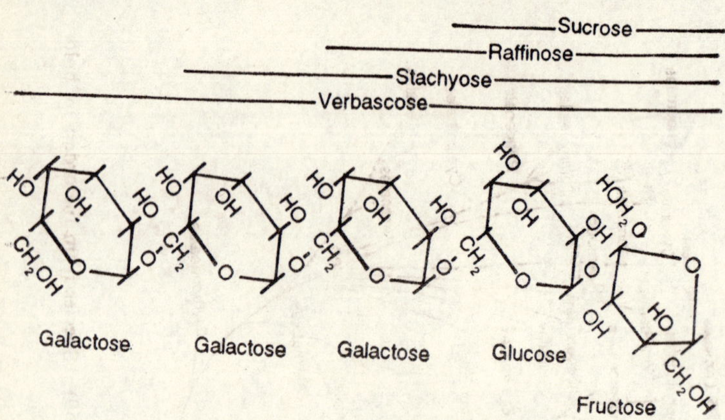

Fig. 7.2. The raffinose family of oligosaccharides.

These transport substances reach the conducting elements after their formation during phytosynthesis or after the mobilisation of storage material (starch, fat). As far as is known to date, in the chloroplasts themselves at best only sucrose is synthesised among all the sugars and sugar alcohols mentioned above (Fig. 7.3) Several authors assume that in chlorenchymatous cells also the cytoplasm is the main site of sucrose synthesis. There are also doubts, however, as to whether sucrose, formed in the chloroplasts during photosynthesis or after mobilisation of the transitory starch, can pass unchanged through the chloroplast envelope in order to reach the cytoplasm. According to experiments of Heldt and Sauer (1971), in spinach chloroplasts only the outer membrane is unspecifically permeable to sucrose, as it is to other molecules of low molecular weight. Sugarcane chloroplasts, on the other hand, are reported to have a high permeability for sucrose but in these experim ents no differentiation was made between the inner and the outer membrane of the chloroplast envelope.

Dihydroxyacetone phosphate (DAP) and 3-phosphoglycerate (3PGA) are considered to be the main photosynthetic products leaving the chloroplast. For these compounds a specific carrier is working in the inner membrane ('phosphate translocator'), which also facilitates the counter exchange of inorganic phosphate. Another carrier in the inner membrane of the chloroplast envelope is involved specifically in the counter exchange of dicarboxylic acids, *e.g.* malate and oxaloacetate. It is possible that this 'translocator' is responsible not so much for the net export of assimilates but is part of a malate/oxaloacetate shuttle between the stroma of the chloroplast and the cytoplasm, which is especially important for the transfer of reducing equivalents from one compartment to the other. In C_4 plants this carrier perhaps plays a specific role in the malate export from the mesophyll chloroplasts.

Sucrose formed in the chloroplast or in the cytoplasm from the low molecular weight products of photosynthesis can be mixed with sucrose from the vacuole or with sucrose synthesised after the mobilisation of starch, as was shown in kinetic analyses of the export of ^{14}C-assimilates from leaves. Since the export rate does not change very much between day and night, the supply of sucrose from these two additional sources is especially important during the dark phase. Sucrose, which is applied to the phloem via cambium or xylem, is split at least partially during the transfer to the conducting channels and is resynthesised again. Application of uniformly labelled ^{14}C-sucrose results in the formation of ^{14}C-sucrose labelled mainly in the glucose moiety, even in the presence of excess ^{12}C-fructose. Also supply of ^{14}C-labelled D-fructose, glucose, maltose, D-galactose, UDPG, fructose-1,6-diphosphate, and fructose-6-phosphate to the xylem of the inflorescence stalks of *Yucca flaccida* yielded sucrose as the main labelled constituent in the sieve tube sap.

In the intact leaf a part of the sucrose formed in the chlorenchymatous cells may reach the conducting elements unchanged.

The other main transport substances in the carbohydrate fraction are easily formed in a few steps from the compounds involved in the sucrose synthesis (Fig. 7.3). The sugars of the raffinose family are synthesised by the transfer of the galactosyl moiety of galactinol (1 L-(+)-O-α-D-galacto-pyranosyl-myo-inositol) to the corresponding receptor molecules Galactinol is regenerated by transfer of a galactosyl residue from UDP-galactose to free inositol. Myo-inositol itself is also trans-

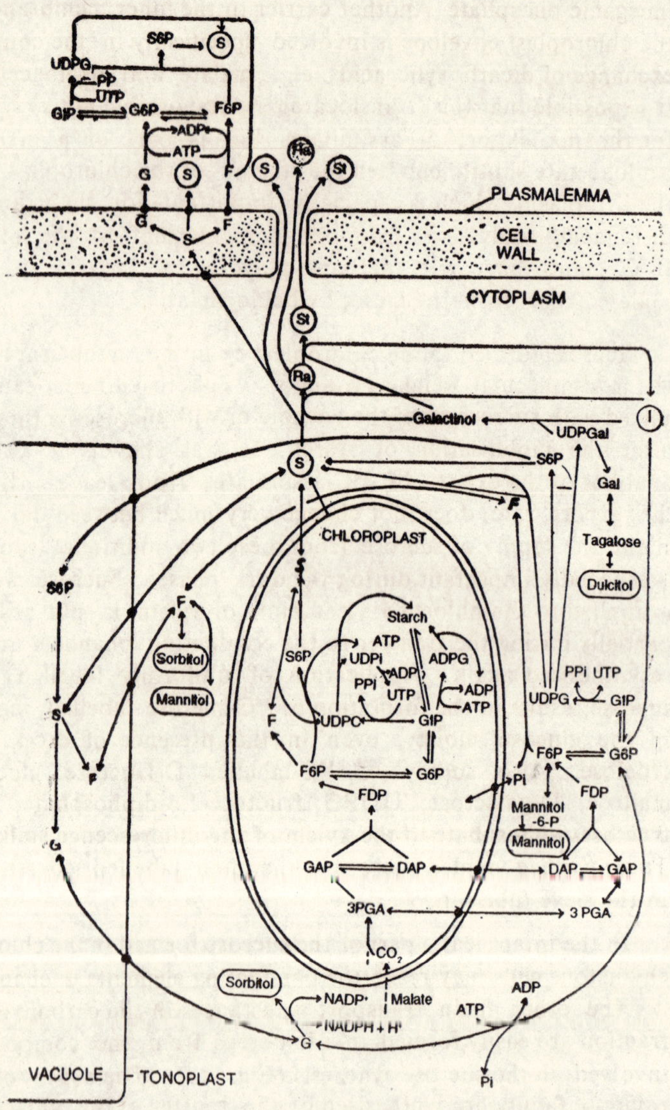

Fig. 7.3. For legend see facing page. Biochemical processes involved with the production and the transfer of transport sugars and sugar derivates in a chlorenonymatous cell. ● Supposed carriers in the membranes. S, sucrose ; S6P, sucrose-6-phosphate ; G, glucose ; F, fructose ; 3PGA, 3-phosphoglycerate ; GAP, glyceraldehyde-3-phosphate;

located in the sieve tubes in significant amounts ; it is formed from glucose-6-phosphate. It is unknown whether galactinol itself enters the sieve tubes, but only very little of this compound is found in the stem and root. A kinetic study of the labelling in the leaves of *Lamium maculatum* following photosynthesis in $^{14}CO_2$ revealed that the galactinol pool is saturated with ^{14}C more rapidly than the sugars of the raffinose family. Also label accumulates initially in the galactose moiety of galactinol, then in the galactosyl part of raffinose and stachyose. Under conditions of continued photosynthesis, label accumulates in sucrose, raffinose and stachyose and decreases in galactinol. In stachyose, the highest specific activity during a short period of photosynthesis is found in the terminal galactosyl residue.

The sugar alcohols, too, can by synthesised quite easily from the common intermediates of carbohydrate metabolism (Fig. 7.3). There are, however, no special proofs of the presence and operation of the responsible enzymes and of the proposed intermediates in the leaves of those species which translocate the different sugar alcohols.

Very little is known about the sites of synthesis or liberation of the other compounds occurring in the sieve tubes (e.g. the amino acids and the organic acids). For these substances, too, the composition in the sieve tube sap is different from that in the leaf extract. This may be due to the fact that the leaf extract contains also the contents of compartments which do not directly supply the conducting channels, or to the existence of specific carriers in the phloem loading sites.

As far as the amino acids are concerned, it is striking that in many species, besides glutamine/glutamate and asparagine/aspartate, serine also has a higher concentration than the other amino acids in the sieve tube sap (Table 7.1). This may be due

---

DAP, dihydroxyacetone phosphate ; F6P, fructose-6-phosphate ; G6P, glucose-6-phosphate ; G1P, glucose-1-phosphate ; UDPG, furidinediphosphate-glucose ; UTP, uridinetriphosphate ; UDP, uridinediphosphate ; ADPG, adenosinediphosphate-glucose ; FDP, fructose-1, 6-diphosphate ; UDPGal, uridinediphosphate-galactose ; M6P, mannitol-6-phosphate ; I, myo-inositol ; Ra, raffinose ; St, stachyose. Main transport substances encircled.

Table 7.1. Concentration of free amino acids in the sieve tube sap of several tree species, measured by an amino acid analyser.

	Asp	*Thr*	*Ser*	*Glu*	*Pro*	*Ala*	*Val*	*Heu*	*Leu*	*Phen*	*Lys*	NH_3	*Hist*	*Arg*	*Gly*	*Tyr*	*Whole µmol* ml^{-1}
Quercus bicolor Willd., Fagaceae	17.7	2.1	13.8	9.8	tr	0.9	6.6	3.6	1.7	2.5	0.3	2.9	0.8	0.1	—	0.5	58.4
Q. palustris Munchh., Fagaceae	0.1	tr	0.1	0.1	—	0.0	0.1	0.1	tr	0.1	0.1	—	0.0	tr	tr	tr	0.8
Q. imbricaria L., Fagaceae	4.4	0.6	6.0	3.9	—	0.2	1.3	0.7	tr	0.6	tr	1.9	0.3	tr	0.1	0.3	20.3
Q. borealis Michx., Fagaceae	1.2	0.2	7.3	4.0	—	0.2	0.4	0.3	—	0.4	0.3	1.1	tr	tr	—	—	15.5
Q. cerris L., Fagaceae	3.0	tr	8.0	7.6	—	0.3	1.6	3.2	1.6	1.4	—	6.7	0.4	tr	—	0.5	34.1
Fagus engieriana Seem., Fagaceae	5.0	3.5	10.0	10.0	—	1.0	4.2	3.2	1.8	1.9	2.7	10.0	1.5	2.9	—	0.8	58.4
Tilia × euchlora Koch, Tiliaceae	2.3	4.0	10.0	6.8	12.5	1.03	6.7	5.7	4.4	2.9	—	1.0	0.4	—	—	—	57.7
T. petiolaris DC, Tiliaceae	0.8	1.1	6.6	1.2	—	1.2	1.0	1.1	0.8	0.7	0.8	1.2	0.4	—	—	—	17.0
T. vulgaris Hayne, Tiliaceae	15.9	4.9	17.3	10.9	—	1.5	5.0	4.1	3.2	3.7	1.8	3.7	1.2	0.4	—	0.6	74.3
Tilia × blechiana Dieck, Tiliaceae	20.0	3.6	10.0	10.0	tr	0.9	3.8	2.6	2.5	2.2	0.6	4.3	0.6	tr	—	—	61.3
T. oliveri Szysz., Tiliaceae	10.7	0.9	3.3	9.7	tr	0.2	0.8	0.8	0.4	0.6	—	1.6	—	tr	—	—	29.0

Fraxinus americana varr. *acuminata* Wesm., Oleaceae	tr	0.8	21.5	0.5	—	0.1	3.4	2.9	3.3	—	0.5	1.4	0.1	tr	—	0.5	88.6
F. oregona Nutt., Oleaceae	0.8	2.1	28.1	4.7	—	0.3	3.0	2.0	0.7	1.1	—	3.2	0.1	tr	0.3	0.8	47.2
Betula ermani Cham., Betulaceae	7.8	2.9	30.0	3.6	—	0.6	6.4	5.8	2.2	2.2	1.1	1.5	1.4	tr	—	—	65.6
Magnolia kobus DC, Magnoliaceae	0.2	1.1	20.9	3.7	—	5.0	2.2	2.0	2.2	0.6	tr	1.7	0.6	—	—	0.2	40.4
Robinia neomexicana M. Gray, Fabaceae	5.5	3.5	30.9	4.7	68.2	3.0	6.9	4.3	5.9	2.1	—	1.5	1.2	tr	—	—	137.8
Salix × rubens Schrank, Salicaceae	0.1	1.1	10.0	0.5	—	0.1	0.4	0.2	0.1	0.3	—	0.9	0.2	tr	—	tr	14.1
S. pendulifolia Spath, Salicaceae	0.9	9.8	20.0	13.3	—	1.0	10.6	6.5	7.4	6.5	0.5	19.5	2.2	tr	—	1.9	100.2
Liquidambar styraciflua L., Hamamelidaceae	0.5	0.8	20.0	0.7	—	—	1.1	0.9	2.3	0.7	0.5	2.1	—	tr	—	—	29.6
Acer heldreichii Orph., Aceraceae	1.0	2.8	20.0	4.5	6.8	3.9	4.9	3.9	9.0	3.1	0.6	2.7	0.8	tr	—	—	64.0
A. platanoides L., Aceraceae	8.5	6.1	10.0	10.3	—	1.7	2.0	5.6	4.7	1.3	—	6.9	1.4	1.2	—	—	59.8
Buddleia davidii Franch., Loganiaceae	0.2	0.2	0.6	5.0	—	0.3	0.4	0.4	0.5	0.1	0.1	0.6	tr	tr	—	0.1	8.5
Maclura pomifera Schn., Moraceae	1.1	0.7	10.1	1.7	tr	0.4	0.7	0.5	0.1	0.5	tr	4.6	0.2	tr	0.2	0.3	20.4

tr, traces.

to the fact that this amino acid is formed readily from 3PGA, which—as we have seen—is one of the photosynthetic products leaving the chloroplast. Serine is, therefore, a suitable transport carrier for nitrogen which is to be translocated in the phloem. In *Robinia*, which as a member of Fabaceae is especially rich in nitrogen, proline is quantitatively the predominant amino acid in the sieve tube sap, especially at the end of the season, when the nitrogen is transported out of the leaves after mobilisation of the leaf protein. Proline is synthesised from glutamate and carries additional reduction equivalents (in comparison to glutamate) which can be used at the sink sites as such or as a source of energy.

The main transport route from the chlorenchymatous cells to the translocation channels is very probably the symplast. Studies on the leaf structure of sugar beet showed that a 33 μm length of minor vein services 29 mesophyll cells with the result that translocate moves an average of 73 μm or 2.2 cell diameters during transport. During the symplastic transport the assimilates may undergo several biochemical transformations. A special situation is manifested in those plants which show the C_4-dicarboxylic acid type of photosynthetic carbon dioxide fixation. It is assumed that in C_4 plants the primary carbon dioxide fixation occurs in the mesophyll cells and this is followed by a transfer of the fixation product (malate or aspartate, which can pass the chloroplast envelope, see above) to the bundle sheath, where the transfer substances are decarboxylated, the carbon dioxide refixed by the Calvin-cycle in the bundle-sheath chloroplasts, and the remaining C_3 compound (pyruvate from malate and alanine from aspartate) carried back to the mesophyll cells.

In relation to the high photosynthetic activity of the C_4 plants this shuttle traffic should be a very effective process. It may be facilitated by the strong concentration gradients of the moving substances as well as by the numerous plasmodesmata between the mesophyll and the bundle sheath cells.

There are indications that the export of assimilates from the leaves of C_4 plants, in comparison to the normal C_3 plants, is more effective both in the rate of export as well as in the total percentage of assimilates which move out.

It is unknown whether an apoplastic transport plays an important role in the transfer of assimilates from the photosynthesising cells to the conducting elements. It does, of course, if isolated bundles or parts of the secondary phloem are supplied with external substances. Some experiments indicate that apoplastic transport may also proceed in the intact plant. In this case, the function of the specific transfer cells (which are developed in some species especially at the sites of phloem loading ; would be understandable. The transfer cells could be responsible for the re-entrance of the substances from the apoplast into the symplast and for the final transformation into the transport compounds.

During the movement of assimilates in the cell walls there could be again a biochemical transformation, e.g. a hydrolysis of sucrose to the hexoses ; invertase was shown to be active in this compartment. Sucrose has then to be resynthesised following the entrance of the hexoses into the cytoplasm.

An apoplastic transport of assimilates, however, would have very pronounced disadvantagaes. In spite of the fact that leaving the symplast and entering it again are both energy-consuming processes, the translocation in the cell walls is not under plasmatic control. It is subject rather to water streaming, the direction of which is governed by differences in the water potential. Most of the substances would not reach the conducting bundles but get accumulated at the transpiring surfaces.

Very little is known about the biochemical reactions involved with the entrance of the substances into the sieve tubes. It is very probable that it proceeds mainly in the companion cells or in their morphological equivalents : these cells are connected with the sieve tubes or the sieve cells by numerous plasmodesmata of a special type. It is, however, unknown whether the main loading proceeds via these plasmodesmata or via the whole plasmalemma of the sieve elements. At the last step, anyway, the final selection and perhaps the final transformation in the transport form happens. Since the sieve tubes do not have a boundary between the cytoplasm and the vacuole (the tonoplast disappears in the mature cells), it is hard to say whether the lumen of the sieve tubes belongs to the symplast or forms its own system, which is specifically separated from the remaining symplast.

The increase in sugar uptake in the conducting elements by addition of ATP can be explained in several ways ; ATP can enhance either the biosynthesis of the transport substances (Fig. 7.4) or an active secretion of the compounds into the sieve tubes. Light may act by promoting ATP formation when it facilitates the export of assimilates from leaves. In sugar cane, it is reported that darkening does not inhibit the entrance of sucrose into the bundle sheath, but does affect the outward movement of the assimilates from the sheath.

THE BIOCHEMICAL CAPACITY OF THE SIEVE TUBES

Considering the special cytological structure of the seve elements (lack of nucleus and of the tonoplast, scarcity and poor development of the mitochondria, special appearance of the cytoplasm) one would assume that these cells would not have very pronounced metabolic activities. The analysis of the enzyme activity in the whole sieve element (including the ectoplasm) can at present be carried out only histochemically, because as yet it is not possible to separate the sieve tubes and the companion cells without damaging the cell content. These experiments, therefore, include all the disadvantages of the histochemical methods. Sieve tube sap, on the other hand, collected by the incision or by the aphid method, can be analysed more exactly, but it only forms a part of the sieve tube, and may perhaps also contain compounds which are originally not found in the sieve tubes (or at least not in the same concentrations as in the exudate), but are secreted in the conducting cells as a consequence of the wounding effect. The following results relating to enzyme activity analyses in the sieve tube sap should, therefore, be interpreted with some caution, inasmuch as the conditions in the sieve tubes are in

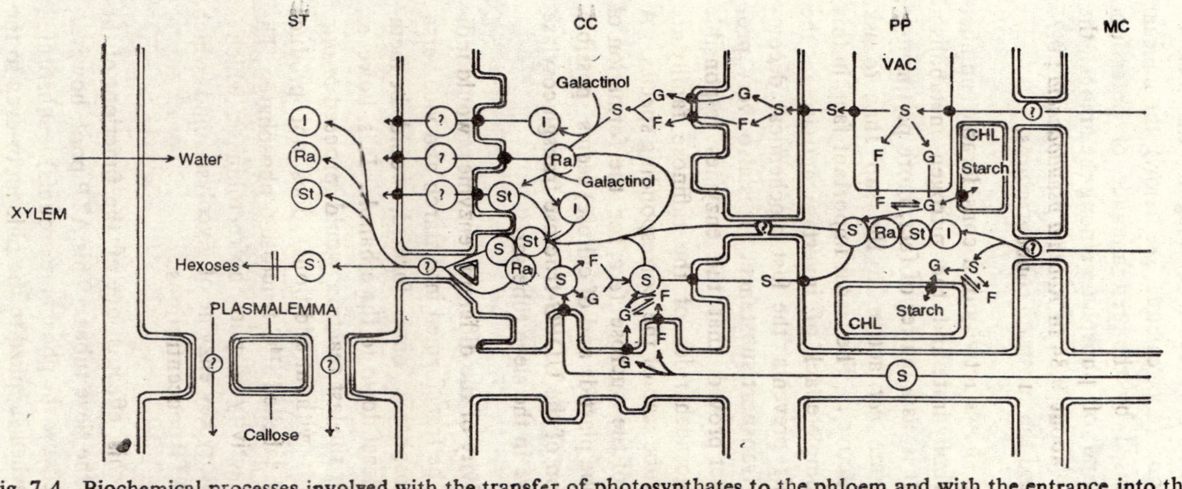

Fig. 7.4. Biochemical processes involved with the transfer of photosynthates to the phloem and with the entrance into the sieve tubes. MC, mesophyll cell PP, phloem parenchyma cell (with amyloplasts and vacuole). CC, companion cell (developed as a transfer cell with wall protuberance) ; note the absence of larger vacuoles and in general of amyloplasts and the specialised plasmodesmos to the sieve tube member. ST, sieve tube member with plasmalemma. ● Supposed carriers in the membranes. (?) Sites, for which active transport is discussed, but not proved. CHL, chloroplast ; VAC, vacuole ; and other abbreviations as in Fig. 7.3.

some respects far from the optimum for many enzyme reactions, viz. the predominance of a few sugars, the lack or scarcity of certain ions (e.g., calcium, see below), the unusual pH (mostly 7-8, caused by the predominance of potassium among the cations and of phosphate among the anions), the high energy charge (about 0.84 in *Robinia pseudoacacia* ; 0.71 in *Tilia platyphyllos* and the substrate concentrations.

The sieve tube sap shows activity of a considerable number of enzymes of the carbohydrate and the nitrogen metabolism (Fig. 7.5 and 7.6). It is, as a matter of fact, more pertinent to ask 'what enzymes are *not* active in the sap' than to ask 'what enzymes are active'. The most important fact, in this connection, is the complete lack of invertase activity in the sieve tube sap. This prevents the fast biochemical decomposition of the main transport substance. It is also very probable, although not yet proved, that the enzymes responsible for splitting the oligosaccharides of the raffinose family are absent in the sieve tube sap of the corresponding species. A metabolic utilisation of the sucrose (*e.g.* for the formation of starch in the sieve tube plastids and of callose) seems possible, however, by the action of the UDPG-fructose transglucosylase, which is weakly active in the sieve tube sap.

The potential activity of the different enzymes would indicate that sucrose can be transformed into starch and vice versa, and that hexose-phosphates, once formed, would be metabolised in the sap at least down to the ethanol. It is, however, very doubtful whether the last reaction would proceed *in situ* : original sieve tube sap (without additions) does not produce $^{14}CO_2$ from uniformly labelled ^{14}C-fructose-6-phosphate. The block may lie in the activity of phosphofructokinase, which is known to be a regulatory key enzyme of glycolysis, and which is inhibited by high ATP concentrations.

It is possible that this effect is one of the functions of the high ATP content in the sieve tubes. This ATP pool, however, is not a static one, because the phosphate group(s) is substituted very fast by $^{32}PO^{3}_{4}{}^{-}$ when supplied to the phloem or even to the sieve tube sap alone. Besides its appearance in ATP, the

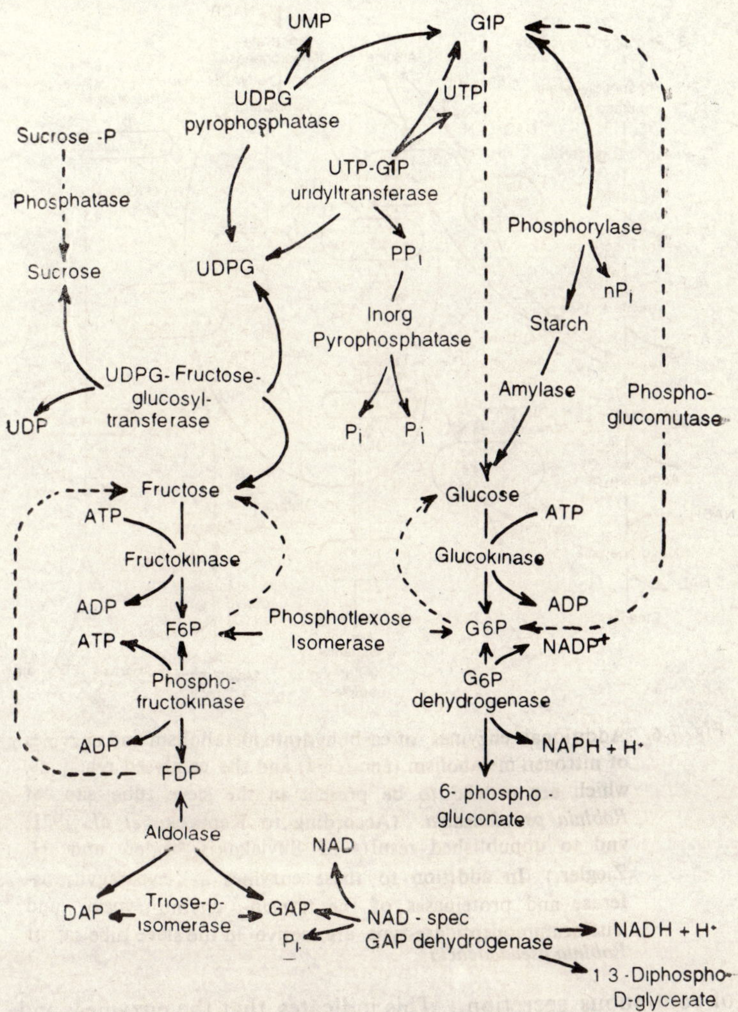

Fig. 7.5. Enzymes of carbohydrate metabolism (italic) and the catalysed reactions, which are shown to be present in the sieve tube sap of *Robinia pseudoacacia*.

$^{32}PO_4^{3-}$ is incorporated into a great number of additional organic substances. The phloem exudate from the inflore-scence stalk of *Yucca flaccida* does not lose this enzymatic ability even when the samples are taken after several hours

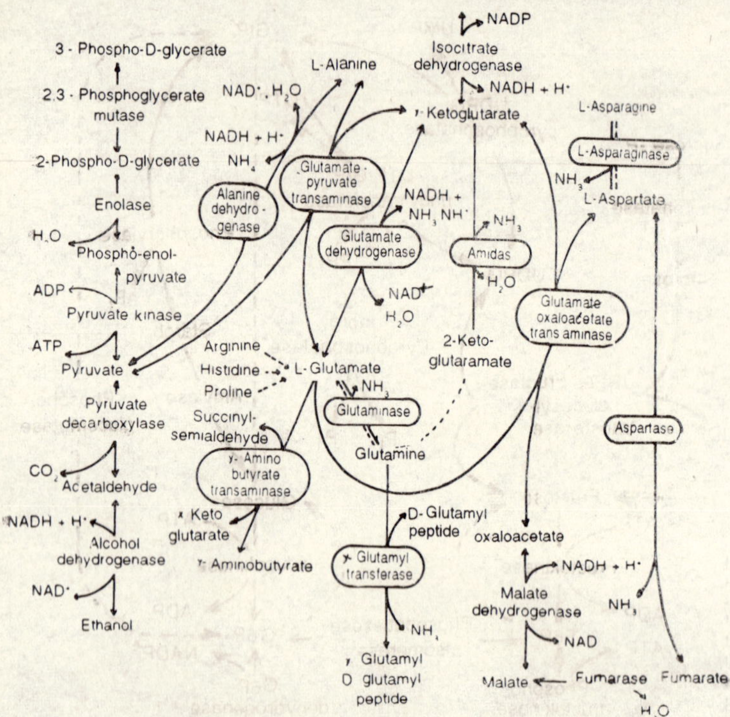

Fig. 7 6. **Additional enzymes of carbohydrate metabolism and enzymes of nitrogen metabolism (encircled) and the catalysed reactions, which are proved to be present in the sieve tube sap of *Robinia pseudoacacia*. (According to Kennecke *et al.* 1971, and to unpublished results of Skylakakis-Ziegler and H. Ziegler.) In addition to these enzymes a γ-asparagyltransferase and proteinases of the trypsin-, chymo-trypsin, and leucineaminopeptidase-type are active in the sieve tube sap of *Robinia pseudoacacia*.**

of continuous secretion. This indicates that the enzymes and the substrates for the phosphorylation are secreted continuously into the lumen of the sieve tubes, having been synthesised, perhaps, in the ectoplasm of the sieve tubes or—more probably—in the neighbouring cells. It is remarkable that the exudate from the conducting cells ('trumpet hyphae') of the brown alga *Laminaria saccharina* does not incorporate $^{32}PO_4^{3-}$ in any organic compound.

Of special interest is the question of whether mature sieve tube members are able to synthesise protein, because they lack nuclei. Neumann and Wollgiehn (1964) demonstrated autoradiographically the incorporation of labelled phenylalanine into the protein of sieve tube members of *Vicia faba*, which had lost the nuclei and had ceased RNA-synthesis. They discuss the presence of a 'long-lived' RNA or a protein synthesis which is independent from nuclear-DNA and perhaps governed by a plasmatic-DNA (which is present in the sieve tubes). Another but perhaps more probable explanation would be the delivery of mRNA to the sieve tubes by the neighbouring cells (*e.g.* companion cells). Actually, if protein synthesis occurs in the mature sieve tube members, it can only proceed in the ectoplasm : sieve tube sap (of *Robinia pseudoacacia*) does not show protein synthesis. This may be due to the lack of ribosomes (which may also be absent in the ectoplasm of mature sieve tube members).

One of the most striking peculiarities of the assimilate transport is an almost complete exclusion of some ions, which are, therefore, not available for redistribution in the plant (*e.g.* calcium, boron). This has a lot of very interesting consequences, which are, however, outside the topic of this review. Epstein (1973) points out that the exclusion of calcium may be the reason for the special cytological development of the sieve tube members with the disintegration of the cell content and especially of various membranous structures, which may be a prerequisite for the longitudinal conduction of a solution flow. The only membrane which is fully functional and the function of which is essential for the phloem transport, the plasmalemma, is exposed to the calcium present in the cell wall and could satisfy its need from there. As a matter of fact, absence of calcium inhibits assimilate translocation in bean plants to a large degree.

The exclusion of boron, on the other hand, could prevent the formation of callose at the sieve pores, which narrows the open conduits necessary for solution flow and which is enhanced by the presence of borate.

Beside these cytological consequences calcium may be absent from the sieve tubes for a very simple reason : the concentration of the inorganic phosphate in the conducting cells is so high that calcium would be almost completely precipitated as calcium phosphate.

Nothing is known about the mechanism of exclusion of calcium and boron.

The metabolic abilities of the phloem, especially the sieve tubes, are of special interest in connection with the question relating to the driving force and the energy requirements of the transport and of the maintenance of the structure in the sieve tubes. There are several reports that respiratory inhibitors block phloem transport. Surprisingly, exclusion of oxygen from the bundles does not inhibit this translocation. Also most of the experiments concerning the effect of temperature on the transport indicate that the translocation in the sieve tubes themselves is practically temperature-independent, while the entrance and the exit show a temperature coefficient like normal metabolic processes.

Most of the measured metabolic activity of the whole phloem should, therefore, be attributed to the companion cells (and—to a lesser degree—to the phloem parenchyma), which show an unusual density of mitochondria and of cytoplasm. Canny (1960) calculated that in stems of *Vitis vinifera* about 0.3—0.5% of the moving sucrose is broken down by respiration (*c.* 0.5 mg cm^{-3} phloem h^{-1}).

BIOCHEMICAL REACTIONS INVOLVED WITH THE UNLOADING OF THE SIEVE TUBES

As is shown by the exudation of sieve tube sap through aphid mouthparts or at the surface of inflorescence stalk sections in *Yucca* a solution flow in the phloem does not demand a metabolically active sink. In the intact plant, however, active removal of the transport substances at the sink sites will determine the direction and the steepness of the concentration gradients. As far as we know, these gradients determine the direction and the rate of transport.

The exit of the substances through the selectively permeable ectoplasm of the sieve tubes (or through the plasmodesmata between the sieve tubes and the neighbouring cells) is controlled metabolically in a way similar to that which operates during their entrance into the conducting cell. Very little is known, however, about the mechanisms involved with this removal process. From studies relating to the concentration changes of the different sugars in the sieve tube sap of defoliated *Fraxinus americana* trees Zimmermann (1958) concluded that there is a formation of sucrose at the cost of stachyose and raffinose. Since free galactose is not detectable in the sap, it is probable that the removal of the galactose residue takes place in the ectoplasm (plasmalemma ?) of the sieve tube, followed by an immediate transfer to the surrounding cells, while the sucrose remains in the sieve tube.

Much more is known about the biochemical processes involved in the utilisation and storage of the substances leaving the sieve tubes. We have most detailed knowledge about the events leading to the accumulation of sucrose in the storage parenchyma of the sugarcane and to the formation of starch in the amyloplasts.

There are many results reporting an influence of plant hormones on the direction and rate of the phloem transport. The most simple mode of influence would be a regulation of the sink activity : an enhancement of the metabolism of assimilate consuming cells at the sink sites would increase the exit of the transport substances from the sieve tubes, a retardation would lower it. Since the sieve tubes themselves have a very restricted storage potential (change in sugar concentration or formation or mobilisation of starch) a change in the unloading of the sieve tubes at one place has to be compensated by a complementary change at another region or at the site of loading. This may be brought about by a simple competition between the different 'sinks' and by feedback inhibition of the loading of sugar into the phloem by an increased sugar content in the sieve tube.

THE MECHANISMS OF PHLOEM TRANSLOCATION

Phloem is a morphologically distinct tissue type in the vascular system of the higher plants which conducts in its sieve tubes long distance translocation of various key substances from one region to another. By far the major constituent (other than water) of the phloem translocation stream is carbohydrate, sucrose in most plants, but raffinose, stachyose and verbascose have also been reported, and in a few cases the sugar alcohols mannitol and sorbitol have been detected. The ionic contents of phloem sap are apparently under membrane regulation ; potassium concentrations are much higher than sodium, magnesium levels are much greater than calcium, and the most common inorganic anions are chloride and phosphate.

In many species the phloem concentrations of these ions, and of various nitrogenous compounds including proteins, markedly increase during leaf senescence as these soluble molecules are exported from the leaf to the body of the plant, prior to leaf drop. It is therefore apparent that the composition of phloem sap is under physiological control, and that phloem translocation may serve the plant in several ways. Of course there is no doubt that the chief function of phloem is to remove photosynthetic sugars from the leaves and to distribute these sugars throughout the plant, and it is with this aspect of phloem translocation that we shall be primarily concerned here. However, we should always bear in mind that translocation does sometimes function in the apparent absence of large osmotic gradients of photosynthetic sugar and we must judge proposals of mechanism accordingly.

PHLOEM STRUCTURE

This section contains only a very brief outline of current ideas on structure. More extensive literature surveys can be found in Esau, Currier and Cheadle (1957), Zimmermann (1966), Weatherley and Johnson (1968) and Wooding (1971). In the angiosperms the phloem sieve tube is made up of a series of sieve elements arranged sequentially end-to-end. During sieve element ontogeny the end walls become perforated by numerous channels, the sieve pores, which connect one sieve element to the next and so form the sieve tube ; in this mature state these end walls are termed sieve plates. Identification of the sieve plates is the chief diagnostic feature for the localisation of sieve tubes and, depending upon the species, the sieve plates may be inclined to the longitudinal axis of the sieve tube at any angle from 30° to 90°.

A second specialised cell type is found in many angiosperm species immediately adjacent to the sieve elements. These are the companion cells, a densely cytoplasmic cell type, connected to the sieve elements through many plasmodesmata across the side walls. In specialised regions the companion cells may be further modified by having prolific wall ingrowths at their common walls with the neighbouring parenchyma. In this state, the companion cells have been termed transfer cells. The adaptation is thought to reflect an enhanced transport capability ; solutes absorbed across the much increased membrane surface of the transfer cell move through the plasmodesmata to the sieve element and so load the phloem. It has been recognised for about eighty years that sieve elements can be plasmolysed, implying that they are bounded by an intact membrane capable of sustaining an osmotic gradient.

These features of sieve element structure are generally accepted. The controversy on anatomy centres upon what structures there might be in the lumen of the sieve element, and in the pores of the sieve plates. One body of opinion holds that the sieve element lumen contains a rather open network of microfibrils of P-protein (phloem-protein), the sieve pores being essentially unrestricted. The P-protein network is thought to be passive during undamaged, normal functioning of the phloem, but to act as an aggregating defence mechanism to minimise phloem bleeding following injury, by blocking the sieve pores. This primary healing process is then made permanent by callose deposition at the sieve plates.

A second school of thought, associated with the observations reported by Lee, Arnold and Fensom (1971) on the phloem of a *Heracleum* species, also believes that the sieve elements contain a fibrillar protein network. However, these workers think that the fibrils run from cell to cell through the sieve pores. Under Nomarski optics, marker particles on the fibrils are seen to oscillate in a non-random manner in living phloem. It is thought that the oscillations are energy dependent and that they may impel translocation along the phloem.

The third point of view is propounded by Thaine (1969), who thinks that the P-protein fibrils reported by other workers are the breakdown products, caused by microscopic preparation, of a much more organised system of tubules which run from cell to cell through the sieve pores. He believes that translocation is conducted chiefly along these tubules.

One can find eager exponents of each of these three points of view ; the topic has been extensively argued elsewhere This section on anatomy has been included merely to allow more intelligible discussion of translocation mechanisms.

PHLOEM LOADING AND UNLOADING

An excellent, quantitative investigation of the processes by which sugar is loaded into the sieve elements from the chloroplasts of the mesophyll cells has been described by Gunning, Pate and co-workers. Further discussion of the points raised by these authors is quite superfluous, but there are several matters of a more general nature that I should like to air.

There are two possible parallel pathways from the mesophyll to the phloem ; first, the sugar may move through the symplasm, chiefly by diffusion. There are many plasmodesmata connecting the cells all the way from the mesophyll to the sieve element, so that symplasmic transport of sugar is an obvious possibility. Second, sugar may move from the mesophyll to the phloem by diffusing through the apoplasm or extracellular free space. Active extrusion of sugar at the mesophyll membranes and active uptake at the companion cell (or transfer cell) and at the sieve element membranes will ensure the concentration gradient to drive the apoplasmic sugar transport. The vascular tissue in the leaves is prolifically anastomised (the main veins split into smaller veins which then split into many individual vascular bundles), so that every mesophyll cell is relatively closely associated with phloem. Thus it would appear that simple physical diffusion along either or both of the parallel pathways described above can provide the sugar fluxes required to load the phloem, without any need for massive and unreasonable concentration differences.

Sugar is probably transported across the chloroplast membranes and the cell membranes by the action of carrier enzymes or permeases. Kursanov (1968) has speculated in some detail on possible biochemical reactions involved in this permeation, but extensive discussion of these seems removed from the main thrust of this paper. Suffice it to say that he believes the membrane penetration requires a successive phosphorylation and dephosphorylation of the sugar with the involvement of ATP, the hydrolysis energy of which provides the metabolic energy input for transport. Glasziou and Gayler (1972) have kinetic evidence for the action of some such enzyme carrier system for sugar efflux from the mesophyll cells.

Some of the published arguments about the pathways of sugar movement to the phloem seem to be too polarised; those who insist that one or other, apoplasmic or symplasmic, is the only operative route, must at least specify the species and conditions with which they are concerned. It seems far more likely that in general each pathway carries a significant proportion of the total flux. However, we may note an important distinction between the pathways. The only energy input to drive the symplasmic flux will be at the chloroplasts since it is generally considered to be unlikely that active transport can occur at plasmodesmata; therefore the high sugar concentrations in the phloem can only be achieved by symplasmic loading if there are even higher sugar concentrations just outside the chloroplasts. This in itself poses no real problem but we may suspect that symplasmic loading is more light sensitive than the parallel apoplasmic loading. Light dependence of fast phloem translocation has been reported in sugarcane by Hartt and Kortschak (1967).

Apoplasmic loading of the phloem will be driven by additional energy transduction processes. Active efflux of sugar across the mesophyll plasmalemma and active uptake of sugar at the companion cell (transfer cell) and sieve element plasmalemma can ensure phloem loading even if the sugar concentration at the chloroplasts falls below that in the phloem. Under these conditions one might imagine there to be a significant diffusive back-flux of sugar through the symplasm from the phloem to the chloroplasts in the mesophyll, but this is probably quite negligible because it will be opposed by a convective flow of water *entering* the sieve element through the plasmodesmata from the companion cells, water which, of course, provides the translocation stream. The active transport processes referred to above will be powered by ATP hydrolysis, presumably by mechanisms similar to those suggested by Kursanov (1968). As such we may expect apoplasmic phloem loading to be less light sensitive than symplasmic loading, because ATP derived from respiration will be available. However, these considerations cannot be taken too far ; the high energy charge in illuminated green cells will tend to accelerate all energy-requiring reactions and the starch-sugar conversion in chloroplasts in the dark may keep the symplasmic flux to the phloem at a significant level well into a dark period. The joint effect of these two factors may mask observable changes and make very difficult such simple resolution of the fluxes.

So far this discussion has dealt with the popular aspect of phloem loading, *viz.* transporting and accumulating sugar into the sieve tubes. However, there is a second major component of the phloem translocation stream, water, and it too must be loaded into the phloem. The one thing common to all present models of translocation mechanism is that all presume there to be a net flow of water along the sieve tubes from the leaves and obviously this water must be supplied to the sieve tubes in the leaves from the xylem. The driving force for water entry to the phloem is the osmotic gradient created by the accumulation of sugar, and we may therefore suppose that a second reason for the proliferation of membrane surface in the transfer cells is to ensure an adequate supply of water. The calculations of Gunning *et al.* show that only very small pressure gradients are required to drive sufficient water flow through the plasmodesmata connecting the transfer cells to the sieve elements to maintain observed translocation rates, and so the total membrane area of sieve elements plus transfer cells is essentially available for osmotic water entry to the translocation stream. Indeed if the symplasmic pathway for sugar loading to the phloem were to prove to be the major one, this enhancement of water permeability by membrane surface area increase might well prove to be the principal *raison d'etre* for transfer cells.

In summary of this section, there are apparently two parallel pathways to load sugar into the phloem, one through the symplasm from the mesophyll and the other through the apoplasm from the mesophyll. They may be to some extent complementary ; with strong illumination and active photosynthesis the symplasmic route may be the more important, because high sugar concentrations in the mesophyll will accelerate symplasmic diffusion. Furthermore, if these high photosynthetic rates are accompanied by high transpiration rates, then apoplasmic sugar diffusion from the mesophyll may be hampered by an opposing water flow through the apoplasm to supply transpirational demands, a factor which will tend to reduce apoplasmic sugar transport. Under the inverse conditions, it seems likely that the apoplasmic pathway, having a more flexible energy input for sugar transport and accumulation, could provide the major sugar loading route into the phloem. It seems certain that no dogmatic statement can be made about how sugar is loaded to the phloem without first specifying species and conditions.

Similar considerations will apply to phloem unloading. In this case sinks for diffusion gradients are created by either metabolism of the sugar or conversion to polymerised, nondiffusible storage forms.

MODELS OF TRANSLOCATION MECHANISM

There are presently four proposals of translocation mechanism, the Munch model, the electro-osmotic model, the peristaltic tubule model and the streaming model. All these models presume that translocation involves a bulk flow of sucrose solution along the sieve tubes. The reason for this particular consensus of opinion is the values of measured phloem translocation velocities over relatively large distances. Mason and Maskell (1928) showed that sugar transport in the phloem does occur down a concentration gradient (as would be expected of a diffusive process), but at such a rate that the 'apparent diffusion coefficient' is ten thousand times larger than the measured value for sucrose in aqueous solution. The conclusion seems inescapable that sugar transport along the phloem requires bulk flow of water in the translocation stream, sweeping the dissolved sugar along. Each model we shall now discuss includes this basic feature of water flow in the translocation stream, and differs from the others only in the mechanism by which the water is driven.

Munch Model

I wish to stress at the outset that this proposal has a different status from all the others. Providing that (*i*) the sieve element-transfer cell (companion cell) complex is bounded by a semi-permeable membrane, (*ii*) sugar is accumulated in this complex, and (*iii*) xylem water tension is not too negative, it is certain that there will be osmotic water entry into the sieve tube at the leaf, along the pathways described in the previous section. In other regions of the plant where sugar is utilised or converted to osmotically inactive forms, there will be a reverse flow of water from phloem to xylem (Fig. 7.7). Under these appropriate conditions, water *must* osmotically enter the phloem at the loading zone and leave at the unloading zone and indeed all the other proposals of translocation mechanism accept, at least tacitly, that water loading and unloading does occur osmotically much in the manner originally suggested by munch (1930).

The only uncertainty about the Munch proposal is whether the hydraulic resistance of the sieve tube is sufficiently low that the turgor pressure in the phloem at the leaf, created by the osmotic entry of water there, can *by itself* drive translocation

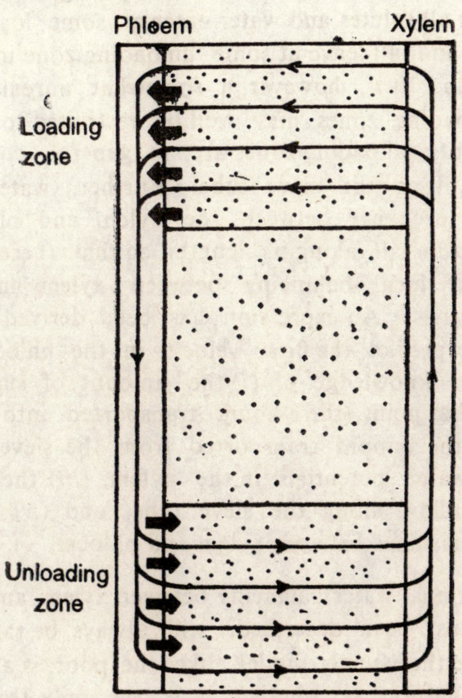

Fig. 7.7. A schematic representation of the circulation suggested on the Munch model. Solutes (represented by the heavy arrows) are accumulated into the phloem in the loading zone and are exported at the unloading zone, while water (represented by the light arrows) circulates around phloem and xylem in the manner indicated.

at the observed velocities. The value of the hydraulic resistance to longitudinal water flow in the sieve tubes in critically determined by the effective open diameter of the sieve pores, and this matter has been extensively and comprehensively argued elsewhere. The crucial area of uncertainty in such deliberation is of course the controversy about the internal anatomy of the normally functioning sieve tube (see section on phloem structure), chiefly to what extent the sieve pores are occluded by P-protein fibrils.

The most common, and rather simplistic statement of Munch-type water circulation 'up the xylem and down the phloem' is similar to the one I have just given where it is implied that all solutes and water enter at some loading zone in the leaf, and all leave at some unloading zone in a fruit or storage organ. It is, however, a somewhat unreal idea, and although loading zones may well be restricted to the leaves in many plants, unloading zones are, in general, continuously distributed along the sieve tube. Further, water may be expected to exchange between the xylem and phloem of a vascular bundle all along its length, so that there will be at every point, local balancing between xylem and phloem water potentials. An expression has been derived which can in principle predict the flow velocity in the phloem at any point, from knowledge of (*i*) the amount of sugar in the phloem at that point (the amount transported into the sieve tube minus the amount transported from the sieve tube), (*ii*) the local water potentials in the xylem, (*iii*) the hydraulic resistance to flow along the sieve tube, and (*iv*) the overall hydraulic resistance between xylem and phloem.

The continual water balancing between xylem and phloem, which has just been described, will always be taking place. In that sense the Munch model, like the poor, is always with us. The question is whether it is the only translocation mechanism or whether there are others operating. Indeed under certain circumstances, the usual argument about whether the sieve pores are sufficiently open to allow Munch flow to maintain observed translocation rates, might have to be reversed. In special conditions we may have to ask whether the sieve pores can be sufficiently occluded to prevent a disadvantageous Munch flow.

Electro-osmotic Model

The first subsidiary (in the sense that Munch flow will always be occurring to some extent) mechanism I wish to discuss is the proposal that translocation is driven by electro-osmosis at every sieve plate. This model was originally suggested independently by Fensom (1957) and Spanner (1958). The basic idea is that sugar solution is driven through the sieve pores, and therefore along the sieve tubes, by the electro-osmotic drag of K^+ ions, accumulated and secreted in turn across every sieve plate by the action of the companion cells (Fig. 7.8). This model has been subjected to much analysis, the

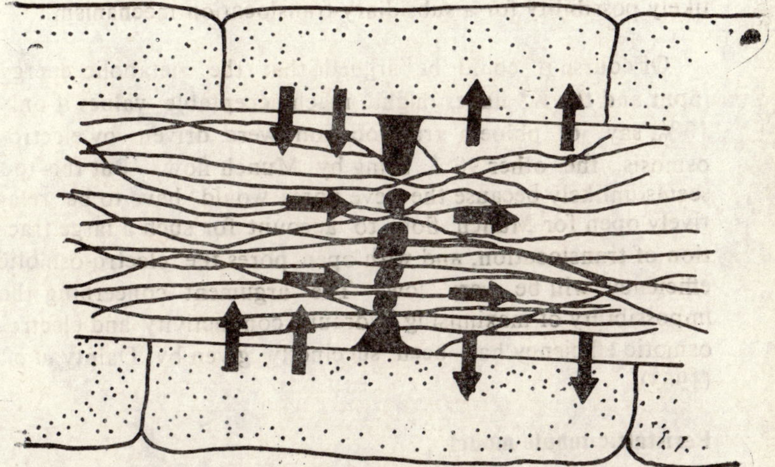

Fig. 7.8 The electro-osmotic model suggests that K^+ fluxes (here represented by the heavy arrows) circulate from the companion cell, through the sieve plate and back into the companion cell as indicated. The sieve pores are thought to contain many P-protein fibrils which cause the only available channels from sieve element to sieve element to be narrow and charged, thus maximising electro-osmotic efficiency. The translocation stream is dragged through the sieve pores as the K^+ flux is maintained.

best probably being in the recent review by MacRobbie (1971), but see also Dainty, Croghan and Fensom (1963). All impartial reviews are agreed that the proposal is untenable as the sole translocation mechanism. The metabolic energy requirements, the magnitude of the K^+ fluxes across the companion cell membranes, and particularly the apparent impossibility of translocating anions in the phloem all cause this proposal to seem highly unlikely. Spanner's (1970) deliberations are incomplete, to my mind, and Bowling's (1969) experimental observations of a sawtooth electric potential profile in the phloem are less than convincing. It would be a matter for comment if the potential difference between two electrodes inserted into plant tissue were zero, rather than if values of a few millivolts were erratically obtained. All the objections raised by MacRobbie (1971) seem pertinent and valid, and the electroosmotic model is thought to be the least likely possibility for a subsidiary translocation mechanism.

Of course it could be argued that the metabolic energy input and the K^+ fluxes might reach acceptable values if only 10%, say, of phloem translocation were driven by electro-osmosis, the other 90% being by Munch flow. But this too seems unlikely because the sieve pores would have to be relatively open for Munch flow to account for such a large fraction of translocation, and with open pores the electro-osmotic efficiency will be very low. The argument concerning the impossibility of maximising hydraulic conductivity and electro-osmotic efficiency has been succinctly given by Dainty *et al.* (1963).

Peristaltic tubule model

Thaine (1964, 1969) has maintained for many years, in the face of much criticism, that the sieve tubes of the higher plants contain a much more highly organised system of transcellular tubules than the majority of anatomists would agree to. His original suggestion of a translocation mechanism based on this tubular system has been modified and quantified by Aikman and Anderson (1971), who showed that this proposal would require apparently acceptable rates of energy supply from phloem respiration to maintain observed translocation rates, even if all the flow were driven by peristalsis. The model basically supposes, following Thaine (1969), that a series of tubules runs from sieve element to element, passing through the pores, and that these tubules are bounded by contractile microfibrils much like those in cilia or flagella. Rhythmic contraction of these microfibrils impels the translocation stream along the tubule (Fig. 7.9). In addition, the turgor pressure gradient generated by the Munch mechanism can drive flow through the tubules, as was indicated, though not very emphatically, by Aikman and Anderson (1971). In this system maximisation of the Munch component of flow

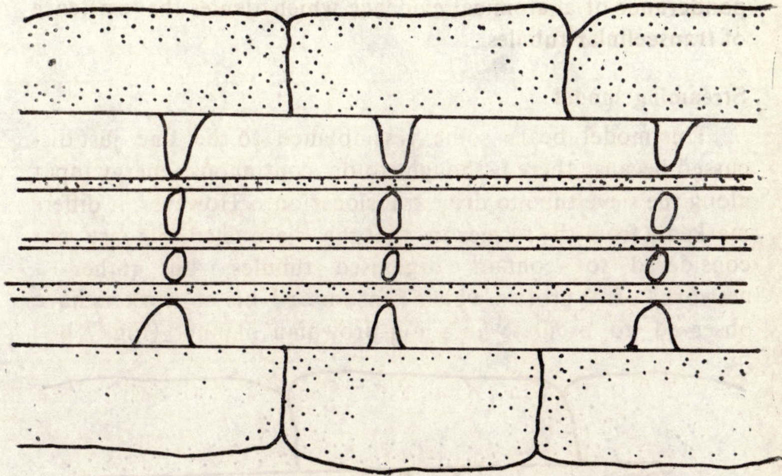

Fig. 7.9. A schematic representation of the structure of the sieve elements as proposed by Thaine (1964). Peristaltic waves of contraction along these tubules could be considered to drive the translocation stream. Calculations show that the process would be energetically acceptable on the basis of measured phloem respiration rates. As is stated in the main text, Munch flow can also occur along these tubules.

is completely compatible with maximising the peristaltic component ; both are achieved by maximising the tubule radius.

There are several other attractive features to this model, which may be summarised as follows : (*i*) contractile systems are the most common and perhaps most efficient mechanochemical transducers ; (*ii*) sugar translocation is by convective solution flow driven by pressure waves within the tubules and is a hydrodynamically and thermodynamically efficient process ; (*iii*) the metabolic energy input is continuously distributed along the sieve tube ; (*iv*) simultaneous bidirectional transport is possible within a single sieve tube ; (*v*) the number of tubules within a single sieve tube transporting in a particular direction may vary to give a very flexible translocation capability.

In summary, the model seems feasible and attractive but these features must be weighed against the overwhelming preponderance of anatomical evidence which denies the existence of transcellular tubules.

Streaming model

This model bears some resemblance to the one just discussed because there is thought to be continuous energy input along the sieve tube to drive translocation. However, it differs markedly from the former in that the sieve elements are not considered to contain organised tubules, but rather a network of P-protein which Fensom and his co-workers have observed to oscillate in a non-Brownian manner (Fig. 7.10).

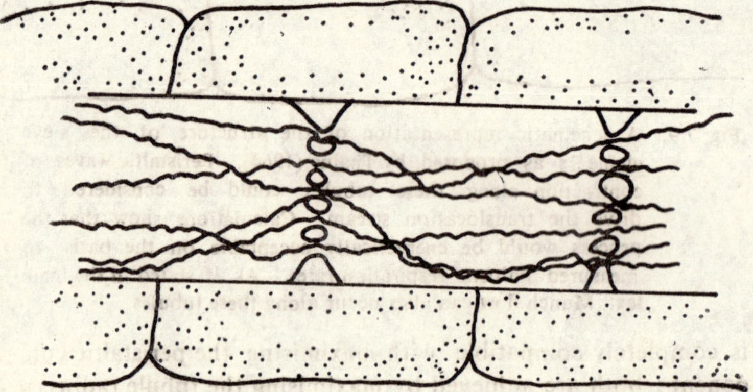

Fig. 7.10. A representation of how synchronised wave motions along the P-protein fibrils may impel translocation flow as an additional component to the flow generated by the Munch mechanism.

MacRobbie (1971) has gone into the possibilities of streaming in sieve tubes as a mechanism for phloem translocation. She has extensively reviewed streaming in slime moulds and algae, and suggests a very interesting comparison between P-protein in the sieve tubes and the contractile fibrils in other cells which produce shearing forces in cytoplasm and thus streaming. This proposal seems to fit the observations of Fensom's school on the nonrandom movement of P-protein in the functioning phloem of *Heracleum*. However, it must be realised that there is unlikely to be any uni-directional flow of sieve element cytoplasm, because there is no obvious mechanism to re-cycle it. What is implied is that the streaming mechanism found in other cells has been modified in sieve elements to drive the translocation stream. The chief modification would appear to be breakdown of the tonoplast-bounded vacuole in mature sieve elements and redistribution of contractile fibrils (P-protein) throughout the space previously occupied by the vacuoles, which is now the lumen of the sieve tube. In this mature state the remaining cytoplasm, usually known as the parietal cytoplasm of the sieve element, is of course not separated from the lumen fluid by a membrane and presumably is maintained undiluted by being in a highly cross-linked gel state. Its function may be to maintain the integrity of the exterior plasmalemma of the sieve element and to provide anchor points for the P-protein fibrils.

The main attraction of this model is that it incorporates the structure agreed by the majority of anatomists into a proposal of translocation mechanism which is biologically plausible and of possibly high efficiency. It may also be noted that this is the most natural mechanism to be subsidiary to Munch flow. In the electro-osmotic model the sieve pores must be packed with P-protein (so that Munch flow is virtually zero) to form the very narrow channels necessary for efficient electro-osmotic coupling. In the peristaltic tubule model Munch flow is restricted to the tubules throughout the length of the sieve element, and the overall hydraulic resistance is therefore higher than in the streaming model where the translocation stream can occupy the total cross-section of the sieve tube lumen. The main disadvantage to the streaming model, which the peristaltic tubule model does not have, is that it cannot accommodate simultaneous, bidirectional, segregated flow within a single sieve tube. The only item of evidence that such a flow occurs is the report by Trip and Gorham (1968) in squash, but it might be said that this evidence seems reasonable hard. Were it to be further substantiated that under certain conditions, bidirectional translocation within a single sieve tube was a common occurrence, then Scientists think that they would have no other option than to believe that there are transcellular tubules. Note here an important distinction.

There is no question that in cells generally there is cytoplasmic streaming in both directions, but in a sieve element having a P-protein network in an essentially open lumen, and therefore with no diffusional segregation of the directions of flow, there would be little likelihood of achieving any sort of efficient net flow of solutes in either direction. This is the type of bidirectional flow required by Trip and Gorham's (1968) result, and it cannot be plausibly imagined on the streaming model.

ENERGY INPUT FOR TRANSLOCATION

In what follows we shall discuss only the energy input required directly to drive translocation ; we shall consider it to be perfectly clear that a certain amount of repair work must be done by all cells in order to maintain themselves and will therefore not refer explicitly to the need for maintenance work in the sieve tubes.

Munch flow is driven by solute accumulation into the sieve element companion cell complex at the leaf, and possibly by solute transport from the sieve tubes at unloading regions of the plant. Thermodynamically the work done will be given by the product of the flux times the force, $J_s RT \ln(C_1/C_2)$, where J_s is the sugar flux (moles cm^{-2} sec^{-1}), R is the gas constant, T is the temperature (°K) and C_1 and C_2 are the concentrations of sugar (moles cm^{-3}) in the sieve tube lumen and in the free space just outside the phloem respectively. The total energy input to the phloem will be $AJ_s RT \ln(C_1/C_2)$ where A is the total membrane area across which the sugar flux is transported. The sugar concentration gradient is then transduced by the semi-permeability of the membrane to an osmotic force on water, which force actually drives the translocation process, so that the work done on water by the Munch model is $\sigma AJ_s RT \ln(C_1/C_2)$, where σ is the reflexion coefficient of the membrane.

Energy is then dissipated by the water fluxes across the semi-permeable bounding membranes of the phloem and along the sieve tubes ; the energy dissipation rate will be given by $(v_1^2 R_1 + v_2^2 R_2)$ where v_1 and v_2 are the water fluxes across the bounding membranes and along the sieve tubes respectively, and R_1 and R_2 are the appropriate hydrodynamic resistances. It seems likely that the energy requirements to drive all the translocation flow on this model are not excessive, providing that the sieve pores are relatively open, although it is somewhat hard to decide on energy availability. Most of the energy input is generally thought to occur from sugar loading at the leaf, and the membrane areas involved in the loading will obviously depend on what proportions of the sugar flux are ascribed to the symplasmic and apoplasmic pathways, as discussed in an earlier section.

On the electro-osmotic model the energy input for translocation will be the energy required to transport K^+ to and fro across the companion cell membranes. (Note that there may be a short-circuit through the many plasmodesmata connecting companion cells and sieve elements which will make such membrane transport in itself rather inefficient.) As the sole driving force for translocation, this mechanism would require quite excessive energy input, many times greater than could be supplied from phloem respiration at the measured rates. Further, the electric current densities across the companion cell membranes to supply the necessary K^+ fluxes are of such magnitude as to cause electric 'punch-through' in other higher plant cells.

Both the peristaltic tubule and the streaming models require continuous energy input all along the sieve tube. The putative contractile mechanism would almost certainly be ATP powered, and there are several reports of ATP concentrations in extruded phloem sap which strongly imply that ATP is indeed available all along the sieve tube. Furthermore, as Ziegler has indicated in the preceding paper, sugar can be metabolised and ATP produced within the sieve tube. Aikman and Anderson (1971) have computed that the peristaltic tubule model will consume the equivalent of 0.04g sucrose cm^{-3} phloem day^{-1} to drive translocation at 100 cm h^{-1}. This is about 2% of the total sugar translocated and falls within the reported values for the respiration rates of phloem. This value will decrease proportionately as the Munch component of translocation is assumed to increase from zero.

Energy dissipation rates on the streaming model have not yet been calculated, and cannot be until some assumption of P-protein geometry within the sieve element is made. However, it is likely to prove as acceptable on energetic grounds as the tubule model. The reason is this ; although transduction of ATP hydrolysis to useful driving pressure may well be less efficient in the P-protein network than it is in the highly organised tubule, the resistance to flow will also be less because flow can occur across the whole cross-section of the sieve tube and is not restricted to a rather narrow tubule.

At this point we might say that we consider the streaming model to be something of a misnomer, but have kept it here because of the likely ontogeny of the mechanism and because MacRobbie (1971) has used the term to include the possibility we believe to be the most probable mechanism by which the P-protein network might impel movement along the sieve tube. Perhaps we can best explain how we imagine this to take place by analogy. A flag will flap because it is driven to do so by the wind ; conversely, if a flag is made to flap by a man waving it, a wind will be produced. The question raised by Fensom's observations is whether the P-protein is being driven by the flow or the flow by the P-protein. Both these motions will appear non-random and indeed quite indistinguishable from one another. But be that as it may, we think it is highly likely that the P-protein is utilising ATP and is producing contraction waves to drive a component of the translocation flow. The local pressure gradients in the fluid are contained by the sieve tube walls and it is easy to see that synchronised contraction waves can drive a net unidirectional flow.

CONCLUSIONS

(*i*) Providing the sieve elements are bounded by a semipermeable membrane and water tension in the xylem is not too high, there must be osmotic water entry at the leaf when sugar is accumulated into the phloem and water loss elsewhere when sugar is exported. Thus the Munch model is of different status from the others. The only question to be asked is whether it is by itself sufficient to drive translocation at the observed rates. The answer depends most heavily on the effective open radius of the sieve pores, but also upon the xylem water potential at any given time and place.

(*ii*) It is plausible that there is a subsidiary translocation mechanism, acting in addition (and perhaps under special circumstances in opposition) to Munch flow, because measured phloem respiration rates are high along its length while the energy input to Munch flow is essentially localised at the chief loading and unloading zones. The use of the word subsidiary does not imply which component of flow is major.

(*iii*) The electro-osmotic model seems to be the least likely candidate for the subsidiary mechanism. As the sole translocation process the energy requirements would be impossibly high, and translocation of anions apparently forbidden. As the subsidiary process, the criteria to maximise electro-osmosis are incompatible with those to maximise Munch flow.

(*iv*) The peristaltic tubule model is a likely proposition on physiological grounds. As the sole driving mechanism it could impel at observed translocation rates within the respirational capabilities of phloem. Therefore as a subsidiary mechanism it can drive with ease a large fraction of translocation. It alone of all proposals can accommodate bidirectional flow within a single sieve tube. Against this model is the stark fact that the anatomical evidence for the existence of tubules comes from only one source.

(*v*) The streaming model (or the peristaltic P-protein model as it might be better named) is by far the most likely subsidiary translocation mechanism. It has all the advantages of the tubule model with the exception of allowing bidirectional flow of diffusion-segregated streams within a single sieve tube. It has been directly observed in operation in the living phloem of one species and the P-protein network is the structure most commonly observed in electron micrographs.

Let us now set up the steady-state conditions of the Munch model for solute and water movement in the vascular system. Consider a vascular bundle containing a single xylem vessel and a single phloem sieve tube. We take the vessel to be a right cylinder of radius a (cm) and the sieve tube, which runs parallel to the vessel, to be a right cylinder of radius b (cm) having sieve plates at regular intervals c (cm) along its length. The sieve plates have thickness d (cm) and contain N cylindrical pores of radius e (cm) and are inclined to the longitudinal axis of the sieve tube at an angle.

The space separating xylem and phloem is of unspecified composition, but will of course contain parenchyma cells, companion cells and possibly transfer cells. There are at least two parallel pathways through this space, but here we shall characterise transport through this space by 'mean values' of the various parameters. This simplification can be justified for parallel permeability barriers, but may cause difficulties if there are barriers in series.

Let us next measure displacement along the vessel and the sieve tube by the variable x (cm) and choose the point $x=0$ to be the leaf end of the vascular bundle. From the ontogeny of both xylem and phloem we know that end wall perforation does occur at some developmental stage and this is the point taken as $x=0$. Perforation may not happen at the same distance from the meristem in xylem and phloem, but we make this simplification ; the point $x=0$ is then the leaf end of both the open xylem vessel and the 'open' sieve tube. The simplification just made is not critical to the argument which follows. Similar considerations will apply at the sink end of these vascular elements, the point we shall refer to as $x=L$.

The model system

The basic processes of the Munch proposal are outlined schematically in Fig. 7.7. We suppose that solutes (principally photosynthetically produced sugar) are accumulated into the sieve tube over a region near the plant leaf and are exported from the sieve tube to be utilised in metabolism or stored in osmotically inactive forms in regions other than the leaves. Concommitant upon these solute movements, water will tend to enter the sieve tube osmotically at the leaf and leave osmotically at the solute unloading regions. If the solute transport rates are constant in time, a standing osmotic gradient system will be established in the sieve tube. The source of water at the leaf and the sink for water in the unloading regions must be the xylem. Thus we can set up a water circulation 'down the phloem and up the xylem'.

The rate of flow of this circulation will obviously depend on the driving forces and the resistances at every stage of the total pathway. In this context, the association of the companion cells/transfer cells with the sieve elements is envisaged as a magnification of the available membrane area, for both solute and water transport, to supply the standing osmotic gradient in the sieve tube.

In order to characterise the flows, we require an expression for the solute concentrations in both xylem and phloem at any distance from the leaf, the difference in which two values will give the osmotic gradient at that point. We also require to know the pressure gradients necessary to drive both phloem and xylem translocation streams. Finally, by insisting on water equilibrium between xylem and phloem at every point distant x from the leaf, we can derive the phloem translocation rate.

Sotute concentration equation for phloem

We write a transport function $T(x)$ (moles $cm^{-2}\ sec^{-1}$) which is taken to describe solute transport to and from the phloem sieve tube, across the side walls of the tube (Fig. 7.11). (The solute is principally sugar, but other solutes, *e.g.* K^+, will also be involved.) We may then write for the sieve tube contents at any distance x from the leaf

$$D\frac{\partial^2 C^p}{\partial x^2}-\frac{\partial}{\partial x}(C^p v^p)+\frac{2T(x)}{b}=0 \qquad (1)$$

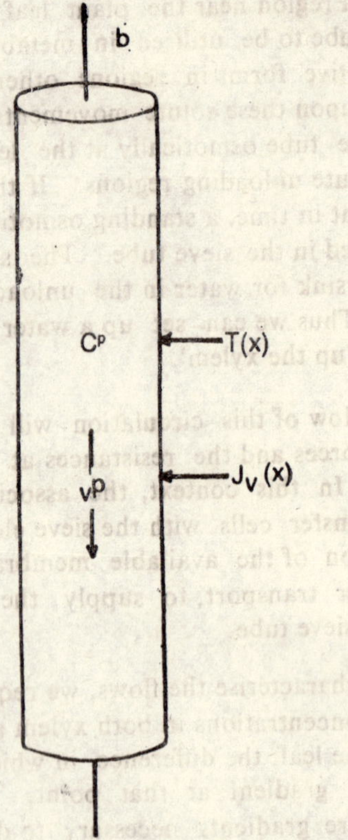

Fig. 7.11. **The action of the solute transport function T (x) (in sieve tube, radius b) causes increase or decrease in the local solute concentration C^p, which in turn disturbs the local water equilibrium and a flux of water $J_v(x)$ will enter or leave the sieve tube. In order to conserve volume the axial flow velocity, v^p will change in accord with eqn. (7).**

as the steady-state conservation equation for the solute concentration $C^p(x)$ (moles cm^{-3}) in the phloem. It is assumed that there is perfect radial stirring in the sieve tube. D (cm^2 sec^{-1}) is the diffusion coefficient for the solute in water and v^p (cm sec^{-1}) is the velocity of convective flow along the sieve tube. We ignore the complications of parabolic or other flow profiles.

We now simplify eqn (I) by approximating the diffusion component to zero as compared to the convective component. This approximation has been used previously and will be valid where transport distances of more than several centimetres (to be very safe) are involved. Transport along a sieve tube, involving hundreds or thousands of centimetres, is obviously such a situation. Eqn. (1) then reduces to

$$\frac{d}{dx}\ (C^{pvp}) = \frac{2T(x)}{b} \tag{2}$$

which equation gives to a very good approximation the steady-state conservation of solute in the phloem at any distance x (cm) from the leaf end of the vascular bundle.

One point should be made. Over the whole length of the sieve tube, L (cm), there can be no net accumulation or loss of solute for the steadystate condition to apply. Thus we require the constraint

$$\int_0^L T(x).\ dx = o \tag{3}$$

We take $T(x)$ as positive for solute transport into the phloem and negative for transport from the phloem.

Solute concentration equation for the xylem

No transport function is used to describe solute movement to or from the xylem, not because there is no such movement, but because the osmotic pressure changes so caused are relatively minor and can probably be safely neglected. We thus refer to xylem solute concentration as the constant value C^x (moles cm^{-3}).

Axial hydrostatic pressure gradient along the phloem

The phloem translocation stream will be maintained by the axial hydrostatic pressure gradient acting across the hydrodynamic resistance to flow along the sieve tube. This resistance is not uniform because of the intervening sieve plates, but nevertheless it can be handled by adopting the following procedure. Consider a segment of sieve tube from the centre of one sieve element through the sieve plate to the centre of the adjacent sieve element. There are two flow velocities in this segment, v^p in the lumen and v_{pore} in the sieve pore. If we take the flow in the lumen to be Poiseuille, then we may write for the pressure drop over the entire lumen length of a single sieve element

$$\Delta P_1 = \left\{ \frac{8(c-d)\eta^p}{b^2} \right\} v^p,$$

where $(c-d)$ is the length (cm) of a single sieve element lumen and η^p (poise) is the dynamic viscosity of the flowing solution.

In spite of possible arguments for the occurrence of flow other than Poiseuille in the sieve pores, especially if there are P-protein fibrils running through them, we again assume Poiseuille flow. The velocity of flow in the pores will have to be

$$v_{pore} = \frac{b^2}{Ne^2} v^p$$

to conserve volume flow in the sieve tube.

The pressure drop across a sieae plate will therefore be

$$\Delta P_{pore} = \frac{8d\eta^p}{e^2} v_{pore}$$
$$= \frac{8d\eta^p b^2}{Ne^4} v^p$$

and we are now in a position to compute the total pressure drop from the centre of one sieve element through the sieve plate to the centre of the next adjacent sieve element as

$$\Delta P^p = \Delta P_1 + \Delta P_{pore}$$
$$= \frac{8(c-d)\eta^p}{b^2} v^p + \frac{8d\eta^p b^2}{Ne^4} v^p$$
$$= 8\eta^p \left\{ \frac{c-d}{b^2} + \frac{db^2}{Ne^4} \right\} v^p,$$

where c (cm) is the sieve element centre to centre distance.

Over some distance x (cm) where $x > c$ we will have (x/c) such centre to centre segments, so that the total pressure drop over x (cm) of sieve tube length will be

$$\Delta P^p = \frac{8\eta^p x}{c} \left\{ \frac{c-d}{b^2} + \frac{db^2}{Ne^4} \right\} v^p. \quad (4)$$

Axial hydrostatic pressure gradient along the xylem

The pressure gradient along the xylem to drive the flow is straightforwardly available from Poiseuille's law on the assumption that the vessel end walls are fully perforate.

$$\triangle P^x = \frac{8x\eta^x}{a^2} v^x \qquad (5)$$

will give the pressure difference over a distance x (cm) of xylem when the flow velocity in the xylem is v^x (cm sec^{-1}).

Water movement between phloem and xylem

We assume that the only forces driving water between phloem and xylem are the hydrostatic pressure gradient and the osmotic pressure gradient ; no matric potential component is considered. The well-known formalism of Kedem and Katchalsky yields a water flux at any point along the vascular bundle x (cm) from the leaf end of J_w (cm sec^{-1}) equal to

$$J_w = L_p(x)[\triangle P(x) - \sigma(x)\, RT\triangle C(x)], \qquad (6)$$

where $L_p(x)$ is the hydraulic conductivity for the total pathway from xylem interior to phloem interior and may be a function of distance from the leaf, $\triangle P(x)$ is the hydrostatic pressure difference between xylem and phloem at any point x (cm), $\sigma(x)$ is the reflexion coefficient for the pathway from xylem to phloem, R is the gas constant, T is the temperature (°K) and $\triangle C(x)$ is the solute concentration difference between xylem and phloem at any distance x (cm) from the leaf. It is appreciated that we may have here a multi-membrane problem which could cause difficulties with the formalism just given, but for the sake of keeping this treatment within reasonable limits the simplification has been carried through.

The volume flux, $J_v(x)$, will enter or leave the phloem at any point x, across the side walls (and possibly through the companion cells, as earlier discussed). In order to conserve volume (xylem and phloem being taken as incompressible) we require the following condition relating the change in axial flow velocity to the volume flux across the side walls.

$$\frac{dv^p}{dx} = \left(\frac{2}{b}\right)\{J_v + \overline{v}_s\, T(x)\} \qquad (7)$$

where most symbols have been defined previously, and $\bar{v}_s$ is the solute partial molar volume. Note that we can still define J_v in terms of the sieve element surface area, even if the companion cell/transfer cell complex is in fact the membrane area over which the osmotic water flow occurs.

The equivalent condition for the xylem is

$$\frac{dv^x}{dx} = \left(\frac{2}{a}\right) J_w^x, \tag{8}$$

where J_w^x is the normalised water flux across the xylem vessel side walls.

Phloem-xylem circulation

We now have available all the factors necessary to formally describe the circulation 'down' the phloem and 'up' the xylem as envisaged in the Munch model.

At any point along the vascular bundle x (cm) from the leaf end the exchange of water between xylem and phloem will be given by eqn (6)

$$J_w = L_p(x)\ [\triangle P(x) - \sigma(x)\ RT \triangle C(x)],$$

where J_w is the water flux normalised over unit area of sieve tube side wall (which is taken to include companion cell or transfer cell contributions to water flow, if appropriate). Thus in unit centimetre length of sieve tube the volume rate of water entry will be $2/bJ_w$ ($cm^3\ sec^{-1}$), which will also be the rate of volume gain or loss in the xylem if we assume no net water gain or loss in the parenchyma cells.

Our basic aim is to relate the velocity of axial flow, v^p (the phloem translocation velocity), to the other parameters of the system, which we can do immediately by substituting from eqn (7) into (6),

$$\frac{dv^p}{dx} = \frac{2L_p(x)}{b}\ [\triangle P(x) - \sigma(x) RT \triangle C(x)] + \frac{2}{b}\bar{v}_s\ T(x). \tag{9}$$

We may now make further substitutions from eqns (2), (4) and (5) if we remember that

$$\triangle P(x) = \triangle P^p(x) - \triangle P^x(x),$$
$$\triangle C(x) = C^p(x) - C^x,$$

to obtain as the final result

$$\frac{dv^p}{dx} = \frac{2L_p(x)}{b}\Bigg[8x \left\{ \frac{\eta^p}{c} \left(\frac{c-d}{b^2} + \frac{db^2}{Ne^4} \right) v^p - \frac{\eta^x v^x}{a^2} \right\}$$

$$-\sigma(x)RT \left\{ \int_0^x \frac{2T(x)}{b}\, dx - C^x \right\} \Bigg] + \frac{2}{b}\, v_s T(x). \qquad (10)$$

This equation then gives the relationship between the axial velocity of the translocation stream in the phloem (actually the flow velocity in the sieve element lumen) and the solute transport into and from the phloem, the mechanism originally proposed by Munch (1930) and here given the modern dressing of being considered as a standing gradient osmotic flow along the sieve tube. The equation cannot be integrated analytically but can be readily evaluated by computer ; details are available for integration of a similar form Note that v^p will depend on v^x, the xylem flow velocity, as well as on x, the distance from the leaf, and the solute transport function, $T(x)$.

At times of rapid transpiration, v^x may be taken as constant with small error because it will principally represent transpirational loss and the small adjustments to the xylem flow velocity as a result of Munch-type exchange between xylem and phloem will have a negligible effect. Otherwise the xylem velocity will be modified by this exchange in accordance with eqn (8), where $J_w{}^x = (b/a)\, J_w$. In general, as the xylem flow velocity increases the phloem translocation velocity will decrease, other things being equal.

8

Symplasmic Transport in Plants

Theoretically, transport of molecules through plant tissues could take place either by release into and diffusion through the extracellular spaces (apoplast) followed by uptake at a distant point or by diffusion through the system of protoplasts and protoplasmic connections (plasmodesmata) known as the symplasm. Although transport via both pathways probably does occur, it is clear on general grounds that the symplasmic pathway would have advantages for the plant, firstly because it avoids use of the energy that would probably be expended in the final uptake step and, secondly, it would prevent loss of materials to the environment through the free space in organs, such as roots, that lack a cuticle.

Plasmodesmata were first observed in endosperm tissue by Tangl (1879) who suggested that they might be involved in transport. Work on plasmodesmata from these early beginnings has been reviewed thoroughly by Meeuse (1957). There has, in fact, been very little experimental work on the function of plasmodesmata and it is possible for the author of a recent *Plant Physiology* text to conclude, with more candour than many authors, that '...the extent and importance of the role of plasmodesmata in intercellular transport is not known'. To assess the present status of our knowledge we shall review some of the older evidence together with that made available recently by the use of electrophysiological techniques. Since phloem transport has been treated separately we shall not include it in our consideration of the symplasm, though Arisz (1969) frequently makes no distinction between phloem and other tissues in this respect.

STRUCTURE OF THE PLASMODESMA

Although the existence of plasmodesmata was inferred from light microscopy, the dimensions of plasmodesmata, as revealed by electron microcopy, show that they are well beyond the resolving power of the optical microscope. Taiz and Jones (1973) have shown, however, that the cell wall surrounding the plasmodesma in barley aleurone tissue has a modified structure which forms a 'wall tube'. They suggest that it might be this that was detected optically. In other cases swelling and staining treatments were used.

Ultrastructural studies show that plasmodesmata typically have an overall diameter in the range 20-80 nm and the membrane lining the pore is continuous with the plasmalemma of both cells. There has been considerable controversy over structures in the central core and whether or not the plasmodesma provides as open pathway between adjacent cells. Recently, however, there appears to be a growing recognition that a variety of plasmodesmatal structures probably occur in different plants and tissues and that the function of these structures cannot be predicted on the basis of structural evidence alone. It is also important to recognize that the structures appearing in the plasmodesma could be due to a repair or blocking mechanism similar to callose and slime formation in cut phloem. Studies with the electron microscope have, however, removed any doubt that, with the exception of guard cells and cells in certain reproductive and senescing tissues, plasmodesmata are present between all cells in higher plants.

SYMPLASMIC TRANSPORT

Assertions that the symplasm plays a role in transport have, for the most part, been based on structural evidence or physiological evidence of an indirect nature. To those familiar with plant physiology it may appear self-evident that plasmodesmata provide a transport pathway. If it were not so, cells in the cortex of the root, for instance, would be dependent on diffusion through the cell walls for their supply of carbohydrate from the phloem. It is now well established that there is no diffusion barrier between cortical cell walls and the solution outside the root. This situation would, therefore, lead to massive losses of carbohydrate to the soil solution and, since such losses are normally small in the mature parts of the root it seems reasonable to conclude that transport takes place in the symplasm. The demonstration that the Casparian strip in the endodermal cell wall provides a barrier to solute diffusion strengthens this conclusion. The problem of proving the existence of symplasmic transport may, therefore, appear trivial. However, the technical problems are certainly not trivial, particularly if a quantitative description of the situation is required. Since the problems associated with the movement of water and solutes are quite different, we shall treat them separately.

WATER

The study of the pathways of water movement in plant tissues presents the most serious technical difficulties. In principle, it might be possible to use isotopically labelled water to trace the pathway. The fluxes of water across plant cell membranes are so great, however, that the label almost immediately becomes uniformly distributed throughout the tissue, the rate being controlled by unstirred layers rather than by membrane permeability. Our knowledge of the pathways of water movement is, therefore, based on other types of experiments. We know from the existence of exudation and the osmotic behaviour of root systems that there must be an osmotic barrier at some point in the root. Whether this barrier is at the endodermis or nearer the xylem is a point of controversy at present. However, the presence of a barrier means that the protoplasm must form a major element of the pathway in this region.

The importance of the symplasm as a pathway in other tissues has not been established. A kinetic method that purportedly demonstrated the importance of the cell wall pathway in leaf tissues was later shown to be invalid due to the misinterpretation of growth effects though these authors report that it is valid for mature leaves.

As a result of these technical problems it is only possible to make an educated guess as to the relative importance of the various pathways. It seems likely that the cell walls provide the major pathway except in those cases where they are impermeable to water. A detailed analysis of the contribution of the symplasm to water and solute transport has been made by Tyree (1970) taking the best data available for plasmodesmatal frequency, area and constitution. His calculations indicate that the presence of plasmodesmata greatly increases the hydraulic conductivity for passage of water between cells. However, he did not compare the symplasm pathway with the cell wall pathway. Furthermore, it is clear from the work of Clarkson, Robards and Sanderson (1971), in which they combined a detailed examination of plasmodesmata in the endodermis of barley roots with measurements of water and phosphate fluxes, that calculations of the pressure difference necessary to support the observed flow of water through the plasmodesmata depend critically on the exact area of the plasmodesma available for water movement and the viscosity of the material in the pore. These parameters are not known with any accuracy at present.

SOLUTES

Tracer methods. Tracers have been used to investigate symplasmic transport at least since 1909 when Bierberg used dyes to investigate the effect of cytoplasmic streaming on intercellular transport in several species of plants, including members of the Characeae. The experiments on *Chara* were confirmed by Littlefield and Forsberg (1965) using radioactive phosphate. From the precautions they took it is clear that transport took place through the symplasm. In general, however, tracer methods suffer from certain disadvantages, among which a lack of spatial resolution at the cellular level in higher plants is the most serious. A combination of micro-injection and microautoradiography might overcome this problem but at present it is difficult to determine the exact cellular pathway by which the tracer moved, let alone whether it moved through the plasmodesmata.

In most tissues this situation is complicated by the presence of a cuticle or other extracellular barrier that would permit passage through and uptake from the apoplast. Some of these problems have been circumvented by Arisz in his extensive work on *Vallisneria spiralis*. As a submerged aquatic plant, *Vallisneria* lacks a cuticle and does not transpire. The vascular bundles do, however, contain sieve tubes and, unfortunately, Arisz usually uses the term symplasm to include the phloem. It is by no means certain that there is protoplasmic continuity between the parenchyma and phloem cells. Indeed, his experiments on the effect of azide on the movement of chloride from the vascular bundles to the parenchyma suggest that transport across a membrane may be involved. However, he has provided convincing evidence that chloride can move through the symplasm of parenchyma tissue. The simple anatomy of *Vallisneria* leaves makes it possible to remove short sections containing the vascular bundles, leaving only 'parenchyma bridges' connecting the two parts of the leaf. When sufficient time had elapsed to allow for recovery from wounding and consequent leakage of electrolytes, it was possible to demonstrate movement of chloride through the parenchyma bridges using autoradiography and a two-compartment chamber for loading and transport. Since there was no leakage of chloride from the tissue into the transport compartment and since such leakage would have occurred due to the lack of a cuticle if the apoplast had been part of the pathway, it must be concluded that the chloride moved through the symplasm. Close examination of the autoradiographs suggests that once the chloride has passed the parenchyma bridges it enters the phloem and is transported rapidly over long distances. However, these experiments appear to offer the best proof from tracer studies that transport takes place through the symplasm of higher plants. Besides chloride, other ions and a variety of small organic molecules, including indole acetic acid, are translocated in *Vallisneria.*

The tentacles of *Drosera capensis* have also been used by Arisz in transport studies. These structures contain no phloem so transport must be through the two layers of living cells. Since there is a cuticle it is conceivable that transport could take place through the apoplast. However, inhibition by anaerobic conditions and a variety of inhibitors suggested that uptake into the cells was an essential step in transport which, therefore, took place through the symplasm.

A large body of evidence concerning lateral transport of salts across roots to the xylem is also consistent with the symplasm theory.

Electrophysiological techniques. For ions there is an alternative method of tracing transport pathways. In general, passage of current through a biological preparation will produce potential gradients that are proportional to the resistances of the pathways through which the ions are moving at any given point. With multiphasic tissues containing various mixtures of ions, the interpretation of such data is extremely difficult. Nevertheless, Lou (1955) used external electrodes to make such measurements. Using a preparation of two internodal cells of *Nitella* immersed in oil, he measured the potential gradient along the system when a current was passed between the two ends. There were sharp gradients where the current entered and left the interior of the cells near the current electrodes and a smaller gradient along the length of the cells. At the node there was also a steep gradient but it was not as great as would be expected if all the current flowing inside one internodal cell had to cross the membrane before entering the adjacent internodal cell. He concluded that there were direct low-resistance connections between the internodal cells. Lou also performed similar experiments on roots but, since the external electrodes are only capable of measuring the gradient along the root rather than along the length of individual cells and across cell junctions, it is more difficult to interpret the results. He found a greater spread of the potential gradien near the current electrodes in living than in dead roots. He interpreted this as penetration of current into the symplasm in the living roots. However, it might merely reflect impeded spread into the apoplast or the stele in the presence of intact cells. A better comparison would be between roots with intact and disrupted plasmodesmata. Unfortunately, there appears to be no way to selectively disrupt plasmodesmata while leaving the cells intact.

A better method with good spatial resolution involves the use of microelectrodes. Basically the change in potential in one cell is measured when current is injected into a neighbouring cell. If there is no connection, all the current will flow out through the intercellular space, since it has a relatively low resistance, and no change in potential will be detectable. However, a direct connection will permit current to flow into the cell and out across its membrane with a resulting change in membrane potential. This method was first used for the study of non-excitable animal cells by Loewenstein and Kanno (1964) and Kuffler and Potter (1964). It has since been used widely in animal physiology.

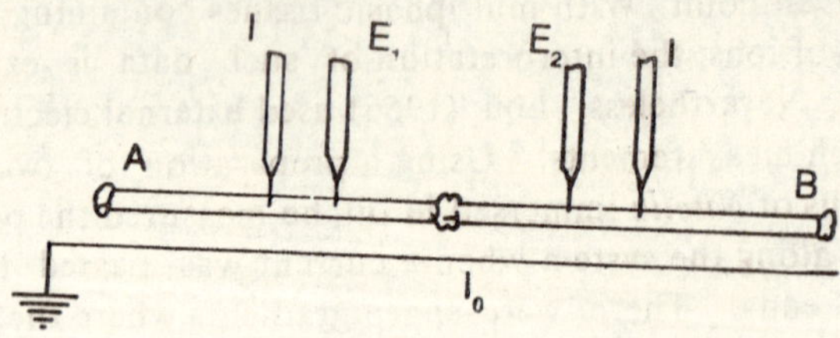

Fig. 8.1. Diagrammatic representation of the system of two internodal cells of *Nitella translucens* (A and B) and the arrangement of the current microelectrodes (I), potential microelectrodes (E_1 and E_2) and the external current electrode (I_0). The external reference electrode is not shown.

It occurred that the same technique could be used to determine whether plasmodesmata did in fact provide a direct pathway between plant cells. The method was initially tested on *Nitella translucens* using a system consisting of two internodal cells joined by a node (Fig. 8.1). By using *Nitella* it was possible to overcome one of the problems pecular to plant cells, *i.e.* the presence of the tonoplast membrane. Because current is injected into the vacuole, there is a change in potential across both the tonoplast and the plasmalemma. However, since current flowing to the adjacent cell through the plasmodesmata does not have to cross the tonoplast, measurement of the changes in potential between the vacuoles and the external solution would give an overestimate of the attenuation of the pulse. By inserting electrodes into the cytoplasm it was possible to measure the attenuation across the node accurately and calculate the nodal resistance. These experiments showed that the specific resistance of the node, which contains at least one layer of small internodal cells, was only 1.7 kΩ cm^2 as compared to about 90 kΩ cm^2 for the plasmalemma of the internodal cells. Although the nodal resistance is low and clearly indicates that the plasmodesmata can carry current, and therefore ions, between the cells, it was about 300 times greater than the resistance calculated from the plasmodesmatal dimensions and frequency. It is, therefore, likely that plasmodesmata contain structures that hinder free diffusion of ions. Electrical coupling between characcan internodal cells has also been demonstrated by Sibaoka (1966) and Skierczynska (1968).

Application of this method to higher plants is not easy. Spitzer (1970) demonstrated coupling between cells in the developing lily anther. Goldsmith, Fernandez and Goldsmith (1972) were unable to detect electrical coupling between parenchyma cells of *Avena* coleoptiles. However, this appears to have been due to technical difficulties since coupling in this tissue can be demonstrated. With higher plant cells it is not possible to insert microelectrodes directly into the cytoplasm and it is, therefore, difficult to determine the attenuation across the plasmodesmata since an unknown fraction of the potential change in the cell into which current is injected is across the tonoplast. However, by measuring the potential changes in three cells in a row on the edge of an *Elodea* leaf and using an approximate equivalent circuit for the tissue, it was possible to make an estimate of the plasmodesmatal resistance as well as the separate resistances of the plasmalemma and tonoplast. Again, the resistance of the plasmodesmata appeared to be greater than would be expected if they were completely open. The structural information required for an accurate comparison is not yet available. In the same paper, coupling was also demonstrated between the cortical cells of maize roots.

Macromolecules and viruses

Scientists have been unable to find any experimental evidence for the movement of macromolecules through plasmodesmata. Reports that viruses can move through plasmodesmata would appear to indicate that macromolecules should be able to do so also and that the plasmodesmata are open. However, plasmodesmata which contain viruses have a larger diameter than those in control preparations and this suggests that the virus may penetrate by modifying the structure of the plasmodesma.

TRANSMISSION OF ACTION POTENTIALS

1973 marks the centennial of the discovery of action potentials in plants. In 1873 Sir John Burdon-Sanderson received some Venus Flytraps (*Dionea muscipula*) from Charles Darwin and on 12 September recorded action potentials from the leaves. At the beginning of this century there was considerable interest in the transmission of stimuli which nowadays is usually considered to involve the movement of hormones. However, there has recently been a resurgence of interest in the role of action potentials in plants and a recognition that they may be involved in processes other than rapid movements.

We would like to take this opportunity to point out the probable role of plasmodesmata in the transmission of action potentials. *Nitella* provides a convenient model system in that the internodal cells are easily stimulated with intracellular microelectrodes. An action potential generated in one cell is not always transmitted to the neighbouring cell and this suggests that the action potential is not propagated directly along the membranes of the intervening plasmodesmata and/

or the intervening layer of nodal cells. Consistent with this is the delay observed when transmission does take place (Fig. 8.2). The simplest explanation of this would be that

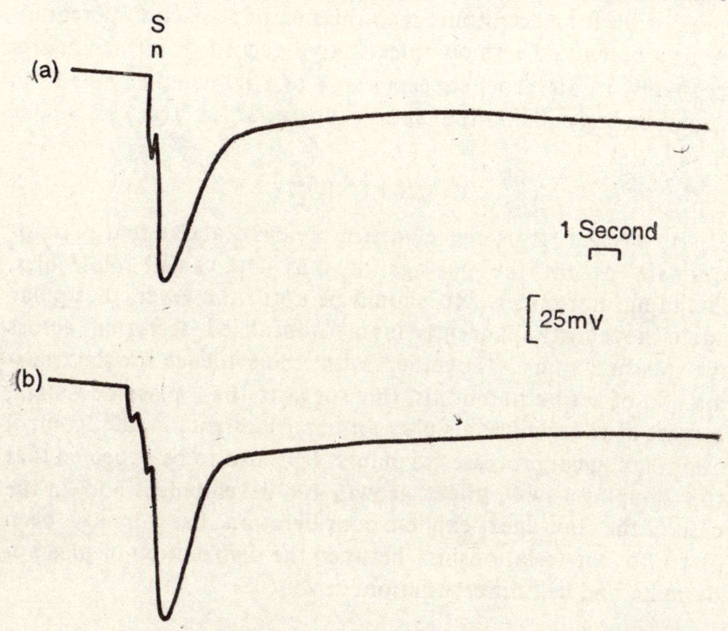

Fig. 8.2. Recording of action potentials from two adjacent internodal cells of *Nitella translucens*. The recording microelectrodes were placed near the node. (*a*) Recording from the cell into which current was injected to produce the stimulus, S, which was followed by an action potential. (*b*) Recording from the adjacent internodal cell. The attenuated component of the stimulus is followed by the attenuated component of the action potential from the first cell which, in turn, triggers off a separate action potential.

depolarization in one cell causes flow of current through the plasmodesmata and hence across the membrane of the neighbouring cell. If this attenuated signal is of sufficient magnitude and duration it will stimulate a separate action potential.

Action potentials in the *Drosera* tentacle are triggered by a receptor potential in the tentacle head and are propagated down the stalk which contains only two layers of living cells. The end walls of the stalk cells contain numerous plasmodesmata and it seems reasonable to assume that these are responsible for electrotonic transmission of the action potential. Action potentials with complex shapes recorded intracellularly probably result from superposition of attenuated signals from neighbouring cells on the action potential in the cell under study.

CONCLUSIONS

It is clear from the evidence reviewed above that plasmodesmata permit the passage of ions and small molecules, including hormones. It should be noted, however, that polar auxin movement apparently involves localized secretion across the plasmalemma. Together with the evidence for the transmission of action potentials, this suggests that plasmodesmata, as part of the symplasm, play an important role in the control of physiological processes in plants. It is also to be expected that the symplasm will affect growth and development and, in the case of the root apex, explicit consideration has already been given to the relationship between the distribution of plasmodesmata and cell differentiation.

9

Translocation in Fungi

Since we know that certain fungi can grow for considerable distances over non-nutrient surfaces and since growth of fungi only occurs at the hyphal tips, nutrients must be able to move within the hyphae to the growing points. *Serpula lacrymans*, the fungus causing 'dry rot' of timber, is able to penetrate stone walls two feet thick, while Findlay (1951) records the presence of two rhizomorphs of *Armillaria mellea*—one about 30 m long and the other about 90 m—in a water tunnel, 200 feet below the ground surface. The rate of growth of these rhizomorphs (2.5 cm week^{-1}) appears to be little different from what one might expect to find in pure culture. Clearly in these cases nutrients must readily move through the hyphae.

Whatever the mechanisms of translocation, the process, unlike the analogous process in higher plants, has to occur in many instances within non-specialised hyphae in which other highly significant physiological processes must be taking place. This makes translocation within fungal hyphae a difficult process to analyse. Indeed, as we shall show there are at least four recognisable ways in which material may be moved within hyphae.

MOVEMENT OF NUCLEI

A nucleus, because of its genetical material, can produce detectable morphological and biochemical changes which are ascribable only to the presence of that nucleus or other nuclei derived from it at a certain locus in the mycelium. Thus these morphological and biochemical changes at different parts of a mycelium can be used as markers for the movement of a nucleus through that mycelium. The classic observations were made by Buller (1931) who showed for *Coprinus lagopus* that, when a plate containing initially only primary mycelium was inoculated on one side with primary mycelium of the opposite mating type, nuclei migrated across the mycelium. The movement was detected by the formation of clamp connections which are characteristic only of secondary raycelium. Since then the phenomenon has been studied in several other homobasidiomycetes and in a few ascomycetes.

There are a number of key features about the process. Snider and Raper (1958) studied nuclear migration in *Schizophyllum commune* using biochemical mutants as markers and two mycelial geometries. In one instance the recipient mycelium had grown out radially from a point source (radiate); in the other, the recipient mycelium had grown from hyphal fragments distributed uniformly over the plate, such that there were numerous anastomoses between hyphae (reticulate). The mycelium producing the migrating nuclei (migrant) was a small inoculum placed at the centre of the recipient. Sampling was either along four radii at right angles or by a grid. At each sampling time, the positions of the four radii or the grid were varied by a few degrees.

The results obtained from these studies, while complicated in detail are in general clear. First, the rate of migration in radially growing recipients was at least ten times faster than hyphal tip growth and the Q_{10} for the rate of migration over the range 22-32°C was about three times that for tip growth. Thus the movement of nuclei cannot be explained on the basis of hyphal growth from the migrant inoculum.

The migrating rate in reticulate mycelium was found to be about 0.5 mm h^{-1} but there was some doubt about the rates in radiate mycelium. The range was from 1.0—5.4 mm h^{-1}. The variation here may be related to a lag in nuclear division and/

or migration. The difference between the rate for reticulate mycelium and that for radiate may be because nuclei follow a more direct route in the latter.

In interpreting their data (for reticulate mycelium), Snider and Raper pointed out that for migration to occur the migrant and resident must first grow a little and anastomose. Once initiated, migration in reticulate mycelium will still require continued growth and anastomosis. This means that the above measurements of Q_{10} relate as much to the rate at which channels are opened up as to the rate of migration *per se*.

We know now that the rate of migration is under the control of B-incompatibility factors, which apparently regulate migration at the septa by causing enzymes to dissolve away part of the complex pore, thus facilitating nuclear passage. In support of this there is a significant rise in the specific activity of a particular cell-wall gluconase in those matings of *Schizophyllum commune* in which the B-sequence is operative. Two interesting points arise from these latter observations. First it is clear that any concept of opening up of channels must include the process of septal dissolution as well as that of anastomosis. Second, the dissolution of septa will facilitate not only nuclear migration but presumably also translocation of other material will be very much faster in those hyphae with dissolved septa than in those containing complex septa.

Nuclear migration might be brought about by either a specific force acting solely on the nucleus or a more generalised force bringing about both nuclear and cytoplasmic movement. Two tests have been carried out to investigate these possibilities. One reported by Snider (1968) using *Schizophyllum commune* indicated that the receptor of light-induction of fruiting did not travel with the nucleus. This does not necessarily mean that there is no movement of cytoplasm since the receptor may be bound to the outer membrane. In the other test, Casselton and Condit (1972) used mitochondrial genes as cytoplasmic markers and here too there appeared to be no movement of cytoplasm.

The only other information concerning the physiological basis of nuclear migration is that obtained by Snider (1968) who showed that the process requires oxygen but the rate in air is reduced by only three times when the oxygen concentration is lowered to 0.1% O_2.

In spite of the paucity of physiological information, electron microscopical observations have provided suggestions regarding a possible mechanism. The relevant information comes from studies on the 'shortrange' nuclear movements which occur in basidiomycetes. Two types of movement have been distinguished. One occurs after nuclear division, when the two daughter nuclei move apart and the other maintains the central position of the nuclei in the cells. Elegant studies by Girbardt (1968) using a combined electron and light microscopical approach have demonstrated that microtubules are involved in short-range movements within the cells of *Polystictus versicolor*. Girbardt also showed that oscillation of nuclei occurs independently of cytoplasmic streaming. Raudaskoski (1973) has demonstrated the presence of microtubules scattered in the hyphae around interphase nuclei in mycelium in which intracellular migration is occurring. Further, during hyphal fusion, microtubules close to an interphase nucleus were found to be connected with electron-dense structures, thought to be centres for the assembly of microtubules. The assumption that the microtubules participate in the movement of nuclei is based on the findings described above plus the observations that microtubules have been most frequently detected by Raudaskoski in those hyphae involved in migration. It is of relevance here that in the imperfect fungus *Drechslera sorokiniana*, fascicles of microtubules have been found running parallel to the long axis of the conidiophore. They are presumed to be involved in the orientation of cytoplasmic streaming and not nuclear movement. Nevertheless they are consistently located close to the nucleus.

THE ABILITY OF FUNGI TO TRANSLOCATE

Using the simple test of whether a fungus will grow from a medium supporting luxuriant growth onto one lacking nutrients, Schutte (1956) and Thrower and Thrower (1968) have provided information that indicate that the ability of a fungus to translocate material along its hyphae is by no means universal. Thrower and Thrower found that the fungi tested could be divided into three groups colonising, non-colonising and indeterminate. All the basidiomycetes tested were colonisers. The phycomycetes showed a range of abilities but none of those from predominantly aquatic habitats (*Achlya*, *Allomyces*, *Apodachlya* and *Saprolegnia*) could colonise non-nutrient media.

It is also interesting to note that cytoplasmic streaming was known to occur in 16 of the fungi tested but only 10 were capable of colonising deficient media. The inability of a fungus to colonise deficient media may be due to the absence within the hyphae of any mechanism capable of bringing about translocation. But one also needs to keep in mind the possibility that the so-called deficient media may not be inert and may in fact be inhibitory to growth. In septate fungi, failure to colonise deficient media may be a consequence of septal blocking.

Although there have been many studies of the ultrastructure of fungal septa little consideration has been given to the relationship between blocking and other physiological processes. Fortunately, a recent study by Trinci and Collinge (1973) has done something to rectify the position. They have shown from studies of several strains of *Neurospora crassa* that the percentage of unplugged septal pores at different distances from the periphery of a fungal colony shows very considerable strain differences, with variation from 0% unplugged at 1 mm in one strain to 100% at 12 mm in another stram. The results certainly show that there may be a direct relationship between hyphal density and the plugging of septal pores. Septa of those strains which produce very dense colonies have their septal pores plugged closer to the periphery of a colony than those which produce sparse colonies.

This sort of study needs to be paralleled by one in which translocation is measured. Since there is also a relationship between septal plugging and the peripheral growth zone (a parameter which can be readily measured for plate cultures), the relationship between ability to translocate and septal plugging can be investigated without recourse to electron microscopy. *Geotrichum candidum* merits investigation since the septa are entire except where traversed by plasmodesmata-like structures.

TRANSLOCATION STUDIES, EMPLOYING PETRI DISH CULTURE

There have been many attempts to study translocation by using petri dish cultures. Essentially, the procedure has been to allow a fungus to grow across a diffusion barrier and examine the movement of a radioactive tracer across that barrier. The results are difficult to interpret, mainly because few attempts have been made to separate the mycelium from the substratum and the labelled compounds used have been those which were also metabolisable by the fungus. This means that one has no measure of the concentration or specific activity of the moving compound at any point along the path of translocation.

Further, in nearly all instances mycelium has been allowed to colonise the entire dish before movement of isotope has been studied. The reason for this is not clear but it means that many of the older hyphae present may be highly vacuolate and indeed, if nutrients are at a low level in the substratum, conceivably without contents because of autolysis. This latter phenomenon has been shown to occur in shake culture over a similar period of time.

In view of this possibility, it is perhaps not surprising that, where an attempt has been made to calculate the rate of translocation, it has been found to be little different from the rate of diffusion in aqueous solution.

We have attempted to surmount these problems by using a labelled compound which is not metabolised and by causing the fungus to grow in a specific direction over the source of the labelled compound. At the same time we have made an estimate of the number and size of the hyphae in which translocation is occurring. The results to date indicate that in

Dendryphiella salina the non-metabolised sugar 3-O-methyl-D-glucose is translocated at a rate faster than diffusion.

There is a body of evidence that the formation of sclerotia and fruiting bodies have a strong directional effect on translocation which a my clian. Surprisingly, the reasons for this have not been clearly established, though the belief is that the above structures act as very efficient sinks for translocated material.

SHORT-DISTANCE MOVEMENT AT THE HYPHAL TIP

There is now plenty of electron microscopical evidence which strongly suggests that vesicles move to the extreme tip from the Golgi apparatus which is located some distance back along the hyphae. It is believed that these vesicles contain wall precursors which are inserted in the wall where extension growth is occurring.

Two hypotheses have been proposed for the way these vesicles move to the tip. Both hypotheses rest on the observations of Slayman and Slayman (1962) that the membrane potential of *Neurospora crassa* decreases as one moves from the mature parts of a hypha to its tip. Bartnicki-Garcia (1973) proposed that the vesicles move to the tip by electrophoresis driven by the potential gradient along the hypha, jennings (1973) believes that the drop in potential as one moves to the tip is a consequence of the fact that there is also a drop in the number of potassium pumps (ATPases) at the outer membrane. He produced in support of this hypothesis evidence of changes with time or potassium/sodium ratios in mycelium of *N. crassa* and of *Dendryphiella salina*. The net effect of this presumed distribution of potassium pumps could be the presence of a standing gradient osmotic flow in the hypha, so that there will be a bulk flow of liquid to the tip. This bulk flow of liquid would have a directional effect on the movement of the vesicles.

We now have evidence in support of the latter hypothesis. Hist ochemistry at the light microscopical level has demonstrated that there is no detectable ATPase activity at the hyphal tips of *D. salina*, activity only becoming apparent at 200 μm from the tip. There is of course a need to show a more exact location of the enzyme (*i.e.* that it is associated with the outer membrane) as well as isolate it and characterise it, especially with respect to activation by monovalent cations.

It is also interesting to note that, using the same techniques, ATPase activity is only detectable at the base of sporangiophores of *Phycomyces blakesleeanus* which is in keeping with the greater rate of potassium uptake by this region of the isolated sporangiophore.

STUDIES ON PHYCOMYCES BLAKESLEEANUS

Cowan, Lewis and Thain have made an intensive study of potassium translocation in *P. blakesleeanus.* They made three types of investigation.

(*i*) The measurement of the rate of steady-state movement of potassium through mature sporangiophores from a solution of higher potassium to one of lower potassium concentration, both solutions having the same osmolarity.

(*ii*) The same type of measurement for potassium movement through mycelium grown originally on cellophane and then separated from it prior to experimentation.

(*iii*) A study of the relationship between potassium and water movement in developing sporangiophores.

It is important that one is clear about what Cowan *et al.* were measuring in the first two studies. In both instances, the rate of translocation was taken as the rate of increase of potassium in the solution of lower concentration. This means that Cowan and his colleagues were examining the sum total of processes whereby potassium is loaded onto the fungal tissue, translocated within it and released into the external medium. The parallel between phloem translocation in higher plants and this process is clear. In both cases, there is loading of solute, translocation and release of the solute into a 'sink'. But the parallel is misguided, for under natural conditions the release of the translocatory compound by a growing fungus into the external medium is hardly to be expected. Further, the rate of translocation was measured when there was a steady state of potassium release. This took place after 45 hours in the first set of experiments and after 20-24 hours in the second set. Bearing this in mind, it is not surprising to find that the movement of potassium in both systems can be attributed to diffusion, for it can be argued on theoretical grounds that potassium would move through the system in this way.

We can think of both systems as a series of resistances. If one resistance is much greater than the others it will essentially govern the rate of flow through the system. At a steady state, therefore, the characteristics of the movement occurring within the system will be that of movement through that part of the system offering most resistance. We know from other studies that potassium is freely mobile in the cytoplasm of *P. blakesleeanus.* We have evidence (see above) that there are no potassium pumps at the hyphal and sporangiophore apices. Potassium ions must pass through these regions by diffusion, although the ions are most probably actively pumped in elsewhere. In any case, there exists no evidence that potassium ions are pumped out of fungi. Therefore the plasma membrane which the potassium ions must cross in order to reach the external medium will be the major barrier to movement through the two systems studied by Cowan *et al.* By the time equilibrium with respect to the rate of efflux of potassium out across the plasmalemma has been achieved, it is highly likely that the rest of the system will also be in diffusion equilibrium.

The more interesting part of the work of Cowan *et al.* are those studies concerning the uptake of potassium into developing sporangiophores under the influence of different water activities ($\omega_a = P/P_0$, where P is the vapour pressure over the medium and P_0 is that over pure water at the same temperature and pressure). It is clear from these studies that water movement caused by evaporation brought about potassium movement into the developing sporangiophore from the mycelium. The potassium concentration at the early stages of development was increased by 50% as water activity was decreased from more than 0.98 to 0.60. Parallel studies on isolated sporangiophores, using a microporometer, showed that water uptake at this stage was increased by about the same amount.

Microscopic investigation of sporangiophores showed two distinct types of streaming—strand streaming of constant velocity which is unaffected by external ω_a and transpirational streaming occurring in the inner regions of the sporangiophore.

As the sporangiophore develops it becomes less permeable to water loss due to cuticularisation. However, throughout the development of the sporangiophore, the apical portion maintains virtually the same concentration of potassium (125 mm l^{-1}). Cowan *et al.* suggest that since the velocity of movement of radioactive potassium to the apex (2.1×10^{-6}m sec^{-1}) is very similar to strand streaming (2.8×10^{-6}m sec^{-1}), it is the latter process which brings potassium to the apex. Diffusion is unlikely to be important since the expected boundary profile was not observed. In any case, the appearance of ^{42}K at the apex is more rapid than would be expected if diffusion were bringing about the movement of potassium. It is implicit in the consideration given by Cowan *et al.* to the mechanism of potassium movement to the apex that this mechanism can only bring about the net movement of a certain amount of potassium, irrespective of the amount of potassium at the base of the sporangiophore.

A constant potassium concentration at the apex could of course be due to the fact that the internal potassium concentration regulates the rate of growth occurring there. If this were so, there might be a variable rate of transport, the constant concentration of potassium being maintained by the growth rate. Under these circumstances, transpiration might be causing some of the movement. The point is difficult to test, since, although water activity had no significant effect on the rate of volume growth of sporangiophores, the standard deviations were large.

PROTOPLASMIC STREAMING

The first major studies on protoplasmic streaming were made by Arthur (1897) and Schroter 1905). They both studied streaming in *Rhizopus nigricans* and *Phycomyces nitens* amongst other fungi, for the most part members of Phycomycetes.

Streaming appeared to be brought about by differences of water potential, *e.g.*, evaporation from one part of the mycelium, or local applications of high concentrations of salt. No streaming was observed in submerged culture or saturated air. Changes in the direction of streaming could be made by appropriate local application of sugar solution. Both Arthur and Schroter describe the psesence of movement in both directions along the hyphae—the central protoplasm and vacuole moving in one direction and the outer mass in the form of a cylindrical mantle moving in the other.

As always with visual observations on fungi, one turns to Buller for information. In volume 5 of his *Researches on Fungi* there is a wealth of information concerning protoplasmic streaming, particularly within hyphae of *Pyronema confluens, Fimetaria fimicola* and *Rhizoctonia solani*. It is quite clear that there can be considerable bulk movement of protoplasm as evidenced by the movement of vacuoles, which with considerable distortion are able to traverse septa. The source of the protoplasm appears to be hyphae which have ceased to grow in length and are beginning to vacuolate. Buller believed the situation to be similar to tributaries feeding a river but with the difference that the streams are, in all instances, about the same width. His analogy holds for the relative velocities since the main stream is faster than the tributaries feeding it and also faster than the branches further along the flow. Buller believed that movement was caused by increased pressure brought about by growth of vacuoles and by increase in the amount of protoplasm resulting from synthetic processes.

Buller provided some confirmation that pressure may be the motive force, by demonstrating temporary septal bulging in *Rhizoctonia solani* in the direction of streaming. Isaac (1964) examined streaming cytoplasm with a micro-interferometer and showed that the density of cytoplasm increases towards each septum in the direction of the stream. For instance, on the side of the septum against the stream, the cytoplasm contained 24% dry matter, whilst on the side away from the stream, the amount of dry matter was only 8%.

These facts argue for a pressure-driven flow. Whether or not the pressure is produced by increased vacuolation is open to question. There is no doubt that vacuolation is an important morphogenetic development in hyphae as they age and the process can be triggered chemically. It is a reversible process. When vacuolation does occur there appears to be some dehydration of the cytoplasm.

Protoplasmic streaming has also been studied in sporangiophores of *Phycomyces blakesleeanus.* It appears to be an active shearing type movement along thin cytoplasmic strands and there are often a great many strands per cell. Flow seems to take place in both directions along a strand. The rate of streaming is about 2.5×10^{-6} m sec^{-1} upward and about 3.0×10^{-6} m sec^{-1} downward. In an atmosphere of nitrogen, streaming in the upper part of the sporangiophore stops in 1—2 min but streaming in the lower part may continue for 30—40 min. There is clearly a need for detailed biochemical investigations to be under-taken here.

TRANSLOCATION IN SERPULA (MERULIUS) LACRYMANS

Dixon and Ball (1922) showed how a consideration of simple data concerning the growth of a potato tuber can provide some meaningful answers about the process of translocation in higher plants. It should be possible to do the same for fungi growing over inert surfaces. Butler (1957) has provided data for the growth of *S. lacrymans* over acid-washed brick which are susceptible to such an analysis. The relevant observations are a growth rate of 35 mm $week^{-1}$ and a hyphal diameter of 3.3 μm. To these, we can add a fresh weight/dry weight ratio of 2.54.

Essentially, we can think of the growth of *S. lacrymans* in terms of the extension of a bundle of hyphae, each of which is growing from a food base. Now, if we can calculate the rate of production of dry matter, we can get a measure of the rate of movement of dry matter through a unit cross-sectional area of hyphae in unit time. It is still better to go one stage further and assume that hexose (or hexitol) is the major material translocated. Thus one calculates the flux of such a compound down the hyphae. If we do this, we find that, for the above growth rate, there is a flux of 1.27×10^4 pmoles of hexose cm^{-2} sec^{-1}. If a 10% solution moves, the velocity of movement is 3×10^{-6} cm sec^{-1}.

A considerable number of points of interest are immediately apparent from these two calculated values.

(1) They are likely to be minimum values because they take no account of hyphal branching nor of cell respiration.

(2) The flux of hexose down the hyphae is several orders of magnitude greater than known fluxes of glucose across the outer membranes of cells. This means that there must be a considerable area of loading for the system. Tangible expression of this fact comes from those studies of Garrett (1956) on the growth of rhizomorphs of *Armillaria mellea*, in which he develops the concept of 'inoculum potential', the energy of growth available for infection of a host. He relates inoculum potential to the size of the food base ; the bigger the food base, the more likely are rhizomorphs to infect potato tubers.

(3) Diffusion is unlikely to account for the rate of translocation. Butler (1957) shows that *S. lacrymans* grew for 41.1 cm without any change in growth rate. If the flux which is calculated above is to be maintained by diffusion of hexose in aqueous solution ($D = 0.68 \times 10^{-5}$ cm^1 sec^{-1}) there must be a concentration difference of 77 $moles^{-1}$. If this growth rate is maintained over greater distances, as seems highly likely, *viz.* the growth rate of *A. mellea* rhizomorphs referred to in the introduction, then even greater concentration gradients have to be maintained if movement of material along hyphae is by diffusion. We can therefore safely discount diffusion as the mechanism of translocation.

(4) If bulk flow of liquid is occurring—a point which we shall return to later—it is possible to obtain some idea of the pressure gradient. Most of the resistance to flow occurs at the septa. The septal formation which will cause the most resistance is the so-called dolipore or complex septum. Taking values of 0.25 μm for the diemeter of such a pore and 0.75 μm

for its length, we can calculate, using Poisculle's relationship, the total length of pore which will allow a velocity of 3×10^{-6} cm sec^{-1} to be maintained under a pressure difference of 1 bar. The length works out at 4.13 m, which means that the number of septa which will be traversed is 3.5×10^{6}. This can be converted into the total length of hypha involved, because we know from Falk (1912) that clamp connections, and therefore septa, are 500 μm apart. Thus the calculated length of hyphae is 17400 m. This is a minimum distance ! A greater pressure difference is possible and the septal pores may be much simpler than we have assumed. There have been no microscopical observations made on the strands of *S. lacrymans* since Falk's (1912) classic study. A combined light and electron microscopical study is now badly needed. Motta's (1971) observations on the rhizomorphs of *Armillaria mellea* indicate that the study of the so-called translocating structures of higher fungi is clearly a fruitful field. However, it is clear that much yet needs to be done, particularly with regard to relating structure to any presumed ability to translocate material.

Our calculations therefore indicate that translocation down hyphae of *S. lacrymans* occurs by bulk flow. The driving force may be pressure generated within the hyphae, evaporation from the ends of the hyphae or some mechanical means within the hyphae. As yet there is no evidence for the latter, since nobody has observed the equivalent of the transcellular strands of sieve tubes in fungal hyphae, though there are difficulties in making the appropriate observations on living cells. Although we know that evaporation increases the rate of translocation of material into basidiomycete sporophores this seems unlikely to be the major driving force for translocation in *S. lacrymans*. Growth occurs best at a high relative humidity. Brown, Fahim and Hutchinson (1968) showed that there is substantial growth over glass at humidities between 100% and 95% relative humidity. We have observed growth rates of *S. lacrymans* in the field at 5 °C on an inert surface at 100% relative humidity which are as high as 60% of those in the laboratory on an agar medium at 20 °C.

It is not without significance that the specific name *larcymans* means 'weeping', referring of course to the ability of the fungus to form drops which can reach large size. This indicates that some exudation process is occurring. We have been examining the formation of such droplets and we think that they are formed predominantly at the tips. We cannot be certain about this because droplets are found further back on the hyphae. However, there are indications that the hyphae grow on ahead of the droplet. Many droplets are clear and in keeping with this we find that analysis of the liquid produced contains only 1% dry matter. However, some droplets contain aggregations and these droplets appear to be associated with empty hyphae, tips of which appear to have burst. We know from other studies that hyphal tips can burst in an unfavourable osmotic environment, the hydrostatic pressure within hypha being greater than that which the tip wall can withstand. Presumably, therefore, we have been observing an analogous situation in *S. lacrymans*. Conceivably, when a tip bursts, pressure is released in the system. This might account for the indications that droplet formation is discontinuous.

It is our hypothesis that translocation in *S. lacrymans* and similar fungi is driven by pressure generated within the hyphae. This is much like the hypothesis put forward by Buller (1933) to explain protoplasmic streaming. Soluble carbohydrate is likely to provide the bulk of the osmotically active material. There can be as much as 15% dry weight, the predominant component being arabitol. Whether or not increased vacuolation is involved we do not know.

This hypothesis offers a new interpretation with regard to the function of strands. We have indicated that for translocation to occur there must be a large area of loading of the compound being translocated. The growth habit of the fungus, namely surface spread over wood, should allow this Strands grow from the mycelial felts onto essentially non-nutrient media. We presume that a large number of hyphae feed into strands. We have already referred to the point made by Buller that, as such hyphae feed into one main hypha, so the velocity of flow within the hypha will increase. This velocity will, however, decrease if the walls distend in any way.

Strands contain vessel hyphae 10—20 μm in diameter which possess 'beam' and 'ring' wall thickenings in the midst of smaller thick-walled 'fibrous' hyphae. It is not unreasonable to assume that the role of these large amounts of rigid wall material is to maintain pressure gradients in the hyphae in which translocation is occurring. Presumably these are the vessel hyphae but there is a need to show this experimentally. The use of a labelled compound and autoradiography of sections of strands is likely to be the best approach. The presence of thickened cells in rhizomorphs can be viewed in a similar manner. Strand formation in *S. lacrymans* is stimulated by low nutrient concentrations in the external medium and by nitrate. The process therefore has many points in common with other morphogenetic processes such as conidiation.

There seem to be four distinct ways in which material may move within a fungal hypha :

(1) By biffusion ;

(2) By cytoplasmic streaming which appears to be similar to the cyclosis observed in other plant cells ;

(3) By nuclear migration. It is of interest that the rates of nuclear migration are much like those of fast transport in axons. This reinforces the idea that nuclear migration even over long distances is brought about by a contractile system ;

(4) By osmotic flow. We use this term to describe bulk flow through hyphae brought about either by a pressure generated within the hyphae or by evaporative water loss at particular places in the mycelium. Osmotic flow may occur over short distances and we have instanced the movement of vesicles to the tips of hyphae as a possible example of short-distance movement. Osmotic flow certainly occurs over long distances. The work with sporophores leaves no doubt that evaporation provides the driving force whereby nutrients are drawn from the substratum into the fruiting body. The evidence for movement brought about by pressure generated osmotically is more circumstantial. We have little doubt that it occurs and indeed there is evidence for its occurrence not only in strands and rhizomorphs, but also for its involvement in the movement of material into sporophores.

Although the situation has to be described in quantitative terms, it is appropriate before finishing to look at the physical situation with which we are dealing. If there is bulk flow of liquid down the hyphae, this volume flow must be dissipated either as an increase in volume of the hyphae at their tips, or by movement of liquid out of the tips, as is the case for *S. lacrymans.* These relationships can be represented by the following equation :

$$J_{V,S}=X\pm J_{V,T},$$

where $J_{V,S}$ and $J_{V,T}$ represent movement of water into the hyphae at the source of the nutrients and at the tip respectively and X represents the rate of volume increase of the hyphae. If water is extruded from the tips $J_{V,T}$ is positive, otherwise it is negative.

We can take it that water will move into the hyphae at the source of nutrient, across a semi-permeable membrane. In *S. lacrymans*, it would seem from the analysis of the liquid which is coming out of the hyphae, that a semi-permeable membrane must also be traversed in the process of fluid exudation.

For the movement of water across a semi-permeable membrane,

$$J_V=L_P\,(P-\pi),$$

where L_P is the hydraulic conductivity, P the pressure difference and π the osmotic pressure difference across the membrane. For a hyphal tip, the water flux out of the tip is given by

$$J_V=L_P(P_F-(\pi_F-\pi_M)),$$

where P_F is the hydrostatic pressure and π_F the osmotic pressure within the hypha and π_M the osmotic pressure of the medium. If we are considering *S. lacrymans* growing on an inert surface in an atmosphere of 100% relative humidity, π_M can be put at zero.

$$\therefore\quad J_V=L_P\,(P_F-\pi_F).$$

Therefore, for extrusion of water, the fungus must generate the requisite internal hydrostatic pressure. This, as we have indicated, must mean that the fungus has a very efficient system for accumulating solute within the hyphae from the medium on which it is growing. By doing this, the requisite hydrostatic pressure is generated osmotically. As a result, solution is driven along the hyphae. Carbohydrates will be removed from solution at the tip by incorporation into insoluble material such as protein and cell wall material. But of course many inorganic ions cannot be dealt with in this way. Since water is extruded from the hyphae inorganic material will tend to accumulate within the tip. The immediate consequence of this will be an increase in π_F. Water extrusion will decrease. It would therefore seem that *S. lacrymans* has either a very effective system at the hyphal tip for removing carbohydrates from solution or that the quantity of ions entering from the nutrient medium is much reduced compared with other fungi. It does mean that we need to know more about the metabolism of this fungus and about its ionic relations. Jennings (1973) has already emphasised our great ignorance about the ionic relations of basidiomycetes.

The final point to make about bulk movement of liquid along a hypha is that, unless some corrective action is taken, many of the organelles might end up at the hyphal tip. Cytoplasmic streaming in the outer mantle may be one way by which this is prevented.

10

Transport of Proteins into Chloroplasts

Most chloroplastic proteins are encoded in the nucleus and synthesised as higher molecular weight precursors in the cytoplasm. The extra amino acids that are present at the amino terminus of precursors are called transit peptides. Work in the past few years has demonstrated that a transit peptide is necessary for transport of precursors into chloroplasts. Moreover, it has been demonstrated in several cases that a transit peptide is sufficient to direct the transport of foreign passenger proteins into chloroplasts. However, the mechanisms responsible for decoding the targeting information present in the transit peptide are less well characterised.

A key development that allowed important advances in understanding protein transport into chloroplasts was the demonstration that import could be reconstituted *in vitro*. In these early studies a mixture of radioactive precursor proteins, obtained by *in vitro* translation of polyadenylated mRNA, was incubated with isolated intact chloroplasts. The precursor to a small subunit of ribulose-1, 5-bisphosphate carboxylase (prSS) was imported into chloroplasts, processed to its proper size and assembled with large subunit to yield holoenzyme. This *in vitro* assay has subsequently been used to examine the import of numerous precursor proteins including precursors to light-harvesting chlorophyll *a*/*b* protein (LHCP) plastocyanin (PC) ferredoxin (Fd) and 5-enolpyruvylshikimate-3-phosphate (EPSP) synthase.

The utility of the *in vitro* import assay was further enhanced by the development of *in vitro* transcription systems. These systems allow the synthesis of precursor proteins from cloned genes *via* sequential *in vitro* transcription and translation reactions. One advantage of this technique is that it allows experiments to be performed with radiochemically pure precursors, rather than with the mixture of precursors that results from the translation of total polyadenylated mRNA. This not only eliminates the possibility of competition between precursors, but also simplifies data analysis by eliminating the need for immunoprecipitation of imported proteins. An additional and probably more significant advantage is that it allows the relatively rapid production of altered precursor proteins *via* modification of precursor genes by recombinant DNA techniques. The ability of the modified precursor proteins to be imported and properly localised can be examined using reconstituted import assays.

Studies using *in vitro* assays have led to several important insights into the import process that would have been difficult or impossible to obtain from *in vivo* experiments. For example, the *in vitro* assay has been used to demonstrate that import occurs after synthesis of the precursor protein is complete. A second important insight derived from *in vitro* studies is that the import process can be divided into several discrete steps. The first step is the binding of precursor proteins to the chloroplast surface. It has been suggested that this binding is mediated by a receptor protein present in the outer envelope membrane although evidence for this suggestion remains incomplete. The second step in the import process is translocation of the precursor protein across the two envelope membranes. This step requires energy in the form of ATP. A third step, which occurs during or immediately after translocation, is proteolytic processing of the precursor to remove the transit peptide. For many proteins, additional steps are required to reach their final location and achieve their active form. For example, the small subunit (SS) of ribulose-1, 5-bisphosphate carboxylase must assemble with the chloroplast-synthesised large subunit to form the active holoenzyme. Other imported proteins must have prosthetic groups or metal ions while still others must be inserted into or across another membrane before reaching their final location. In this brief review we will focus primarily on the binding and translocation events.

Binding of Precursors to Chloroplasts

The first step in protein import is the binding of precursors to the surface of chloroplasts. Cline *et. al.* demonstrated that precursor binding can be assayed separately from translocation across the envelope membranes. They showed that binding of either prSS or prLHCP without transport could be achieved by incubating the chloroplasts with the precursors in the absence of ATP. Chloroplasts treated in this manner could be washed to remove unbound precursor, and then assayed for transport of the bound species upon addition of exogenous ATP.

The initial characterisation of the binding of prSS to chloroplasts was extended by Friedman and Keegstra. They found that binding was both saturable and specific. Under the conditions used for the assays, binding approached saturation at a precursor concentration of 5-10 nM, with approximately three thousand precursor molecules bound to each chloroplast. The binding was specific in that p-SS was bound, but SS was not. These results are consistent with the idea that prSS binding is mediated by specific receptor proteins located in the outer envelope membrane.

The early studies of Cline *et al.* on precursor binding assumed that the binding step did not require energy because binding could be measured at levels of ATP that did not support import. This assumption has proven to be incorrect. Recent studies, using precursor preparations with ATP completely removed by gel filtration, have demonstrated that precursor binding requires ATP. Binding is maximal at an ATP concentration of approximately 100 μM, a level which cannot support protein translocation. The function of ATP in precursor binding is not known, but non-hydrolysable analogs will not substitute for ATP, which suggests that ATP hydrolysis is needed.

Membrane Receptors for Precursor Binding

A number of attempts have been made to indentify the receptor(*s*) involved in protein import. Pfisterer *et. al.* examined the binding of a mixture of precursor proteins to isolated chloroplastic envelope membranes, and more recently, Bitsch and Kloppstech provided evidence that precursors could be bound to solubilised envelope proteins reconstituted into lipid vesicles. The latter authors also attempted to cross-link prSS to proteins on the surface of chloroplasts but no clear receptor protein emerged from that study. Cornwell and Keegstra also used cross-linkers in an effort to identify the receptor for prSS. Utilising a heterobifunctional, photoactivatable cross-linking reagent, they found a specific interaction of prSS with a 66 kD protein on the surface of chloroplasts. This protein was postulated to be the receptor, or part of the receptor complex, that mediates the binding of prSS to the chloroplastic surface.

Recently Pain *et. al.* used a very different approach to identify a putative receptor protein. They prepared anti-idiotypic antibodies that act as an analogue of prSS and bind to the putative receptor, thereby blocking the import of prSS. These antibodies were used in several types of studies. One was to demonstrate that the putative receptor, identified by this antibody preparation, was localised at contact sites where the two envelope membranes are tightly oppressed. They also used antibodies for Western blotting in an effort to identify the putative receptor. The antibodies reacted with two chloroplastic proteins. The first was identified as the large subunit of ribulose-1, 5-bisphosphate carboxylase. The reason for a positive reaction with this protein remains unexplained. The second protein which reacted with the antibody was the major 30 kD protein of the chloroplastic envelope membrane. They concluded that this protein is the putative receptor protein.

If one assumes that the level of precursor binding at saturation gives an accurate estimate of the amount of receptor protein, then it is possible to calculate the relative abundance of the putative receptor protein. Assuming 3000 receptors per chloroplast and a molecular weight of 66 kD, the receptor protein would be approximately 0.003% of the total chloroplast protein. If one assumes that envelope proteins constitute approximately 1% of the total chloroplast protein, then the receptor protein would be approximately 0.3% of the total envelope protein. Thus the putative receptor protein should be a relatively minor protein in envelope polypeptide profile. This conclusion is in conflict with the observations of Pain *et. al.* who identified the major 30 kD protein of the envelope membrane as the putative receptor.

PROTEIN TRANSLOCATION ACROSS THE ENVELOPE MEMBRANE

Energy Requirements for Import into the Stroma

The energy requirement for the import of nuclear-encoded proteins into chloroplasts was first addressed by Grossman *et. al.* They demonstrated that chloroplastic protein import was stimulated in the light over that observed in the dark, and that uncouplers were effective inhibitors of that stimulation. DCMU, which inhibits linear, but not cyclic, photophosphorylation, did not eliminate protein import. Exogenously added ATP was able to support protein import into dark-incubated chloroplasts, either in the absence or presence of uncouplers. These experiments led the authors to propose that protein import into isolated chloroplasts requires ATP as the sole energy source. As mentioned above, the latter approach, wherein exogenously added ATP was used to overcome the uncouplerinduced inhibition of import, was later employed by Cline *et. al.* to separate the binding and translocation steps.

Several additional reports in the energy requirements for protein import into chloroplasts have appeared recently. The hydrolysis of ATP was found to be necessary for import of prSS into spinach chloroplasts, pea chloroplasts and castor bean leucoplasts. We have recently extended these observations to demonstrate that ATP is required for import of the precursors of ferredoxin and plastocyanin into pea chloroplasts. Other nucleotides (GTP, CTP, UTP and AMP) were found to be ineffective substitutes for ATP in supporting protein uptake by plastids in the dark, as were a number of non-hydrolysable ATP analogs. Moreover, it was found that the ATP-dependent import of precursors into dark-incubated chloroplasts was unaffected by agents which collapse either or both the pH and electrical components of the protonmotive force. Thus the consensus from all these studies is that protein translocation across the envelope membranes requires ATP, but not an electrochemical potential.

Another important question is the location of ATP utilisation during protein import. One strategy for examining this problem is to generate ATP inside plastids by the addition of glycolytic or Calvin-Bensen cycle intermediates. Import of prSS into chloroplasts is supported by internally-generated ATP. These results are not unexpected since light can stimulate import and ATP produced by light is also generated in the stroma. Although internally generated ATP supported protein import into chloroplasts, it was unknown whether this ATP was actually utilised internally or after it had been transported to the external medium. To address this question, ATP was generated in the stroma while external ATP was consumed by exogenous membrane-impermeable ATP traps. Presumably, ATP-mediated protein import would occur under these conditions only if it was driven by ATP inside the chloroplasts. However, the results of the experiments seem to depend upon the choice of enzyme used to consume ATP. Pain and Blobel used either glucose/hexokinase or apyrase and observed that import of prSS into pea chloroplasts occurred normally, leading them to conclude that ATP was hydrolysed in the stroma. Schindler *et. al.* performed a similar experiment with glucose/hexokinase, and observed that protein import was inhibited. They concluded that ATP was required externally for protein import. Flugge and Hinz performed the most careful study of this question, measuring the actual ATP levels in the chloroplast under conditions of the import assays. Furthermore, they reduced possible complications which could result from rapid transport of ATP across the envelope by using spinach chloroplasts, which display lower adenylate translocator activity than do pea chloroplasts. These authors found that when either glucose/hexokinase or fructose-6-phosphate/fructose-6-phosphate kinase were used as external ATP consuming systems, the import of prSS was not inhibited. However, when alkaline phosphatase was used as the external ATP trap, the import of prSS was inhibited and correlated with the external, rather than the internal ATP concentration. They concluded that ATP outside the stroma appeared to be required for import, although an alternative explanation could involve the participation of an externally accessible phosphorylated intermediate, if this intermediate was dephosphorylated by the alkaline phosphatase.

We have recently re-examined the problem of the location of ATP utilisation using pea chloroplasts and several different precursors. Our strategy was similar to that employed by the other groups, except that we examined the rate of precursor import under the various conditions. Because pea chloroplasts have an active adenylate translocator, thereby allowing internal and external pools of ATP to rapidly equilibrate, it was possible that external ATP traps could deplete internal ATP levels. This emphasises the importance of measuring the ATP levels produced by various treatments. However, by monitoring the ATP levels and correlating them with the rates of protein import, we have tentatively concluded that the ATP utilised for protein translocation is needed inside chloroplasts.

The recent evidence that ATP is also needed for precursor binding may complicate the issue of the location of ATP utilisation during import. At present we have not examined the location of the ATP requirement for protein binding, which is approximately one order of magnitude lower than that needed to support protein translocation. However, it seems reasonable to speculate that this ATP is needed outside chloroplasts. If correct, this leads to the conclusion that ATP is required on both sides of the envelope membranes ; on the outside to support precursor binding and on the inside to support protein translocation. This tentative conclusion needs further experimental support.

Although initial studies indicated that ATP was required for protein import into mitochondria subsequent studies indicated that instead, the electrical component of the proton electrochemical potential gradient was required. More recently, the energy requirements of protein import into mitochondria were reinvestigated, and a requirement for ATP was observed. Thus the emerging consensus is that both ATP and a transmembrane electric field are necessary to drive the import of proteins into mitochondria. A requirement for ATP hydrolysis may be common to all protein translocation systems; such a requirement has also been reported for the transport of proteins across the endoplasmic reticulum membrane, movement of secretory proteins through the Golgi apparatus, and for protein secretion from E. coli. In the last case a protonmotive force also appears to be involved.

Mechanism of Protein Translocation

The actual mechanism by which proteins are translocated across the two envelope membranes remains obscure. Eilers and Schatz addressed this question for mitochondria and have proposed that protein unfolding must precede translocation. This hypothesis is based largely on the observation that the import of a chimeric precursor containing the transit peptide from subunit IV of cytochrome oxidase fused to dihydrofolate reductase was inhibited by the folate antagonist, methotrexate. Methotrexate also imparted some protection against dihydrofolate reductase degradation by thermolysin. Additional evidence for the requirement of unfolding prior to translocation was recently provided by Verner and Schatz, who showed that a prematurely terminated chimeric precursor, which presumably was incompletely folded, was imported into mitochondria without the usual ATP requirement. While these results are consistent with the notion the protein unfolding occurs during import into mitochondria, other interpretations of these data cannot be rigorously excluded.

In chloroplasts, recent evidence suggests that, if proteins are unfolded during import, they are not exposed to a hydrophobic environment during translocation. Lubben *et. al.* fused two different hydrophobic stretches of amino acids that function as stop-transfer sequences to the carboxyl terminus of prSS. Neither chimera was halted during translocation by an interaction of the hydrophobic stop transfer region with the membranes, which indicates that the regions were probably never exposed to the lipid layer. These results argue against the threading of an unfolded protein directly through the lipid bilayer or other hydrophobic environment. They also suggest that if translocator proteins are a universal feature of protein transport, then a putative chloroplast translocator protein must be sufficiently different from the endoplasmic reticulum translocator protein such that it could not recognise the stop-transfer domains of either chimera.

Although such studies begin to provide some insight into the mechanism of protein translocation across the envelope membranes, a great deal of work still needs to be done. At least two fundamental but related questions need to be answered. First, are there translocator proteins present in the chloroplast envelope membranes that mediate the movement of proteins across the two membranes ? Such a hypothesis has been described in detail most recently by Singer *et. al.* The second question has to do with the conformation of the precursor during translocation. Specifically, is the protein unfolded during transport so that it can be threaded across the membrane or is the protein translocated in a folded state ?

Protein Translocation Across the Thylakoid Membrane

The internal thylakoid membrane system of chloroplasts contains several proteins that originate in the cytoplasm. These proteins not only must be transported across the envelope membranes, but also must be inserted into or transported across the thylakoid membrane. The transport of PC, a thylakoid lumen protein, has been used as a model system to study this pathway. The precursor to plastocyanin has a complex transit sequence that consists of two structural domains. Smeekens *et al.* provided evidence that the transport of plastocyanin occurs in two steps. The first step, transport of prPC across the two envelope membranes, is similar in many respects to the import of stromal proteins discussed above. This step is mediated by the amino terminal domain of the transit sequence. After import, this region of the transit sequence is removed by a stromal processing protease to form an intermediate-sized form of PC. The second step, transport of the intermediate across the thylakoid membrane, is thought to be mediated by the carboxyl terminal domain of the transit sequence. At present many details of the thylakoid translocation process are still unknown. The best way to investigate this step will be to reconstitute it *in vitro* using isolated thylakoid membranes and either prPC or intermediate PC. Efforts to accomplish this *in vitro* reconstitution are under way in our laboratory. In the interim, studies utilising intact chloroplasts have yielded some insights. Experiments with ionophores have demonstrated that an electrochemical potential across the thylakoid membrane is not needed for PC transport to the lumen. The tentative conclusion at present is that transport across the thylakoid membrane requires ATP, just as transport across the envelope membranes does. However, until the second transport step can be studied independent of the first, it will be difficult to confirm the exact energy requirement. Many other questions also remain unanswered. For example, are stromal proteins or other factors required to initiate translocation across, the thylakoid membrane, as has been demonstrated for insertion of LHCP into the thylakoid membrane ? Does the thylakoid membrane have receptors or other proteins which mediate the translocation of proteins ? These questions indicate that this will be a fruitful area of investigation for many years.

11

Transport from Symbiotic Algae and Symbiotic Chloroplasts to Host Cells

The nutrition of symbiotic organisms is of two main types, 'necrotrophic' or 'biotrophic'. Necrotrophic organisms first destroy host cells before absorbing nutrients; biotrophic organisms cause little or no destruction but obtain nutrients in a continuous supply from living host cells. Biotrophic nutrition is characteristic of mutualistic associations, where the flow of substances between the symbionts is commonly two-way; it also occurs in some parasitic associations (*e.g.* rust infections of higher plants) where the flow is predominantly one-way. The successful operation of biotrophic nutrition necessitates the development of transport systems between living cells of the separate symbionts.

This chapter concerns such transport systems. It will be limited to associations in which one component is autotrophic either an alga or a chloroplast—and the other heterotrophic, and will consider only transport from autotroph to heterotroph (although transport in the reverse direction can undoubtedly occur). These autotrophic symbionts exist in varying degrees of intimacy with the host cells: (*a*) extracellular, as in all lichens and a few algal symbionts of invertebrates; (*b*) intracellular, as in most algal symbionts of invertebrates ; and (*c*) only an organelle, the chloroplast, becomes intracellular, as in certain sacoglossan molluses.

It is convenient to consider the various associations in terms of this increasing degree of intimacy, beginning with lichens. This order is also logical because more information is available about the lichen system than any other.

In all systems, most investigations have been concerned with the movement of products of photosynthesis from autotroph to heterotroph, concentrating upon attempts to assess the quantitative amount of transport and to identify the molecules involved. Only a small amount of information about the types of membrane transport system involved is available.

LICHENS

Each lichen consists of an association between an alga and a fungus. Lichens contain more than a quarter of all described species of fungi, with representatives of both discomycete and pyrenomycete forms of ascomycetes, as well as a few basidiomycetes. About 27 different genera of algae have been reported, either green or blue-green. It is obviously important not to assume that results from one lichen species are necessarily applicable to all others.

Most experimental investigations have concerned transport of ^{14}C fixed by photosynthesis from alga to fungus, but these have recently been supplemented by studies of the movement of nitrogen fixed by *Nostoc* symbionts of lichens to their associated fungi. Although the transport of other types of substances (minerals, growth factors, etc.) has not been investigated, it would be surprising if it did not occur.

Movement of photosynthate from alga to fungus in Peltigera polydactyla Transport of photosynthetically fixed carbon from alga to fungus has been extensively studied in *Peltigera polydactyla* and a detailed account of this work will illustrate the variety of experimental techniques which can be used to study the lichen symbiosis.

The fungus is an ascomycete, and the alga *Nostoc*. In symbiosis the mucilaginous sheath of *Nostoc* is severely reduced and the alga occurs in short chains so contorted as to appear as small packets. As in the majority of lichens, the algae (which comprise about 5% of the dry weight) are confined to a distinct layer about 50μm thick just beneath the surface of the plant ; beneath the algal layer is a fungal medulla 400-1000 μm thick. The fungus does not normally penetrate the alga, but is closely appressed to the algal cells.

Movement of photosynthetically fixed ^{14}C from the algal layer (which contains both symbionts) to the medulla (which contains only fungus) was demonstrated by incubating samples of intact lichen in the light on solutions of $NaH^{14}CO_3$, and then dissecting off the medulla at intervals after the start of the experiment. Fixed ^{14}C was first detectable in the medulla after 10 minutes (Fig. 11.1), and in the next four hours, 40% of the fixed ^{14}C accumulated in the medulla.

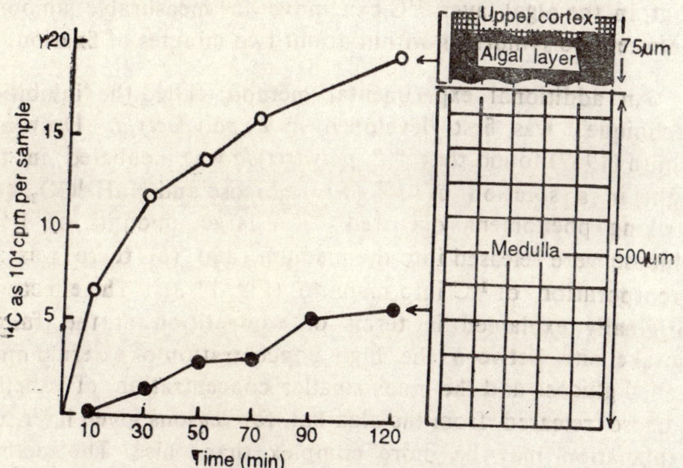

Fig. 11.1. Movement of photosynthetically fixed ^{14}C from the algal region (○) to the fungal medulla (●) in the lichen *Peltigera polydactyla*. Data of Smith and Drew (1965). Samples (10×7 mm diam. thallus discs) floated on aqueous media, pH 6.5, 20°C, 4000 lx ; medullae dissected off at intervals shown.

The fixed ^{14}C appearing in the medulla was almost entirely incorporated into mannitol. Algae isolated from homogenates of the thallus contained no mannitol but released [^{14}C]glucose to the medium during photosynthetic ^{14}C fixation. When [^{14}C]glucose was supplied to isolated portions of fungus, it was absorbed and immediately converted to [^{14}C]mannitol. Glucose therefore appeared to be the principal form in which fixed carbon moved from alga to fungus, and this was confirmed in short term photosynthesis experiments with intact lichens when fixed ^{14}C appeared in glucose within 90 seconds, but none appeared in mannitol until after two minutes (Fig. 11.2). As photosynthesis continued, the amount of ^{14}C in glucose remained very small, while that in mannitol continued to increase, representing 70% of all the fixed ^{14}C after one hour.

Because mannitol is confined to the fungus, incorporation of fixed ^{14}C into mannitol during photosynthesis by the intact lichen is a reliable indicator of the rate and extent of transport between the symbionts. It can be inferred from Fig. 11.2 that, in the algal layer, ^{14}C can move in measurable amounts between the symbionts within about two minutes of fixation.

An additional experimental method, called the 'inhibition technique', was first developed in *P. polydactyla.* Drew and Smith (1967) found that if *P. polydactyla* was incubated in the light in a solution of 2% (w/v) glucose and $NaH^{14}CO_3$, two striking phenomena occurred : (*a*) large amounts of [^{14}C] glucose were released into the medium, and (*b*) there was no incorporation of ^{14}C into mannitol (Fig. 11.3). The effect was originally explained in terms of competition at the fungal uptake site between the high concentration of external unlabelled glucose and the much smaller concentration of labelled glucose released from the alga but, for reasons given later, the explanation may be more complex than this. The method became called the 'inhibition technique' because it involved complete abolition of incorporation of ^{14}C into mannitol. It was specific in that certain other sugars such as fructose were completely ineffective. Because of its specificity, the 'inhibition technique' subsequently proved a useful tool to help identify the mobile carbohydrates in many other lichens.

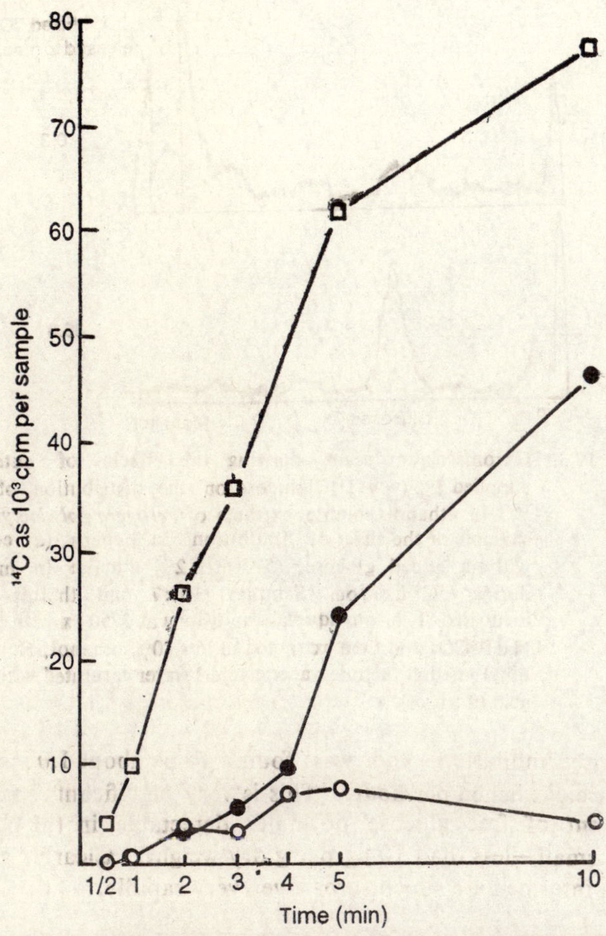

Fig. 11 2. Short term photosynthesis in the lichen *Peltigera polydactyla* illustrating rapid transfer of fixed ^{14}C to the fungus. The alga is believed to release glucose (○), which the fungus absorbs and converts to mannitol (●); □ = total fixation. Samples (6×7 mm thallus discs) incubated at 20°C and 23000 lx.

In *P. polydactyla*, 2-deoxyglucose was found to be as effective as glucose in 'inhibition'. Since glucose and 2-deoxyglucose are easily separable by gas chromatography, the amount of glucose released could be estimated during 2-deoxy-

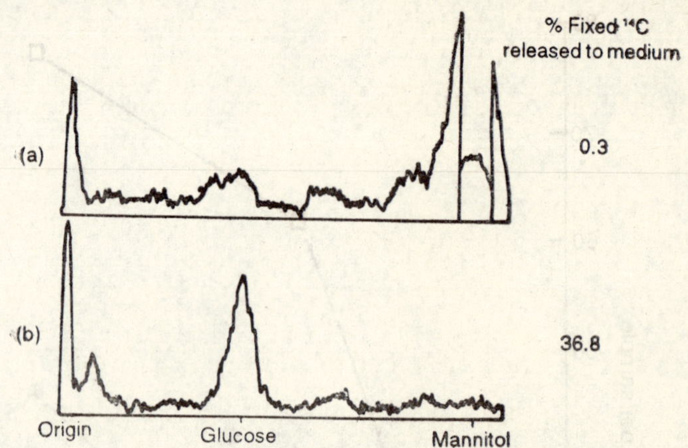

Fig. 11.3. Chromatogram scans showing the effects of externally supplied 2% (w/v) [^{12}C]glucose on the distribution of fixed ^{14}C in ethanol soluble extracts of *Peltigera polydactyla* : an example of the effect of 'inhibition' in lichens. (*a*) control, without added glucose, (*b*) with 2% glucose in medium during ^{14}C fixation. Samples (10×7 mm thallus discs) incubated 1 h on aqueous solutions at 3750 1x with 10 μCi $NaH^{14}CO_3$ and then extracted in hot 80% ethanol. Solvent= ethyl methyl ketone : acetic acid : water saturated with boric acid (9 : 1 : 1).

glucose 'inhibition' and was found to be about 1.0 mg per g dry weight lichen per hour. This is very significant since the amount of free glucose normally detectable in the lichen is very small—less than 10 μg per g dry weight. Clearly, glucose is an intermediate which turns over very rapidly.

Although glucose is involved in transport between the symbionts, there remains the question of how it is produced. Hill (1972) found that in short term photosynthesis, ^{14}C enters sugar phosphate and glucan pools before appearing as free glucose. He suggested that it may be the glucan which is released from *Nostoc*, and that glucose is produced from this by the action of fungal enzymes ; he speculated that the glucan might be a part of the cell-wall/sheath complex of the alga, and that its continued removal stimulated its continued production by the alga. He showed that isolated portions of fungal medulla could break down the glucan to glucose, and he cited the experiments of Richardson, Hill and Smith (1968) showing that glucose 'inhibition' of some lichens with blue-green algae (but not *Peltigera*) resulted in the production of short chain oligosaccharides as well as glucose into the medium.

On the other hand, Drew and Smith (1967) found that freshly isolated *Nostoc* released free glucose rather than a glucan. Although it could be argued that their algal preparations were demonstrably contaminated with fungal fragments, Green (1970) found that clean preparations of *Nostoc* from *P. polydactyla* still released substantial amounts of free glucose during photosynthesis. Smith and Sussman found that 72 hours pre-treatment of *P. polydactyla* with actinomycin D did not significantly reduce the amount of [^{14}C]glucose released during 'inhibition', so if it is produced by fungal enzymes, they must turn over very slowly. Finally, evidence to be given later suggests that at least some of the glucose release in 'inhibition' may occur by exchange with the algal cells, indicating the existence of a cell membrane system capable of transporting substantial amounts of glucose. *Nostoc* in the lichen lacks the polyglucoside granules of free-living *Nostoc*, so it is possible that the glucan pool of Hill could be internal and represent the vestige of these granules.

Information about the composition and mode of synthesis of the mucilaginous sheath of *Nostoc* would clearly be valuable in helping to decide between these two alternative possibilities. Unfortunately very little is known about the composition of the cell envelope of blue-green algae in general.

Movement of photosynthate from alga to fungus in other lichens

Over 35 other lichen species have been investigated using the experimental techniques and criteria pioneered with *P. polydactyla* (Table 11.1).

All lichens with blue-green symbionts release [^{14}C] glucose to the medium during glucose 'inhibition' with a concomitant suppression of ^{14}C incorporation into mannitol. In four

Table 11.1. Identity of carbohydrates moving from alga to fungus in lichens ; a summary of investigations carried out so far.

Type of alga	Algal genus	No. of lichens investigated: Genera	Species	Mobile carbohydrate
Green	*Trebouxia*	7	9	Ribitol
	Myrmecia	2	4	Ribitol
	Coccomyxa	2	3	Ribitol
	Trentepholia	4	6	Erythritol
	?*Phycopeltis*	1 (*Porina*)	1	Erythritol
	Hyalococcus	1	2	Sorbitol
	Stichococcus	2	2	?Scrbitol
	?*Heterococcus*	1 (*Verrucaria*)	2	Sorbitol
Blue-green	*Nostoc*	7	11	Glucose
	Calothrix	1	1	Glucose+? glucan
	Scytonema	1	1	Glucose+? glucan

? = identity tentative only.

lichens, small amounts of an additional compound containing ^{14}C were also released to the medium : in *Lichina pygmaea* (symbiont *Calothrix*) it adhered to the baseline of chromatograms, but in a *Leptogium* and a *Sticta* sp. (symbiont *Nostoc*) and a *Coccocarpia* sp. (symbiont *Scytonema*) it had a chromatographic mobility corresponding to an oligosaccharide. These compounds were not investigated further, and it is not known whether they were produced by the alga, or by the action of surface enzymes using glucose as a substrate.

It is remarkable that all the green algal symbionts of lichens release polyols. Unlike glucose in lichens with blue-green symbionts, the amount of mobile polyol in freshly collected lichens with green symbionts is usually quantitatively significant. The difference in pool size of the mobile carbohydrates may help to explain why ^{14}C is released more slowly from lichens with green symbionts during 'inhibition' experiments (Fig 11.4).

There are a few lichens with two types of algal symbiont, a green symbiont in the main algal layer, and a blue-green in distinct structures, *cephalodia*, which may be either internal or

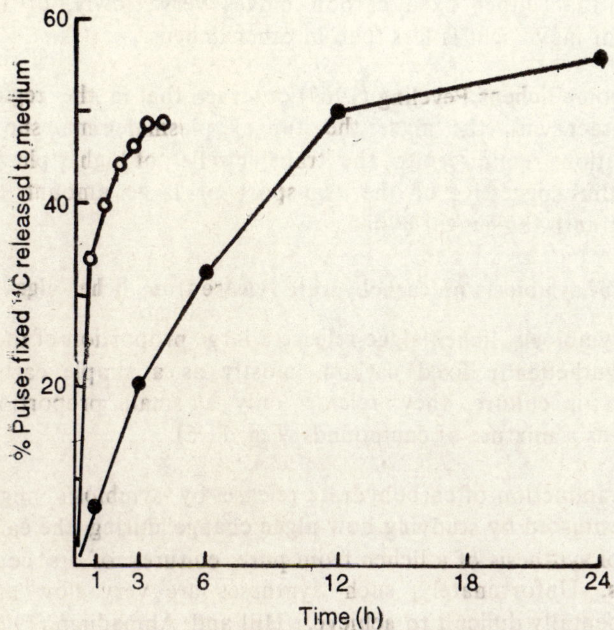

Fig. 11.4. Differences between lichen species in the rate of release of a pulse of fixed ^{14}C during 'inhibition', *Peltigera polydactyla* (○) releases ^{14}C rapidly in the presence of glucose; *Dermatocarpon miniatum* (●) which contains the sorbitol-releasing alga *Hyalococcus*, releases ^{14}C more slowly in the presence of sorbitol. Both lichens given a pulse of ^{14}C before transfer to non-radioactive media containing 1% (w/v) 'inhibiting' compounds.

external to the thallus. In two such lichen species investigated by Richardson *et al.* (1968), glucose moved to the fungus from the blue-green, and a polyol from the green, showing that the fungus does not determine the type of carbohydrate released.

The proportion of fixed carbon moving from alga to fungus was found to be generally large: over half of a pulse of fixed ^{14}C moved from alga to fungus in the majority of lichens investigated by Hill (1970) and Hill and Smith (1972). The main exceptions were a number of lichens containing *Trentepohlia*, where estimates based both upon 'inhibition' experiments and accumulation of ^{14}C into fungal polyols showed that either fixed carbon moves very slowly, or the extent of movement is less than in other lichens.

In some lichens Peveling (1969) observed that in the region of contact with the alga, the fungal plasmalemma shows convolutions reminiscent of the 'transfer cells' of higher plants, and is thus suggestive of the transport of large amounts of material into the fungal hypha.

Effects of symbiosis on carbohydrate release from lichen algae

In symbiosis, lichen algae release a large proportion of their photosynthetically fixed carbon, mostly as a simple carbohydrate ; in culture, they release only a small proportion, mostly as a mixture of compounds (Fig. 11.5).

The induction of carbohydrate release by symbiosis might be investigated by studying how algae change during the early stages of synthesis of a lichen from pure cultures of its components. Unfortunately, such syntheses are very slow and experimentally difficult to achieve. Hill and Ahmadjian (1972) found that 12 weeks after the initiation of a laboratory synthesis of the lichen *Cladonia cristatella,* only a small proportion of photosynthetically fixed ^{14}C could move from alga to fungus, indicating that transport between the symbionts takes a long time to establish. They concluded that there was little evidence to support the theory that the fungus produces a substance affecting the permeability of the algal membrane.

An alternative approach is to study the decline in carbohydrate release by algae after removal from symbiosis, and to investigate whether there are any factors which affect the rate of decline. In a preliminary study, Drew and Smith (1967) found that release of fixed ^{14}C by *Nostoc* from *Peltigera polydactyla* fell off rapidly between 24 and 48 hours after isolation ; at the same time, ^{14}C no longer appeared in the medium as glucose, but as a compound remaining on the baseline of chromatograms. However, the algal preparations were contaminated by fungal fragments, since traces of [^{14}C] mannitol were also detectable.

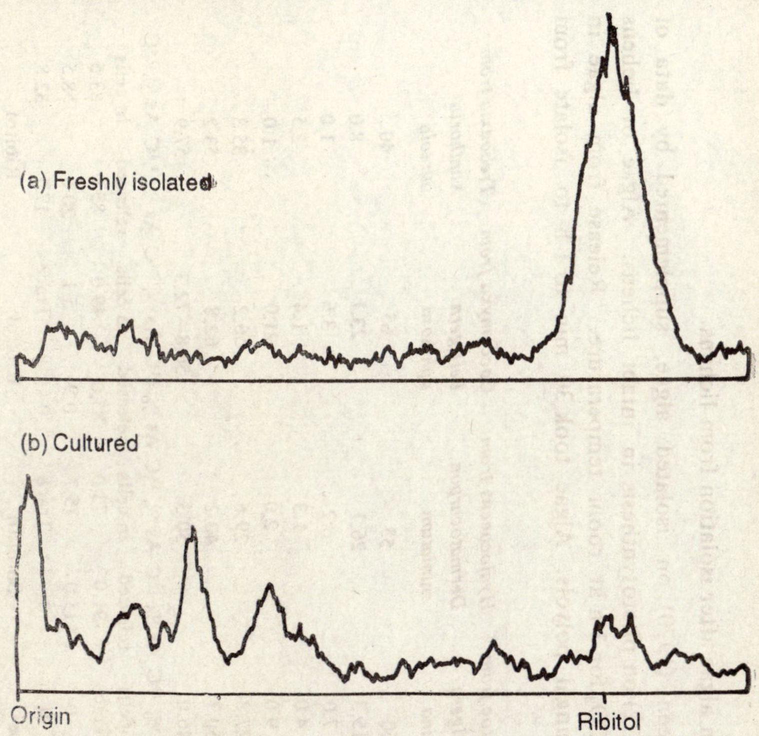

Fig. 11.5. Chromatogram scans illustrating the differences in distribution of fixed ^{14}C released to media by *Trebouxia* (*a*) freshly isolated from *Xanthoria aureola*; and (*b*) after a period of several weeks in culture. Algal suspensions in distilled water at 18 °C, 4700 lx, 1 h photosynthesis. Solvent=ethyl methyl ketone : acetic acid : water saturated with boric acid (9 : 1 : 1).

Green (1970) made a detailed survey of the problem using much cleaner preparations of four different genera of algae (Table 11.2). In all cases, the algae began to change immediately upon isolation. The amount of fixed ^{14}C released dropped abruptly, the proportion of fixed ^{14}C incorporated into the released carbohydrate declined in both medium and cells, and incorporation into ethanol-insoluble material rose. In the absence of the fungus, reversion of the algae from the symbiotic to the cultured condition begins very rapidly-much more

Table 11.2. Changes in algae after isolation from lichens.

A summary of the observations of Green (1970) on isolated algae, supplemented by data of Richardson *et al*. (1968) and Hill & Smith (1972) on photosynthesis in intact lichens. Algae or lichens incubated in distilled water containing $NaH^{14}CO_3$ for 3 h at room temperature. Release from algae in lichens calculated from ^{14}C accumulation in fungal polyols. Algae took 30 min to 1 h to isolate from lichens.

	Condition of alga	*Nostoc from Peltigera canina*		*Hyalococcus from Dermatocarpon miniatum*		*Coccomyxa from Peltigera aphthosa*		*Trebouxia from Xanthoria aureola*	
% Fixed ^{14}C released from alga	In lichen	60		55		65		40	
	0 h isolate	15.3		26.1		23.1		8.0	
	24 h isolate	7.0		6.2		3.6		1.0	
	Culture	4.0		1.3		1.4		2.5	
% Fixed ^{14}C incorporated into ethanol-insoluble	In lichen	9.0		2.0		21.0		1.0	
	0 h isolate	27.8		29.6		56.2		35.3	
	24 h isolate	50.7		40.2		62.8		53.2	
	Culture	46.0		50.3		50.8—72.3		57.9	
		As % ^{14}C released	As % ^{14}C in cells	As % ^{14}C released	As % ^{14}C in cells	As % ^{14}C released	As % ^{14}C in cells	As % ^{14}C released	As % ^{14}C in cells
'Mobile' carbohydrate	0 h isolate	22.0	Trace	91.0	73.0	85.0	49.0	85.0	83.5
	24 h isolate	0	0	15.0	75.7	0.9	5.1	20.0	58.5
	Culture	0	0	0	Trace	0	Trace	17.1	32.8
Identity of carbohydrate		Glucose		Sorbitol		Ribitol		Ribitol	

rapidly than had been indicated by the experiment quoted above of Drew and Smith; since the algal suspension of Drew and Smith contained fungal fragments, it raised the possibility that lichen fungi produce 'factors' affecting algal membrane permeability. Green therefore made numerous attempts to prepare extracts or isolate 'factors' from the lichen *Xanthoria aureola* which would stimulate its isolated alga (*Trebouxia*) to release ribitol, but failed. He concluded that there was no evidence that the fungus of this lichen released a substance which altered the permeability of the algal cells. Although shortage of material prevented a search for 'factors' in the lichen used by Drew and Smith (*P. polydactyla*), Green found that clean preparations of *Nostoc* from this species lost their ability to release glucose much more rapidly. Possibly, some of the algae in Drew and Smith's preparation still had fragments of hyphae attached.

Green also studied the effect of the following external factors upon *Trebouxia*: pH, kinetin, benzylaminopurine, coconut milk, the 'lichen acids' usnic acid and atranorin, and various sugars (including sucrose and ribitol). None had any marked effect on carbohydrate release except pH. Each alga had a distinct optimum for release (below pH 4 for *Trebouxia*, about pH 6.0 for *Nostoc* and 7.2 for *Coccomyxa*), although the extent of release was still less than in the thallus, and only with *Nostoc* was the optimum sharply defined. The proportion of fixed ^{14}C released (mostly as ribitol) by freshly isolated *Trebouxia* tended to increase somewhat as the density of the suspension decreased, although there was no simple relationship between density and release. This effect was very much less evident after 24 hours isolation, and completely absent in cultured algae. However, in *Coccomyxa*, suspension density had no effect on ribitol release by freshly isolated algae.

If isolated *Trebouxia* is exposed to a pulse of $NaH^{14}CO_3$ and then transferred to distilled water, Green found that release of [^{14}C]ribitol virtually ceased, even though nearly 57% of the ^{14}C in the ethanol soluble fraction is in ribitol. This implied that ribitol is available for release only within a short time of its synthesis, and thereafter it enters unavailable pools.

It must be concluded that experiments with isolated lichen algae have so far given no clear and positive indication of why symbiosis induces a massive carbohydrate release. Many more experiments are needed to confirm the impression that lichen fungi do not release 'factors'.

Finally, there remains the suggestion of Smith, Muscatine and Lewis (1969) that the mobile carbohydrate was only formed by the lichen alga when it was in the symbiotic state. This was based upon the observation that, except for *Trebouxia*, incorporation of fixed ^{14}C into the mobile carbohydrate ceased soon after isolation (*e.g*, cf. Table 11.2). However, chromatography of ethanol extracts of cultures of four genera of green algae of lichens by Hill and Ahmadjian (1972) showed the presence of the mobile polyol in each. Presumably, the polyol ceases to be a destination for fixed carbon soon after isolation of the alga. It is as if, in symbiosis, the continual drain from the pool of mobile carbohydrate results in a continual input of carbon fixed by photosynthesis ; as soon as the drain ceases when the alga is removed from symbiosis, fixed carbon is diverted elsewhere, and the pool itself shows little turnover (*e.g.* it is possible that in the non-symbiotic state, the prime function of the polyol pool is the maintenance of an adequate osmotic pressure to counteract the water stress in the terrestrial environment of such algae).

Movement of fixed nitrogen from Nostoc *symbionts to the fungus in* Peltigera Millbank and his colleagues have studied nitrogen fixation in a number of *Peltigera* species containing *Nostoc*. They found that the rate of fixation per mg algal protein by *Nostoc* in the thallus was approximately two to three times greater than that recorded in culture, but that the frequency of heterocysts (3% of cells) in the lichen was much lower than values recorded for blue-green algae in culture (up to 15%). Although heterocysts are generally believed to be the site of nitrogen fixation in free-living *Nostoc*, Millbank concluded that ordinary vegetative cells might carry out fixation within the lichen. This was of interest since nitrogen fixation proceeded in the light, when there is a net photosynthetic oxygen production, yet nitrogenase is particularly oxygen sensitive.

As soon as *Nostoc* was isolated from thallus homogenates, the ability to fix nitrogen was immediately lost and could not be restored. Even cutting the thallus into small pieces caused a sharp reduction in fixation. It was as if close, undisturbed physical contact between alga and fungus is necessary to enable both nitrogen fixation and oxygen evolution to be carried out in the light simultaneously.

Movement of fixed nitrogen to the fungus was first studied in *P. aphthosa* using ^{15}N. This lichen has a green alga (*Coccomyxa*) in the main algal layer of the thallus, but *Nostoc* in surface cephalodia. The cephalodia can be excised from the thallus, and Millbank *et al.* developed techniques for isolating algae quantitatively from either excised cephalodia or residual thallus. They showed that virtually all the nitrogen fixed by the *Nostoc* passed to the fungus, and hardly any moved to the green alga, even over a period of 25 days. There had been a pronounced influx to the fungus, but scarcely any efflux.

General considerations about the lichen symbiosis and the interaction between fungus and alga

In a lichen such as *Peltigera aphthosa*, experiments have shown that the same fungus can have the following effects on its algal symbionts :

(*a*) Cause an efflux of glucose from *Nostoc* in cephalodia.

(*b*) Cause an efflux of ribitol from *Coccomyxa* in the thallus.

(*c*) Cause an efflux of fixed nitrogen from *Nostoc*.

(*d*) Probably enable ordinary vegetative *Nostoc* to fix nitrogen.

In seeking to explain these effects, an obvious possibility is that the fungus produces chemical substances affecting specific processes in the algal cells. Presumably, several such substances would be needed in *P. aphthosa* since the fungus is affecting both eukaryotic and prokaryotic algae, and since a variety of metabolic processes are involved.

In so far as transport processes are concerned, it is a general conclusion from lichen studies that it is the transport of specific molecules which is affected rather than the general permeability characteristics of the alga. It was concluded independently by Green and Smith (1974) working with *Trebouxia* isolated from *Xanthoria aureola*, by Hill and Ahmadjian (1972) on synthetic cultures of *Cladonia cristatella* and by Millbank working on nitrogen fixation and transport in *Peltigera* that the effects of symbiosis on the alga cannot be wholly explained in terms of the production of substances from the fungus.

Alternative mechanisms must therefore be considered. Both Millbank and Green and Smith suggest that *physical* contact between the symbionts is of great importance. One possibility is that, through physical contact, the fungus affects some membrane property of the alga such as the membrane potential or proton movement. Specific transport processes dependent for their directional or vectorial characteristics upon these membrane properties might then be altered—either by an actual reverse in the direction of transport across the algal membrane, or by abolishing or reducing the inward flux relative to the outward. This might happen if the fungus could reduce or reverse the potential of the algal membrane (Fig. 11.6). Membrane potentials of lichen algae and lichen fungi have never been measured, but it is of interest that the membrane potential of the alga *Chlorella* has been measured at−40 mV and of the ascomycete fungus *Neurospora* at−200 mV. Although the algal and fungal membranes are not in contact in the lichen, charge might be carried along protein, polysaccharide or glycoprotein links in the material which separates the membranes of the symbionts. Alternatively (or additionally), it is possible that the fungus might extrude cations—H^+ or Na^+—and these may affect sugar and amino acid transport in the algal membrane. The responses, albeit weak, of excretion by freshly isolated algae to external pH might reflect such a mechanism.

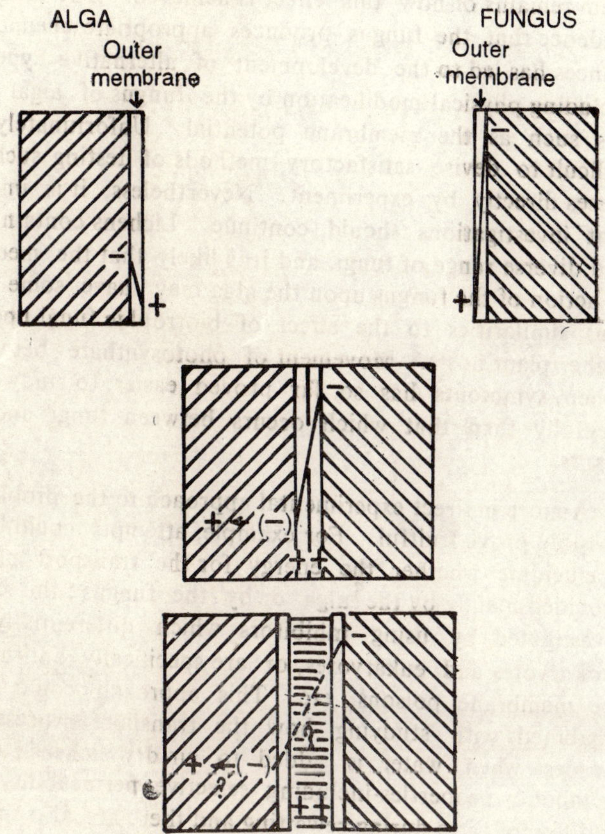

Fig. 11.6. Diagram illustrating a hypothesis to explain how physical contact between symbionts might cause efflux from lichen algae if the fungus can alter the membrane potential of the alga.

Specificity in both the lichen and other fungal symbioses might depend partly upon the ability to make physical links between the symbionts for the conduction of these effects. The delay in the development of a transport system in the synthesising *Cladonia cristatella* culture found by Hill and Ahmadjian might represent the time taken for the manufacture of such links.

In conclusion, while transport from alga to fungus obviously involves a marked effect of symbiosis upon the alga, the problem remains of how this effect is achieved. The lack of any evidence that the fungus produces appropriate chemical substances has led to the development of alternative hypotheses, including physical modification by the fungus of algal properties such as the membrane potential Unfortunately, it is difficult to devise satisfactory methods of testing such hypotheses directly by experiment. Nevertheless, it is important that investigations should continue. Lichens contain a large and diverse range of fungi, and it is likely that the mechanism of action of the fungus upon the alga may have some important similarities to the effect of biotrophic fungi upon their higher plant hosts. Movement of photosynthate between the lichen symbionts has so far proved easier to study experimentally than that which occurs between fungi and higher plants.

A more indirect experimental approach to the problem may possibly prove fruitful. For example, attempts could be made to elucidate whether the energy for the transport process is provided mainly by the alga or by the fungus ; this could be investigated by using inhibitors which differentially affect prokaryotes and eukaryotes, or are specifically antifungal, or are membrane poisons, etc. This approach could also be combined with studying how the transport process recommences when water is added to air-dry lichens ; different metabolic properties-including selective permeability-recover at different rates during rewetting and there are also some very distinctive respiratory phenomena which are probably associated with the reactivation of cell function.

ALGAL SYMBIONTS OF INVERTEBRATES

Of the many and varied associations between algae and invertebrates experimental studies on transport have been carried out on three main types : *Chlorella* symbionts of *Hydra viridis* and *Paramecium bursaria* ; dinoflagellate zooxanthellae in various coelenterates and giant clams ; and the green alga *Platymonas convolutae* in the platyhelminth *Convoluta roscoffensis*.

Chlorella *symbionts of* Hydra viridis *and* Paramecium bursaria. Each alga occurs inside a vacuole in the host cell, and the number of algae per cell is constant for a particular strain of host and particular set of environmental or cultural conditions. In *Hydra,* the algae are restricted to the gastrodermis and average 15-25 per cell ; in *Paramecium,* as many as 1000 algae per cell may occur.

In a general survey of symbiotic *Chlorella* isolated from various strains of *Paramecium* and *Hydra* as well as *Spongilla,* Muscatine, Karakashian and Karakashian (1967) showed that they liberated either glucose or maltose to the medium in amounts ranging up to 86% of the entire photosynthetically fixed carbon. Free-living *Chlorella* did not liberate free sugars, only small amounts of glycollic acid Liberation of maltose by symbiotic *Chlorella* was markedly sensitive to the pH of the medium, but glucose release was much less so.

A strain of *Hydra* containing maltose-liberating *Chlorella* has been extensively studied by Muscatine and his colleagues. *Hydra* is easily cultured in the laboratory, and all the algae can be quantitatively isolated from them by centrifuging homogenates of animal tissue. Eisenstadt (1971) found that 30-40% of the ^{14}C fixed in photosynthesis moved to the animal tissue, and Roffman and Lenhoff (1969) showed that the animal can metabolise products received from the alga. Algal photosynthesis is of nutritional importance, since under conditions of starvation, symbiotic *Hydra* survives much better than aposymbiotic, provided that they are illuminated.

The maltose-liberating *Chlorella* from *Hydra* has so far proved impossible to culture, but it is unlike most other symbiotic algae in that its ability to release free sugar does not decline during the 24 hours after isolation (indeed, its inability to revert to 'non-symbiotic' behaviour may help to explain why it has proved unculturable). Studies of the mechanism of maltose release by pure suspensions of the *Chlorella* were made by Cernichiari, Muscatine and Smith (1969) and it was found to be very sensitive to the pH of the medium. At pH 4.0, 40—50% of the ^{14}C fixed by photosynthesis is liberated to the medium mostly as maltose, together with small amounts in other compounds (alanine, glycollic acid, glucose and an oligosaccharide provisionally identified as maltotriose). Fig. 11.7 shows that as the pH rises, excretion progressively diminishes until at pH 7.0, only 1% of the fixed ^{14}C is released, about half as maltose.

Only traces of maltose ever occur within the cells and sucrose is always the predominant intracellular sugar. At pH

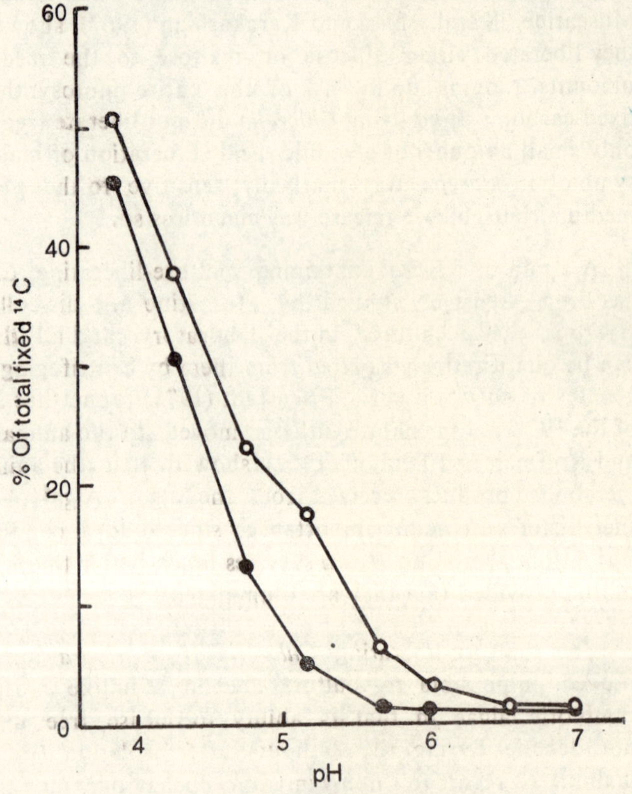

Fig. 11.7. *Chlorella* symbionts isolated from *Hydra viridis* : effect of medium pH on percentage of total fixed ^{14}C released (○), and the percentage of total released as maltose (●). 20 min photosynthesis at 21 °C, 4700 lx illumination. Data of Cernichiari *et al*. (1969).

7.0, external glucose can enter the cells, where it is rapidly metabolised. If cells previously labelled with ^{14}C at pH 7.0, by either photosynthesis or glucose feeding, are transferred to non-radioactive media in the dark at pH 4.0, they immediately begin to excrete [^{14}C]maltose which is apparently synthesised from the internal hexose monophosphate pool. In short term photosynthesis experiments at pH 4.0, fixed ^{14}C appears in sucrose within 20 sec, but none appears in maltose until 60 sec. This, together with the near absence of intracellular maltose and the marked sensitivity of maltose synthesis to external pH, suggests that the mechanism of synthesis is at or near the cell surface. Excretion of photosynthetically fixed ^{14}C at pH 4.0 is somewhat diminished by external maltose, although it could still continue at an appreciable rate even when the external maltose concentration was as high as 10%(w/v). In 10% maltose media, some of the excreted ^{14}C was still in maltose, but most was in maltotriose and maltotetrose, suggesting that a mechanism for transglycolisation exists on the cell surface.

Figure. 11.8 summarises the mechanism proposed by Cernichiari *et al.* to explain maltose release. Glucose liberation by symbiotic *Chlorella* has not been studied in the same way, so it is not known if this is derived from the mechanism shown in Fig. 11.8, or from an entirely different system. Since it is

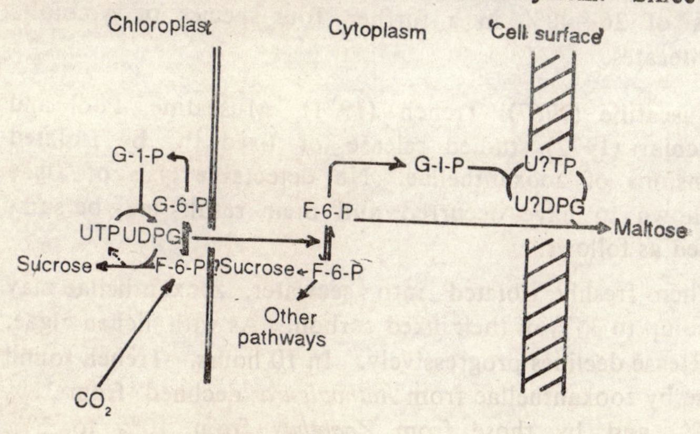

Fig. 11.8. Hypothetical mechanism proposed by Cernichiari *et al.* (1969) to explain maltose excretion by *Chlorella* symbionts of *Hydra*.

relatively insensitive to pH, a different system might well be involved.

It would be valuable to investigate whether a system such as that shown in Fig. 11.8 can be isolated from symbiotic *Chlorella*. Appropriate comparisons with free-living *Chlorella* (which does not manufacture detectable maltose) might then show how the maltose-excretion mechanism evolved.

Dinoflagellate zooxanthellae

A range of lower marine invertebrates contain symbiotic dinoflagellates especially in the Tropics. As with *Hydra*, pure suspensions of algae can be isolated from homogenares of animal tissue, but the cells are less robust and lyse easily. It is therefore important in studies of release of products by isolated zooxanthellae to demonstrate that no lysis has occurred in algal suspensions.

Transport from alga to animal is extensive. In intact associations, Muscatine and Cernichiari (1969) showed that over a 24 hour period, 40% of the ^{14}C fixed in photosynthesis by the symbionts of the coral *Pocillopora damaecornis* moved to the animal tissue. Trench (1971) obtained values of 45-50% in the anemone *Anthopleura*, and 20-25% in the zoanthid *Palythoa townsendii*. Von Holt and Von Holt (1968) obtained values of 26—40% in a further four species of symbiotic coelenterates.

Muscatine (1967), Trench (1971), Muscatine, Pool and Cernichiari (1972) studied release of fixed ^{14}C by isolated suspensions of zooxanthellae. No detectable lysis of algae was shown to have occurred and their results may be summarised as follows.

When freshly isolated into seawater, zooxanthellae may release up to 30% of their fixed carbon. As with lichen algae, this release declines progressively. In 10 hours, Trench found release by zooxanthellae from *Anthopleura* declined from 31% to 6%, and by those from *Zoanthus* from 10% to 2%. Decline in release was accompanied by increased incorporation into ethanol insoluble material. The products released by

Anthopleura algae were glycerol and organic acids (having chromatographic mobilities corresponding to malate, citrate, succinate and fumarate) ; there were also traces of glycollic acid.

An outstanding feature in all zooxanthellae so far studied (from ten different hosts) is that an addition of a homogenate of host tissue to algal suspensions causes a marked rise both in the net amount of ^{14}C fixed and in the percentage of it released ; the effect is still seen when host homogenate is added to algae 10 hours after isolation. In freshly isolated zooxanthellae, release of [^{14}C]glycerol is stimulated much more than that of other compounds. In those from *Anthopleura*, for example, net ^{14}C fixation increased 10-fold, the percentage released doubled, and the proportion of fixed ^{14}C released as glycerol rose from 19% to 43%. Thus, the amount of ^{14}C released as glycerol after addition of host homogenate increased by more than 30-fold (Table 11.3).

Table 11.3. Photosynthesis and liberation of ^{14}C by zooxanthellae freshly isolated from the anemone Anthopleura elegantissima.

(Results of experiments carried out at 17 °C ± 1°C, 4700 lx illumination, 1 hour photosynthesis)

Additions to zooxanthellae suspended in seawater medium	*Total ^{14}C fixed (counts min^{-1} ml^{-1})*	*Fixed ^{14}C liberated to medium (counts min^{-1} ml^{-1})*	
		Total	*As glycerol*
None (control)	14392	5320	902
Host tissue homogenate	141166	75400	32430
Boiled host tissue homogenate	45191	15650	3290
Homogenate of aposymbiotic (algafree) anemones	45395	15635	—
Homogenate of aposymbiotic anemones after infection	148901	80208	—

Percentage of fixed ^{14}C incorporated into glucose and glycerol by zooxanthellae incubated in host tissue homogenate

	As % of ^{14}C released to medium	As % of emanol-soluble ^{14}C retained in cells
Glycerol	43.0	3.6
G lucose	0.5	32.5

These observations at first led to the supposition that there were factors in the host tissue with a specific effect on glycerol transport. However, Trench (1971) showed that if addition of host homogenate to isolated *Anthopleura* algae was delayed for ten hours then [^{14}C] alanine—released in only small amounts by freshly isolated algae—became the major product released. Further, Muscatine *et. al.* (1972) labelled freshly isolated zooxanthellae from the coral *Agaricia agaricites* for 30 minutes in the dark with [^{14}C]glucose, and then transferred them to non-radioactive seawater without glucose in the dark. Addition of host homogenate still doubled the percentage of ^{14}C excreted (from 7.5% to 14.0%) but with alanine containing nearly 60% of the released ^{14}C and glycerol only 7.8%.

'Host factors' do not therefore stimulate a single transport system, but have a more general effect on both alanine and glycerol transport. It is noteworthy that hardly any free glucose is released from the cell even though it comprises 32% of the soluble ^{14}C in *Anthopleura* algae. From experiments in which zooxanthellae were exposed to a pulse of ^{14}C before transfer to unlabelled host homogenate, Trench infers that ^{14}C is only available for release a short time after fixation, and that there may be pools of metabolites in the cell unavailable for release. This is similar to the conclusions of Green about isolated lichen algae. The absence of an effect of host homogenate upon glucose release could be due to glucose being 'unavailable' for release. There is little positive evidence as to the nature of the 'factors' in host homogenate. In *Pocillopora* and *Anthopleura* they were thermolabile, but doubtfully so in *Agaricia*. Muscatine found that the effects of host homogenate could not be explained in terms of either pH, nitrate or phosphate. There is lack of specificity since homogenates of giant clam affect coral algae and coral homogenates affect giant clam algae. Trench (1971) found that homogenates of aposymbiotic anemones had no stimulatory effect on release, and only a minor one on photosynthesis. After infection of aposymbionts by algae, the homogenates then developed a stimulatory property (Table 11.3).

Analysis of animal tissues of symbiotic coelenterates show that most of the ^{14}C received from the alga enters lipid, amino acid and protein fractions. This confirms the importance of glycerol and amino acids in transport, especially since Trench showed that it was the glycerol moieties of lipids which became labelled. Von Holt (1968) suggested from his studies of isolated zooxanthellae that they may also release nucleoside phosphates and various other compounds. Unfortunately, he did not show the absence of cell lysis in his algal suspensions.

Convoluta Roscoffensis

Convoluta roscoffensis is a small acoelous platyhelminth containing the alga *Platymonas convolutae*. Larvae do not contain algae when they hatch from the eggs, but they only grow to maturity if they become infected by algae. Once the worms are adult they cease to feed, and the association presumably behaves as a closed system with a 'tight' circulation of nutrients between alga and animal.

Muscatine, Boyle and Smith (1974) showed that in intact animals, the rate of photosynthesis (per mg chlorophyll) was comparable to free-living algae, and the main product of ^{14}C fixation was mannitol also the main photosynthetic product of cultured *Platymonas*. Photosynthetically fixed ^{14}C could be easily detected in animal products—eggs and mucilage—and was prominent in amino compounds in hydrolysates of these products.

Investigations of isolated algae are difficult because the algae lose their cell walls in symbiosis and so are very fragile and can only be obtained in low yield from animal homogenates. The amount of fixed ^{14}C released by such freshly isolated algae was low—4-8% of total fixation-of which alanine was the major radioactive compound released. There was some evidence for the existence of 'host factors' since the proportion of fixed ^{14}C released by suspensions of gently homogenised *Convoluta* rose to 16% the principal radioactive product again being alanine. Although mannitol was the main product of photosynthesis by isolated algae, no release of it could be detected.

Muscatine *et al.* conclude that the principal products moving from alga to animal are amino acids, especially alanine. There was no evidence that mannitol moved. Animals could not metabolise externally supplied [^{14}C] mannitol but [^{14}C] alanine was extensively metabolised. After animals had been exposed to a pulse of ^{14}C, disappearance of fixed ^{14}C from mannitol during the subsequent 24 hours was very slow. Muscatine *et. al.* conclude that mannitol in the alga may constitute a pool of carbon unavailable to the animal and they believed that a major role of the algae is the recycling of nitrogen within the system ; Gooday (1970) found that cultured *Platymonas* can grow on uric acid as a sole nitrogen source and noted that platyhelminths reportedly lack uricase. The low proportion of immediately fixed ^{14}C released by the alga (compared to other symbiotic systems) was believed to be due to the fact that since the adult animal does not grow, its carbon requirements are less. Moreover, some of the energy made available by photosynthesis might be used in conversion of animal nitrogenous waste back to usable amino acids rather than all of it being devoted to export of fixed carbon.

SYMBIOTIC CHLOROPLASTS

Chloroplasts from certain siphonaceous seaweeds (especially *Codium* and *Caulerpa* spp.) occur within cells lining the digestive tract of some species of sacoglossan molluscs. These animals are capable of photosynthetic carbon fixation for prolonged periods after removal from their host seaweed.

Freshly collected specimens of *Elysia viridis*, containing chloroplasts from *Codium fragile*, exhibit rates of photosynthesis per mg chlorophyll of a comparable order to *Codium* plants. Indeed, studies by electron microscopy showed no sign of physical damage to chloroplasts inside the cells of the animal. It should be stressed that chloroplasts from siphonaceous algae such as *Codium* are very different from those of higher plants in both ultrastructure and other characteristics. Especially notable are their remarkably tough outer envelopes and the ability of *C. fragile* chloroplasts to continue photosynthesis for several days after isolation into a simple mineral medium. Within *E. viridis*, chloroplasts continue photosynthetic carbon fixation for at least two months after ingestion.

The amount of fixed carbon moving from chloroplast to animal has been estimated by calculating the proportion of a pulse of fixed ^{14}C which passes through a known animal product. In *Tridachia crispata*, Trench, Trench and Muscatine (1972) found that 30% of a pulse entered animal mucus. In *E. viridis*, Trench *et. al.* (1973) showed that one of the products in which photosynthetically fixed ^{14}C accumulates is a large pool of free galactose. Since free galactose was not detectable in isolated chloroplasts or parent seaweed, it was assumed to be an animal product ; 30 minutes after exposure of animals to a 15 minute pulse of ^{14}C, 36% of it was incorporated into galactose (Fig. 11.9). Transport from chloroplast to animal is therefore substantial. Since animals starved in the light lose

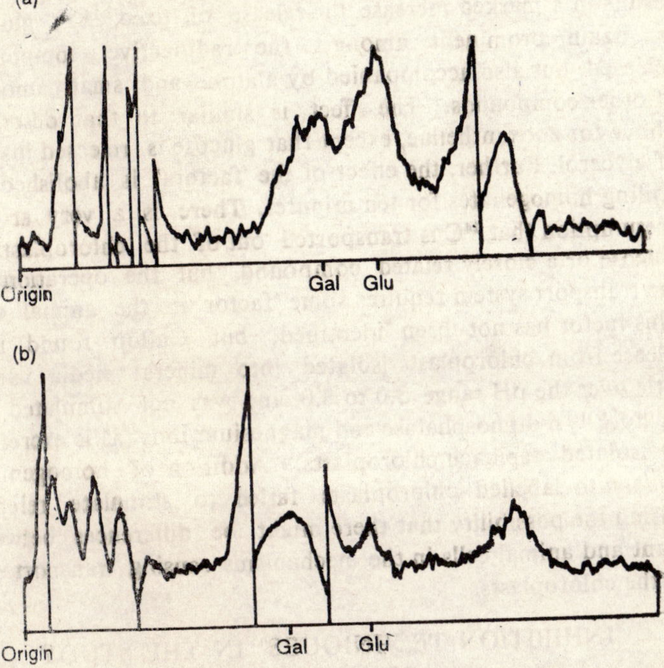

Fig. 11.9. Chromatogram scans of methanol-soluble extracts of *Elysia viridis* (*a*) after 15 min incubation in the light in seawater containing $NaH^{14}CO_3$; and (*b*) 30 min later after transfer to non-radioactive seawater. Gal, galactose ; glu, glucose. Data of Trench *et. al.* (1973*b*).

weight less rapidly than those in the dark, photosynthesis is nutritionally important to the animal.

Isolated chloroplasts from *C. fragile* release only about 2% of their fixed ^{14}C to the medium, mostly as glycollic acid, although their rates of fixation are comparable with or even higher than the intact plant. If *E. viridis* is homogenised, the rate of photosynthesis of the resultant suspension does not decline, but about 40% of the fixed ^{14}C is released to the supernatant mostly as free glucose accompanied by small amounts of glycollic acid ; incorporation into insoluble material is reduced. The existence of 'host factors' stimulating release is confirmed in experiments where addition of animal homogenates to chloroplasts previously labelled with ^{14}C results in a marked increase in release of fixed ^{14}C; glucose is again prominent amongst the radioactive compounds released, but also accompanied by alanine and small amounts of other compounds. The effect is similar to that described above for zooxanthellae, except that glucose is released instead of glycerol. Further, the effect of the 'factors' is abolished by boiling homogenates for ten minutes. There is a very strong presumption that ^{14}C is transported out of the chloroplasts as glucose or a closely related compound, but the operation of the transport system requires some 'factor' in the animal cell. This factor has not been identified, but Gallop found that release from chloroplasts isolated into mineral media varied little over the pH range 3.0 to 8.0, and was not stimulated by fructose 1, 6-diphosphatase and magnesium ions (as is excretion by isolated spinach chloroplasts. Addition of homogenised *Codium* to labelled chloroplasts failed to stimulate release, raising the possibility that there might be differences between plant and animal cells in the mechanisms causing transport out of the chloroplasts.

'INHIBITION TECHNIQUES' IN THE STUDY OF SYMBIOSIS

'Inhibition techniques' were first discovered in lichens. As was described above, when a lichen is incubated in the light in $NaH^{14}CO_3$ and the unlabelled form of the mobile carbohydrate, the labelled form is released and no ^{14}C enters fungal products. The technique proved valuable in studying transport between the symbionts, especially because it was very specific.

When such 'inhibition' experiments are carried out with symbiotic marine animals, fixed ^{14}C is also released by the animals to the medium in the presence of certain specific compounds symbiotic marine coelenterates ; (for *Elysia viridis* ; for *Convoluta roscoffensis*). Table 11.4 summarises results obtained with three different associations, and illustrates that although the animals are from three different phyla and the algae also not closely related to each other, there are nevertheless some broad similarities between them.

Several lines of evidence indicate that in animals, 'inhibition' mostly traps molecules not as they move from autotrophic symbiont to host, but at a later stage after they have entered animal tissue. For example, in *E. viridis*, external [^{12}C]galactose stimulates release of [^{14}C]galactose—an animal sugar not produced free by chloroplasts. There was a substantial time lag of 40-60 minutes before ^{14}C-galactose began to be released—far longer than could be accounted for either by the rate at which ^{14}C is released from chloroplasts or by the rapidity of penetration of galactose or $H^{14}CO_3^-$; the lag was assumed to represent the time taken for ^{14}C to move to the animal pool where exchange by 'inhibition' occurs. There was even a lag of about 20 minutes before ^{14}C is released by glucose 'inhibition' (Fig. 11.10). Muscatine *et. al.* (1974) found similar delays with *Convoluta*, and again concluded that the fixed ^{14}C released by 'inhibition' was mostly not that moving between symbionts but that which had already reached animal cells. In symbiotic coelenterates, Lewis and Smith found glucose generally very effective in 'inhibition', yet freshly isolated zooxanthellae release very little glucose, even in the presence of host homogenates. Delays in the release of ^{14}C can also be inferred from their data. Finally, in none of the animal associations does 'inhibition' prevent substantial incorporation of ^{14}C into animal products.

Table 11.4. Relative amounts of fixed ^{14}C released to the medium by symbiotic marine invertebrates during photosynthetic ^{14}C fixation in the presence of various compounds.

(Consolidated results of 'inhibition experiments' carried out with *Porites divaricata* (coral), *Convoluta roscoffensis* and *Elysia viridis.* In all experiments, animals incubated in seawater containing $NaH^{14}CO_3$ and 1% (w/v) of 'inhibiting compound'. Values in table show relative amount of ^{14}C released to medium (seawater control =1))

Inhibiting compound	*Porites divaricata*	*Convoluta roscoffensis*	*Elysia viridis*
Alanine	20	50 (4.5%)	10
Glycine	7	34	—
Serine	8	22	—
Glutamate	2	9	5
Aspartate	—	6	—
Glutamine	—	15	—
Asparagine	—	16	—
Proline	5	—	—
Leucine	—	7	—
Glucose	56 (10.5%)	2	19 (3.9%)
Galactose	—	4	8
Fructose	1	5	3
Sorbose	1	—	3
Mannose	—	—	17
Glucosamine	3	7	—
Glycerol	46	3	1
Mannitol	—	2	—
Pyruvate	—	20	—
Succinate	2	4	—
Glycollate	—	2	—
Urate	—	1	—
Lactate	—	3	—
Mixtures	54 (glucose+ glycerol) 49 (alanine+ glycerol)	41 (alanine+ pyruvate	16 (alanine+ glucose)
Seawater	1	1	1

'Inhibition' experiments in animals will therefore not give a reliable indication of the identity of molecules moving from symbiont to host, although they have been used to set a lower limit to the overall extent of movement. However, the experiments may yield useful information about transport systems

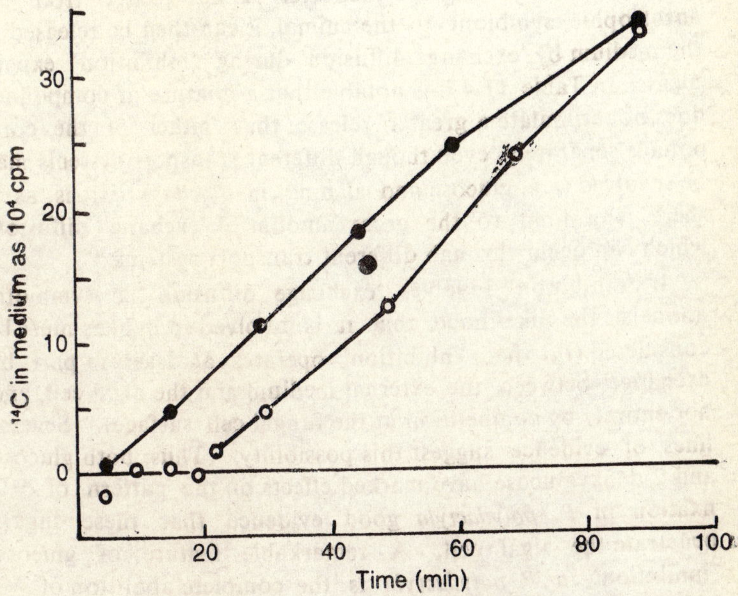

Fig. 11.10. Release of fixed ^{14}C by *Elysia viridis* to media during glucose 'inhibition' either during simultaneous exposure to $NaH^{14}CO_3$ and glucose (1% w/v) (O), or after a previous exposure to a 20 min pulse of ^{14}C before transfer to non-radioactive seawater containing glucose (●).

in the intact animal. In *Convoluta*, unlabelled alanine stimulates the release of labelled alanine, glycine, serine and glutamine but not other amino acids—as if it is operating system A amino acid transport. In *Elysia*, glucose causes release of both glucose and galactose, as if the transport system for one or both of these hexoses is not of very narrow specificity.

'Inhibition' effects in symbiotic animals are strongly reminiscent of 'exchange diffusion'. When Robbie and Wilson (1969) preloaded cells of *E. coli* with ^{14}C-thiomethyl-β-galactoside, addition of the unlabelled compound to the medium markedly stimulated the efflux of the labelled compound. This was explained on the basis that in returning from the inside to the outside of the membrane, the carrier-substrate complex moves more rapidly than the carrier alone. In symbiotic animals, it can be imagined that after ^{14}C has passed from the autotrophic symbiont to the animal, it can then be released to the medium by 'exchange diffusion' during 'inhibition' experiments. In Table 11.4 it is notable that a mixture of compounds does not stimulate a greater release than either of the compounds separately, even though different transport systems may be involved (*e.g.* glucose and alanine in *Porites*). It is as if there is a limit to the gross amount of 'exchange diffusion' which can occur through different transport systems.

If 'inhibition' involves 'exchange diffusion' in symbiotic animals, the likelihood that it is involved in lichens must be considered (*i.e.* that 'inhibition' operates at least in part by exchange between the external medium and the algal cell, and not entirely by competition at the fungal cell surface). Several lines of evidence suggest this possibility. Thus, both glucose and 2-deoxyglucose have marked effects on the pattern of ^{14}C fixation in *P. polydactyla* good evidence that these sugars penetrate the algal cell. A remarkable feature of glucose 'inhibition' in *P. polydactyla* is the complete abolition of ^{14}C incorporation into mannitol (*e.g.* see Fig. 11.3). If 'inhibition' operated simply by competition between labelled and unlabelled sugar at fungal uptake sites, then some [^{14}C]mannitol should be detectable. Hill and Smith (1972) summarise other lines of indirect evidence that 'inhibition' may not operate simply by competition at the fungal uptake sites.

'Inhibition techniques' may therefore be more useful as a tool for studying intercellular transport than transport between symbionts. Its particular value in the lichen symbiosis is that transport between the symbionts is intercellular.

There is one particular respect in which 'inhibition' may be significant for the study of biotrophy. An important property of transport systems which show exchange diffusion is that they can transport molecules *out* of cells. If such outward movement becomes strongly accentuated, then the type of transport system which occurs in biotrophy becomes possible. This idea will be pursued further in the discussion.

DISCUSSION

The two principal features of most of the biotrophic transport systems described here are: (*a*) relatively few types of molecule are released from the autotroph (*i.e.* release is selective); and (*b*) the amount released is usually substantial (*i.e.* measured as a proportion of the total photosynthetically fixed carbon).

In the majority of associations, the autotrophic symbiont loses these characteristics within a short time of removal from the symbiosis. The amount released falls off, the characteristic molecules of biotrophic transport no longer predominate amongst the excreted products and incorporation of fixed carbon into insoluble materials tends to rise. Evidently, existence in symbiosis induces the massive efflux from the autotroph of particular kinds of molecule which are not the normal major components of cell leakage.

Why does existence in symbiosis have these effects ? There are two possibilities, not necessarily mutually exclusive. Firstly, metabolism within the cell may be so greatly affected that a consequence is a profound alteration in both the quality and quantity of products leaked from the cell. Secondly, there may be a direct effect on membrane permeability and transport systems themselves. Hardly any evidence exists in support of the first possibility. Indeed, when isolated autotrophic symbionts of animals are induced to show 'biotrophic' release by 'host factors' (or medium pH for *Chlorella*) no profound changes in internal cell metabolism can be detected. Thus, the more likely—and certainly the more attractive—hypothesis is that existence in symbiosis directly affects cell permeability or membrane transport systems. It is with this hypothesis that the rest of the Discussion will be concerned.

The selectivity of transport in biotrophy has the obvious value that it permits the autotroph to retain essential metabolites for itself. Such selective release might be achieved in three ways.

(1) *The permeability of the outer membrane may remain unaffected, but processes external to it—such as cell wall formation,—become modified.* A general tendency for cell walls to be reduced or disappear in symbiosis naturally leads to supposition that transport systems might arise through the diversion and release of products destined for cell wall

synthesis in the autotroph. In the case of most lichens, it was concluded above that this hypothesis is unlikely ; for chloroplasts, the absence of a cell wall makes the theory untenable. With regard to other symbiotic algae, it should be remembered that some cell wall components are believed to be pre-formed in Golgi bodies, so that breakdown of them at the cell surface should yield a range of carbohydrates and not just the one characteristic of biotrophic transport. Further, it is difficult to explain release of alanine and glycerol in these terms.

(2) *There is a general relaxation of permeability barriers of the outer membrane, and selectivity is achieved by some form of compartmentation in the cell, either inside organelles, or by some form of segregation between the inner and outer regions of the cell.* It has already been shown that this could not explain maltose release from *Chlorella* and it would be difficult to explain release from symbiotic chloroplasts in this way. It is unlikely to be the explanation in zooxanthellae since when Muscatine *et. al.* (1972) supplied [^{14}C]glucose to them in the dark—which should presumably label the outer regions of the cell—subsequent addition of host homogenate caused a highly selective release, mostly of alanine.

(3) *There is a direct effect upon only certain types of transport system in the outer membrane.* This is the most likely possibility and all experimental evidence so far available would be consistent with it, but it is essential to stress that direct and rigorous experimental proof is still lacking. Nevertheless it is valuable to consider the possibility in more detail.

The principal characteristics of biotrophic tsansport in the associations discussed here are summarised in Table 11.5 it illustrates that although the associations are diverse, the kinds of molecule released from the autotroph are restricted to only a few types. There is consequently a temptation to suggest a common explanation for these different associations. This would be that in symbiosis, the membrane transport systems for these molecules are those which are susceptible to modification or reversal so as to produce a substantial net efflux from the cell. Significantly, the same few types of molecule were also effective in 'inbibition' experiments with symbiotic

Table 11.5. Summary of the characteristics of transport of photosynthate fro n symbiotic algae and symbiotic chloroplasts to host cells.

		Lichens: Blue-green symbionts	Lichens: Green symbionts	*Hydra viridis*	Dinoflagellate zooxanthellae	*Convoluta roscoffensis*	Symbiotic chloroplasts in molluscs
Principal form in which photosynthetically fixed ^{14}C moves to heterotroph	Glucose	+++	—	+++ (maltose and glucose)	—	—	?+++
	Polyol	—	+++	—	+++	—	—
	Neutral amino acids (esp. alanine)	—	—	—	++	+++	+
Proportion of total fixed ^{14}C moving to heterotroph		25-70%		30-40%	25-50%	<16%	36-40%
Stimulus required to cause autotroph to release photosynthate		? Physical contact ?		Acid pH	? 'Host factors' ?		

animals, showing that under certain experimental conditions, the systems can indeed produce a substantial efflux-in this case probably by 'exchange diffusion'. Since the systems can move molecules in opposite directions across membranes, the development of biotrophic transport requires simply that movement across the membrane out of the autotroph becomes strongly accentuated relative to movement inwards. It may be relevant that freshly isolated autotrophs usually show little or no ability to accumulate from external solutions the kinds of molecule they release in symbiosis : this has been shown for the *Trebouxia* symbionts of lichens incubated in ribitel, Hydra *Chlorella* in maltose, *Platymonas* from *Convoluta* in alanine and symbiotic chloroplasts in glucose. It is as if symbiosis caused a complete reversal in the normal direction of transport.

The effects of symbiosis on transport are more likely to be caused by modification of a general membrane property than by specific effects on individual systems ; it would be easier to explain situations where several types of molecule move between the symbionts (*e.g.*, zooxanthellae, blue-green algae of lichens). The observations from 'inhibition' experiments that mixtures of compounds do not cause greater efflux than individual compounds suggests that there may be a membrane property which limits the overall amount of efflux.

Table 11.5 suggests that different hosts might use different types of stimuli to induce autotrophic symbionts to release molecules. Homogenates of symbiotic marine animals undoubtedly contain thermolabile stimulatory 'factors'. These could be artifacts arising simply from homogenisation, but it is remarkable that in *Anthopleura*, they could be induced by infecting aposymbiotic anemones with zooxanthellae. Table 11.3. Although this suggests that they occur in intact animals, the crucial experiment was unfortunately not carried out to show that the 'factors' were not produced by homogenising zooxanthellae. However, *Codium* chloroplasts are not stimulated by homogenised *Codium*, only homogenised *Elysia*. Closer characterisation of these factors in sombiotic animals would clearly be helpful. The failure to detect such factors in lichen homogenates should be considered in relation to their environment ; they mostly exist in conditions of water shortage, so it is obviously advantageous to them if release from their algae can be achieved by means other than secretion of aqueous solutions of 'factors'. If the host stimulus affects a general membrane property rather than specific transport systems, then it is easier to understand that different hosts may use different stimuli to achieve the same end. If, for example, the general property was the membrane potential, then this could be altered by a variety of means either physical or chemical.

Both in lichens and in animals with zooxanthellae, there is evidence that molecules are only available for transport for a short time after their formation in photosynthesis. Cells of autotrophic symbionts may therefore be able to maintain pools of mobile compounds which are unavailable for transport to the host. This could offer an additional means of regulating the loss of fixed carbon. In this connection, the role of mannitol in many symbiotic associations is of interest : in mycorrhiza and in lichens, it is believed that mannitol forms a convenient reservoir of fixed carbon in the fungus which cannot be transported back to the autotroph. It is interesting that in *Convoluta*, mannitol may have the some function, but this time as a reservoir in the autotroph.

This article has shown that for the majority of associations, experiments have now clearly established both the amount of photosynthate transport from autotroph to heterotroph, and also the principal molecules involved. Future investigations will presumably concentrate upon the nature of the transport mechanisms and how they arise in symbiosis. In lichens, the apparent need for physical contact between the symbionts and the separate contributions of alga and fungus to the transport process are obvious points of further study. In symbiotic *Chlorella*, the maltose-synthesising system needs to be isolated, and the extent to which the component enzymes are present in free-living *Chlorella* investigated-this may well give the clearest example of how a biotrophic transport system can evolve. The 'factors' present in homogenates of symbiotic marine animals might be closer characterised by studying their effects on different types of cells. Finally, there is the remarkable phenomenon of symbiotic chloroplasts. They are unlike other symbiotic autotrophs because their natural and normal function as chloroplasts is presumably the substantial export of fixed carbon. Nevertheless, once isolated from the cell, they cease to do this. A detailed study of isolated *Codium* chloroplasts and of external factors which induce release of photosynthate from them is obviously called for-and since these organelles are remarkably robust, it should not present some of the technical problems which beset such studies of higher plant chloroplasts. The apparent similarity between the different systems in Table 11.5 could of course be misleading and superficial. Nevertheless, since so few types of molecules seem to be involved, a study of their movement in a range of different systems should give some kind of general picture of the nature of biotrophic transport.

In conclusion, the speculative nature of much of the above discussion must be emphasised. Further, it should be stressed that this chapter has been concerned only with the type of biotrophic transport which is the simplest to study experimentally. In many of the associations discussed here-and certainly in all intracellular symbiotic algae-a two-way flow of substances, which involves movement of ions as well as molecules, must occur between the symbionts; postulated mechanisms of biotrophic transport must take this into account, especially if they involve concepts such as changes in membrane potential and proton-or cation-linked movement of molecules. The experiments illustrate the technical advantage of studying transport between biochemically and genetically distinct cells, and offer the significant possibility that there could be important similarities between, for example, movement of molecules between the alga and fungus of a lichen and movement between adjacent cells in a tissue. Indeed, extending such speculation further, 'biotrophic' fungal parasites of higher plants might obtain nutrients by 'plugging in' to normal intercellular transport processes in the host. On the other hand, because biotrophic transport is easier to study than transport between adjacent cells in a tissue, there may be a real danger of the too ready application of mechanisms of transport between genetically dissimilar cells to the genetically similar.

12

Transport and its Significance in Plant Growth

Transport of water, mineral salts, organic materials, and growth substances within the plant have long been recognized as indispensable integral processes in the continued orderly growth of a plant. At least, the flow of mineral and organic substrates from sites of absorption or formation to regions of utilization is clearly essential; the flow of water is more a consequence of the need to replace the inevitable incessant losses and so maintain a suitable milieu for metabolism; whereas the nature, degree, and significance of the long-distance transport of growth substances remain to be ascertained.

At present, while the qualitative chrysalis of plant physiology is still struggling to emerge to its first quantitative instar, any treatment must be tentative at best-a provisional probing towards the unified numerical representation required for adequate appreciation and assessment. For many reasons, we have to be content here with a very cursory groping towards this goal and with a degree of quantification well below that which now could be achieved. We are aware of procedures and of model systems which could amply repay exploration, and we apologize to the reader for not having had opportunity to do so; we are also conscious of the pitfalls of glib application of such procedures to imperfectly understood systems.

This treatment will also be cursory in the sense that we have not been able to consider all recent relevant contributions.

In general, the quantity q of a substance in solution which moves in time t through an area A normal to the direction of flow is given by

$$dq/dt = -\mathrm{D}A\ dc/dz + fc \qquad (1)$$

where D is a diffusion coefficient, c the concentration, z the distance along the direction of flow, and f the rate of flow of solution in the z direction. Two components of flow are distinguished : a diffusion of the substance in proportion to its own concentration gradient, and a convection (or mass flow) of the substance due to the flow of the solution as a whole. The rate of flow of solution may be described by

$$f = -kA\ (\triangle \mathrm{H} + \triangle(\rho gh))/l \qquad (2)$$

where k is a conductivity and $\triangle P$ and $\triangle(\rho gh)$ the difference in pressure and gravitational potentials of the solvent over distance l in the z direction. Except in tall trees, the gravitational term is usually negligible. [Where a differentially permeable membrane is interposed, the flow rate is related to the difference in water potential as defined by Slayter and Taylor rather than to the difference in pressure potential.]

We must also recognize the possible effects of the movement of other entities—solutes, heat, and electricity—on the transport of the substance being considered. If these effects are substantial, the diffusion term D in Equation 1 should be replaced by

$$\mathrm{D}_{11}dc_1/dz + \mathrm{D}_{12}dc_2/dz + \mathrm{D}_{13}dc_3/dz \qquad (3)$$

where the subscript 1 refers to the substance being considered and the subsequent subscripts to other entities which influence the transport of the first, the coefficients D_{12}, D_{13} .. representing the effect of the gradient of the second, third, ... substances on the movement of the first. Active transport may be defined as that which occurs when D_{12} or D_{13} are large compared with D_{11} ; in effect, where the equilibrium or steady state of one component depends on the lack of equilibrium of one or more other components. Metabolic energy may be expended directly or indirectly in such a system as, for example, in maintaining a flux of ions and thereby influencing the flux of water. In this definition, which is wider than that used by many biologists, expenditure of metabolic energy is not the sole means by which an "active" process is defined; thermo-osmosis, electro-osmosis, and other devices by which flow against an electro-chemical gradient can be maintained should also be included. Moreover, association with metabolism as, for example, a decrease in the flux following inhibition of respiration, does not necessarily imply an "active" process; it may reflect association with the movement of another entity or even a change in structure of the system.

The classical channels of transport are the xylem and phloem. Most investigations and discussions on transport within the plant have been confined to the movement within these tissues. The concentration of inorganic ions in the xylem sap (ca 400 to 2500 μg/ml) is often greater than that of the external medium and contains a greater number of ionic species than phloem sap nevertheless, the concentrations of NO_3^-, K^+, and phosphate ions are comparable in both. Xylem sap also contains amino acids and amides, mainly of the glutamate and aspartate groups—at concentrations 10^{-3} to 10^{-1} that of the phloem (5 to 10 mg/ml)—together with alkaloids ureides, other nitrogenous compounds, sugars—in trace amounts compared with concentrations of 50 to 200 mg/ml in the phloem—organic acids, and growth substances such as gibberellins and kinins. Abscisic acid has been detected in phloem sap.

FLOW OF WATER

The flow of water has been considered in detail in five recent treatises and it is superfluous to give more than the barest of general summaries here. Movement of water in the plant is much better understood than that of other substances—much progress in quantitative analysis has been achieved by accepting only the most essential geometry of the system, making some approximations and regarding the flow as a series of successive steady states of the relations given in Equations 1 and 2. Reasonable estimates of the conductances of the different components in the intricate series/parallel network are available. In general, the rate of flow of water through the plant is determined by the current rate of transpiration, but is modified by the hydraulic properties of the soil, the soil-water content (or water potential), the depth of penetration and concentration of roots, and the internal conductances of root, stem, and leaves. (Of the latter, that of the root appears to be about equal to that of the leaf and possibly half that of the stem.) In soils at field capacity (water potentials ~—0.1 to —0.2 atm), most crop plants can maintain the supply of water to the leaves with potential transpiration rates up to about 7 mm day^{-1}. The major part of the difference in water potential between bulk soil and the leaves is then located in the plant. As the soil dries, adequate supplies of water can only be maintained at lower and lower transpiration rates, the major part of the difference in water potential then being in the soil. Increasing water deficits in the leaf are coupled with partial stomatal closure and a decrease in the actual rate of transpiration below the potential.

The rate of growth of the plant is reduced below its potential rate when the water content of the upper leaves falls to about, say, 90 per cent of that held at full turgor (about−8 atm water potential), and death occurs in many plants when the relative water content falls to 40 to 50 per cent (say, ~—30 to —60 atm water potential). Many physiological processes are affected both directly and indirectly, but it is most uncertain which are the first and key processes involved. Some recent evidence indicates that one of the first effects of a water deficit is a reduction of the transport of mineral ions, sucrose, and other metabolities. The effect seems to be exerted on the transfer from conducting tissue to leaf cells and not on movement within the sieve tubes themselves. Evidence interpreted to the contrary may possibly have been due to retardation of "unloading" at sink sites and other factors. Water deficits lead to enhanced synthesis of sucrose at the expense of polysaccharides and, if large, to the disruption of membranes and irreversible damage to cells.

The unidirectional flow of water provides a means of transport for substances absorbed by and manufactured in the roots to the leaves, fruits, and stem. The extent to which this channel is used, and the influence of other components on transport therein, is considered in relation to particular substances.

TRANSPORT OF INORGANIC IONS

Current evidence indicates that the conductances of the endodermis and stele for the flow of water are not much different from that of the cortex. Moreover, the conductivities of the cell walls and of the symplast are probably not very different, but, associated with its much greater cross-sectional area, possibly about three times as much water moves through the cytoplasm as through the walls. In general, the conductance for water flow across the root would appear to be determined by the conductance of the cell membranes for the entry of water; it may vary as much as tenfold with temperature, ionic composition, and inhibitors, these differences almost certainly arising from variations in structure.

As with water, two pathways of ion transport are normally recognized—the extracellular free-space regions of the cell wall and the symplasm—with transfer surely possible between both at all positions across the root. The measured free space appears to be quite inadequate to allow transport at the rates observed; most of the ions move through the symplast. It is likely that the differential permeability to different ionic species is located mainly in the plasmalemma and that entry, at least of anions, into the cytoplasm involves a carrier-mediated system. Of the two alternative fates for ions within the symplast, accumulation in the vacuole certainly involves "active" uptake systems. Those ions moving directly to the xylem are possibly passively carried, with some local diffusion, by the streaming protoplasm, but movement out of the symplasm into the xylem vessels probably also involves a carrier-mediated system.

The frequently greater concentration of ions in the xylem sap than in the external solution may reflect differing affinities of the carrier-mediated systems for anions and cations. For example, in *Ricinus*, NO_3^-, Cl^-, SO_4^{-2}, $H_2PO_4^-$, and HPO_4^{-2} entered the xylem against their electrochemical potential gradients, K^+ was in equilibrium, and Na^+, Ca^{+2} and Mg^{+2} entered along their gradients. The suggestion that transport of cations requires no expenditure of energy has been challenged, even though many investigators find that much of the transport is not directly associated with metabolism.

Uptake both by excised roots and by intact plants indicate a changing degree of importance of different parts of these pathways according to the external ionic concentration. With low concentrations (<1—10mM) ; the symplasm pathway appears to provide the factor route, possibly because of the high affinity of the transfer mechanism across the plasmalemma and a limitation of the cell-wall pathway by a Donnan exclusion mechanism. With increasing concentration, the latter decreases and finally becomes negligible and the former becomes increasingly saturated. At high concentrations, the rate of transport is linearly related to external concentration : opinions differ concerning the site of the rate-limiting steps under these conditions, some favouring uptake through the plasmalemma and others the sites of transfer to the xylem vessels. We also doubt whether there is yet sufficient evidence to preclude transfer within the symplasm as a rate-controlling component. There is little evidence for the view that ions move through some areas of the plasmalemma by a metabolically mediated mechanism and through others by diffusion, the latter dominating at high external concentrations.

Entry into the vacuoles appears to be rapid and the loss therefrom slow : hence, most of the ions transported to the shoot probably proceed directly, mixing little with the ions already in the roots. Half-times of exchange rates of "symplasm" and "vacuolar" phosphorus in the tomato were found to be about 40 min and 90 hr, respectively. About 30 to 40 per cent of the phosphate ions were transported to the shoot within 1 to 4 hours of entering the root ; about three-quarters of these appeared to be transported directly through the symplasm and cell walls. Of the phosphate absorbed by the root, about one-tenth was in the symplasm and ninetenths in the vacuoles. Consequently, the rate of entry of ions into the xylem may be less than half-the concurrent rate of entry into the roots. Similar results have been found with uptake of phosphorus by maize and barley.

Recent contributions to the extensively studied relationship between transport of ions to the shoot and the rate of water flow indicate that the effects are confined to the transport pathway across the symplasm and along the xylem; uptake and movement through the vacuoles is not greatly affected unless the water potential of the root cells is quite low. When the water potential of the solution surrounding the roots is reduced to about−5 atom (that of a normal culture solution being about−0.4 atm), phosphate transport to the shoots (and presumably flow through the symplasm) is reduced, but uptake and concentration in the root tissues is not affected. Transport of SO_4^{-2} ions to the shoots of wheat appears to be increased by lowering the water potential, possibly as a result of increased permeability. Further reduction to about−7 to −8 atm (about incipient plasmolysis) does lead to substantial reduction in the amount absorbed by the root as well as that transported to the shoot. At water potentials even as low as−20 atm, the mechanisms of ion uptake and integrity of the tonoplast appear to be maintained, but permeability of the plasmalemma and leakage from cells are greatly increased.

The differential transport of ions to the shoot—for example, the relatively greater amounts of K^+ than of Na^+—may well reside in the differential permeability of the plasmalemmas of the cortical or stelar cells or both. [The suggestion that the tonoplast is involved seems unlikely, as it appears to exert little control over transport to the shoot.] Pitman showed that selectivity between the transfers of Na^+ and K^+ to barley shoots was decreased when transpiration was increased and suggested that this effect was exerted on ion movement through the cell walls and symplasm. Much of the exclusion of Na^+ is located at the root surfaces and in the pathway to the xylem some evidence suggests that the retention is due to competition from Na-binding sites outside the transport path and that at low rates of flow, these have to be saturated before it can move to the xylem. There can also be discrimination between Rb^+ and K^+, but whereas the Na-binding sites seem to be Na-specific, Rb^+ and K^+ appear to compete for the same sites. The transport of Sr^{+2} from roots to shoots is relatively less than that of Ca^{+2}. This discrimination probably arises from a mode of transport unique to the alkaline earths ; it involves a series of exchange reactions. The nature of the exchange sites is unknown, although the pectins in the xylem cell walls may be involved. Previously absorbed $^{45}Ca^{+2}$ could be "flushed out" of stems by transferring plants to solutions containing $^{40}Ca^{+2}$, Sr^{+2}, or Mg^{+2} ; transfer to KCl solutions and changes in the rate of transpiration, however, did not lead to any significant movement.

Of the ions reaching a leaf, appreciable amounts are retranslocated via the phloem, the differing degrees of mobility probably being determined by the rate of movement into and out of the phloem. Phosphorus is particularly mobile. Most of that reaching very young leaves comes from older leaves, but most of that imported by rapidly expanding and mature leaves comes directly from the roots. Retranslocation appears to be controlled by the source leaves. Retranslocation of potassium, which has a mobility similar to that of phosphorus, is illustrated in Table 12.1 which refers to barley seedlings grown in a solution of 2.5 meq $litre^{-1}$. The circulation of Na^+ and Cl^- is much slower, at least at high internal concentration. Ions of the alkaline earths are not retranslocated. Analyses of exudate from aphid stylets and honeydew showed no detectable Ca, even when the xylem of *Salix* stem segments was perfused with a 100 ppm solution of Ca. Apparently, calcium ions cannot enter sieve tubes because they are preferentially absorbed on sites outside the phloem, such as in the bundle-sheath cells. Some movement could be detected with high Ca concentrations, when presumably these sites were saturated, or with the addition of hydrogen ions or chelating agents. The evidence of 14 μg Ca per ml of sap supposedly from the phloem of *Yucca* is equivocal because of likely contamination.

Table 12.1. Rates (μeq day^{-1}) of intake of K^+ by leaves of barley.

	Young leaf	*Intermediate leaf*	*Oldest leaf*
Intake via xylem	2.0	2.7	1.9
Intake via phloem	1.3	0.7	—1.6
Net uptake	3.3	3.4	0.3

Although considerable space has been devoted above to discussing details of ion transport within the plant, in almost all ecological (including agricultural) situation the supply of ions to the shoots is primarily determined by the concentration in the soil solution and the ability of the roots to grow in order to reach these ions. Of the three ways in which ions may reach the root surfaces—mass flow of the soil solution, diffusion through the soil solution, the growth of roots into regions not depleted of ions—the first two can only be effective with ions such as NO_3^-, which readily exchange between soil colloids and the soil solution. Some evidence suggests that Ca^{+2} and Mg^{+2} can be supplied by mass flow in sufficient amounts to meet the needs of the plant. Ions such as $H_2PO_4^-$, K^+, Mn^{+2}, Fe^{+3}, and Zn^{+2} move only very short distances, and sustained growth of the roots is necessary to absorb these, for example, ^{32}P is absorbed only from the root hair zone. In almost all field situations, root extension does not keep pace with the increasing demand from new shoot meristems ; it is also possible that new roots formed late in life are less efficient absorbing units than those formed earlier. Consequently, the plant drifts during ontogeny towards a state of increasing internal starvation and with an increasingly greater proportion of the mineral supply to a new growing organ coming from older leaves rather than the external solution.

Both root extension and ion uptake per unit of root length is closely correlated with carbohydrate supply from the leaves, and hence with the intensity of photosynthesis and the size of the photosynthetic surface. For example, defoliation by cutting or grazing transiently reduces root extension and uptake per unit root surface, and ion uptake is intimately associated with the light supply to the leaves. A deficit of water in the upper soil layers where the minerals are in highest concentrations may lead to reduced growth as a consequence of reduced supply of minerals, although ample water is being absorbed from lower layers. The effect of water supply on mineral uptake and distribution has been discussed by Gates.

FLOW OF ORGANIC MATERIALS

Features of the transport system. All workers agree that the sieve tubes are the principal channels for long-distance transport of organic substances from the leaves and that in most plants carbohydrate moves mainly as sucrose. There agreement usually ends ; some workers still favour a simplified mass-flow hypothesis of the Munch type involving water as the transporting medium with or without "pumping stations" across the sieve plates, while others envisage a mass transport involving cytoplasm as the transporting medium. A substantial measure of this disagreement stems from the difficulty of establishing the structure of functioning sieve tubes as they *exist in situ.* The existence of transcellular strands, described by Thaine and at first hotly disputed, has received strong but qualified support for various aspects of structure. There is now reasonable evidence supporting the view that longitudinal transcellular "cytoplasmic" strands extend from sieve tube to sieve tube through the pores, and that the parietal layer of protoplasm found along the vertical walls does not extend across the sieve plates. In many respects this protoplasm differs from that of normal cells, having, in mature sieve tubes, few mitochondria and a degenerate nucleus. (These are probably necessary features of channels in which substances are to be transported appreciable distances without being transformed, provided the structure and functioning of such channels can be maintained). The existence of membranes around these strands, the degree of their permeability to different substances, the significance of the companion cells, and other features are still much in doubt. Nevertheless, it seems likely that there is no mass flow of solutions. Discussion and evaluation of the many uncertain details involved are beyond our present terms of reference.

However, as pointed out in a timely and elegant paper by Canny and Askham, much evidence still supports the original contention of Mason and Maskell that the translocation system behaves as a diffusion analogue with an apparent diffusion coefficient 3×10^4 times greater than that of sucrose in water. It is difficult to find admissable information on concentration gradients, since those usually measured are of bulk tissues, including much that is irrelevant. It is still necessary to establish by acceptable techniques, such as aphid stylets, whether flow along the sieve tubes is proportional to the concentration gradient (diffusion), proportional to the concentration in and linear velocity of the transporting medium (mass flow), or against a concentration gradient as indicated by bulk analyses of sugar beet petioles.

Most of the mathematical models suggested necessarily greatly simplify the system, often ignoring the pathways from the chloroplast surfaces to the sieve tubes and those from the sieve tubes to regions of utilization or storage ; they also fail to take into account the interrelationships between the numerous sources and sinks. They often allow for lateral transport from the sieve tubes but usually assume mass flow of a solution and attempt to assess the velocity of transport assuming a constant velocity, concentration of solution, and unidirectional flow. Possibly these assumptions do not matter in respect to the conclusion that sucrose molecules move along the sieve tubes with velocities of 50 to 100 cm hr^{-1}, but they are unacceptable for a general model of phloem transport in the whole plant.

There is general agreement that the movement of sucrose into the phloem requires a supply of metabolic energy. If a leaf is supplied with ATP, the export of sucrose from the leaf is increased and the uptake of sucrose by a vein immersed in sucrose solution is also increased. There is some doubt with such experiments if the ATP supplied ever enters the cells ; however, it may act on membrane systems. Other indirect evidence indicates that "loading" of sucrose in the phloem involves phosphorylation and chemical transformation of the sucrose, is influenced by respiration inhibitors, and occurs against a concentration gradient —$\sim$10 to 15 per cent in veins compared with $\sim$0.4 per cent in mesophyll. There also appears to be an energy requirement for the movement of ions herbicides and viruses into the phloem.

Estimates of lateral transport ("leakage") from the sieve tubes range from 0.8 per cent of the instantaneous sieve-tube content of ^{11}C-labelled assimilates per centimetre moved to 2 to 11 per cent when ^{137}Cs was used. Part of this laterally transported material moves into the bark and part into the xylem, about 70 per cent of the carbon compounds reaching the xylem being sucrose and the remainder glucose and fructose. There is also some lateral transport from xylem to phloem.

In sugar cane, Glasziou *et. al.* explored some biochemical aspects of sucrose uptake from vascular tissue by stem parenchyma. They found that translocated sucrose moves from the vascular tissue into an outer "free space" which probably includes the cell walls. Here the sucrose is hydrolysed and the glucose and fructose formed move into a "metabolic" compartment, probably the cytoplasm, where they are phosphorylated and then combined to form a sucrose phosphate. This compound then passes into the storage compartment where it is dephosphorylated. The process can be reversed, especially in immature tissue, and the sucrose moved to other regions of the plant.

The movement of amino acids and amines in the phloem may be presumed to show features similar to those relating to sucrose. Few investigations have been made, and it is not known what chemical transformations are involved during the movement into and out of sieve tubes. Some evidence suggests that there is free entry of amino acids with little selectivity into young leaves ; with mature leaves there is little import and only small amounts of highly selected amino acids are exported ; and during senescence there is extensive export of a wide range of amino acids. Although the phloem is the main pathway for export of amino acids manufactured in the leaf in many plants there is extensive manufacture of amino acids and amines in the root and translocation to the growing and storage regions of the shoot. Generally, this appears to be *via* the xylem, although such export *via* the phloem in sugar beet has been reported.

Flow in the plant. While it is difficult to incorporate into a satisfactory quantitative model the influences of the numerous sources and sinks of verying intensities found in a normal intact plant, there is a growing body of data relating to these. The potential sources the leaves, are not importers of sucrose until they are about one-quarter to one-half of their final size and may still continue to import after they have become net exporters. In many plants, the proportion of photosynthate exported from a leaf increases from zero when it is one-quarter to one-third its final size to about 50 per cent by the time it reaches its maximum size. The photosynthetic production of the leaf then commences to decline and the proportion of that production which is exported also declines. During this time, the export of inorganic ions and amino acids usually increases ; it seems unlikely that there is any impairment of transport in the sieve tubes themselves, although the processes "loading" sucrose into them may be affected. [The decline in photosynthetic capacity of the leaf is associated mainly with a reduced intensity of the carboxylation reactions and with reduced activity or some of the enzymes involved.]

In maize and sugar cane, as much as 70 to 90 per cent of current photosynthate may be exported from leaves which also appear to maintain over long periods their maximum capacities for photosynthesis. It is possible that large amounts of export may also be found in other tropical grasses, Amaranthaceae, and other plants in which phosphoenolpyruvate rather than ribulose diphosphate is utilized as substrate for the primary carboxylation ; these plants have a highly efficient photosynthetic system, and much indirect evidence suggests that the rates of photosynthesis and of translocation are closely correlated.

The reduction with increasing age in carbon assimilation by a leaf may proceed until the compensation point is reached, there being no significant resumption of sucrose import. However, flow of some sucrose molecules into mature leaves of wheat and tomato has been reported, and it seems possible for mature leaves to obtain materials moving from the root. The leaf cells of senescing leaves appear to lose the ability to accumulate sugars, although labelled sucrose exported from other leaves is accumulated by active absorbing systems such as aphids feeding on such leaves. The sieve tubes still seem to be functional. There are other systems, however, such as old mother tubers of growing potato plants which, as net exporting organs in an advanced state of senescence, are still able to import and metabolize sucrose exported from leaves.

The regions between which translocation occurs are influenced, not surprisingly, by the location of the sieve tubes and the number of interconnections between them. Much of the flow occurs between leaves on the same orthostichy and adjoining roots ; nevertheless, conductances between orthostichies appear to be relatively large, the proportion of translocated photosynthate moving to differing regions depending largely on the relative activities of the sinks present. For example, photosynthate exported by the third leaf of wheat was found in each successively younger leaf up to the seventh. All leaves of young tomato plants could supply all of the fruit trusses, but as the plants aged, particular groups of leaves tended to supply particular trusses. Vascular connections appear to be more extensive and not to influence translocation as much as previously thought.

FLOW OF GROWTH SUBSTANCES

Evidence relating to the transport of growth substances was fully discussed recently this journal. Here we are concerned with two features only : (*a*) the role of growth substances in the transport of other materials, and (*b*) the significance of the transport of specific growth substances in determining growth.

Effect of growth substances on transport of other substances. The older experiments showing that the application of IAA increased the movement of soluble carbohydrates, nitrogenous substances, and ions into the treated regions extended over several days and did not distinguish between direct effects of IAA on transport mechanisms and those arising from enhanced growth. In more recent experiments, accelerated movement of ^{32}P and ^{14}C-labelled sucrose and photosynthate into internodes of decapitated bean plants on which IAA had been applied was shown within 6 hr of treatment. Further, the accumulation of ^{32}P in an IAA-treated region was reduced by triiodobenzoic acid (TIBA) applied either with the IAA or between the sites of application of IAA and ^{32}P. TIBA is claimed to block the movement of IAA--although it undoubtedly also has other effects, such as inhibiting cell extension—and these authors concluded that movement of ^{32}P was stimulated by the basipetal transport of IAA. Hew *et. al.* showed that when IAA was applied to decapitated soybean plants, the export of photosynthate from the primary leaves was increased. The applied IAA could be recovered only from the stems, and it was concluded that it influenced translocation in the stem, having little effect on processes loading photosynthate into the sieve tubes.

As with IAA, of the few studies on the effect of gibberellins and kinetin on the transport of sucrose and other substances, most fail to distinguish between direct and indirect effects. In experiments ranging from 12 to 36 hr after application, GA_3 and kinetin alone had little effect on transport of ^{3} P to defruited peduncles and decapitated internodes, but they did show a large synergism when applied with IAA. It seemed that a deficiency of IAA had to be remedied before effects of other substances could be demonstrated. In the grape, however, translocation of ^{14}C-labelled photosynthate over a 6-hr period was markedly increased in plants treated 3 days previously with either GA_3 or benzylaminopurine. Transport was reduced when 2-chloroethyltrimethylammonium chloride, which probably prevents gibberellin synthesis, was similarly used.

The above effects may well be similar to the movement of amino acids, segars, and ions into kinetin-treated regions of detached leaves. Mothes and his co-workers believe that kinetin directly influences the movement and accumulation of ^{14}C-labelled amino acids independently of any effect arising from the stimulation of protein synthesis ; for example, the movement of α-aminoisobutyric acid, which is not incorporated into proteins, was also increased by kinetin. Muller Leopold however, consider that the transport of non-metabolites is by mass-flow of solution to the kinetin-treated region, this being generated by the "active transport (of metabolites) through cell membranes at the phloem boundaries". They showed that kinetin stimulated the basipetal (and acropetal) transport of both ^{32}P and ^{22}Na in detached maize leaves, but only the former accumulated in the treated regions.

While the above experiments all substantiate the finding that metabolites move towards regions of high growth-substance concentration, it is not clear whether the recorded stimulation of movement is due to effects on movement within the sieve tubes, entry or exit therefrom, or entry into associated parenchyma. Almost all the effects noted could arise from the stimulation of "metabolic sinks" near or at the point of application ; indeed, the synergistic effect of IAA, GA, and kinetin noted above supports this contention. It seems reasonable to conclude from available evidence that the application of these substances to plant tissues initiates or stimulates many enzymic activities, including the syntheses of protein and nucleic acids, within a few hours.

Long-distance transport of growth substances. Fragmentary evidence shows that growth substances move considerable distances in both the phloem and the xylem. IAA and synthetic auxins such as 2, 4-D have been shown to move throughout the plant and, indeed, the use of many herbicides depends on this. GA_3 appears to behave similarly. There is growing evidence that physiologically active quantities of gibberellins and kinetins are produced in roots and translocated in the xylem. Hoad has shown that abscisic acid (and also IAA and gibberellins) is present in the honeydew from aphids.

Although it is reasonable to conclude from this evidence that there is transport over long distances of auxins in the phloem, upward movement of kinins in the xylem, and possibly both acropetal and basipetal movement of gibberellins, abscisic acid, and the complex involved in the flowering stimulus evidence concerning the effective participation of these transported growth substances in regulating growth is still scanty. A great deal of indirect evidence suggests that the transport of auxins and other growth substances is concerned in the transient inhibition of axillary bud growth. However, there is little evidence that the rate of growth is influenced by growth substances transported over distances greater than a few centimetres and much of the diffuse and indirect evidence scattered through the literature suggests that those molecules of growth substances which do influence the rate of growth are produced *in situ* or in close proximity to the growing tissue.

THE SIGNIFICANCE OF TRANSPORT IN GROWTH

The significance of transport in growth is summarized by considering four particular systems, *viz.* leaves, ears of cereals, potato tubers, and storage roots of sugar beet. Further, information concerning the interrelationships of root and shoot growth mediated through transport is given in a current volume.

Transport in leaf growth. From knowledge of the quantitative development of the leaf it may be inferred that the primordial leaf completely depends on the import of all basic materials. During much of the time between initiation and unfolding from the terminal bud, its growth appears to be limited by the rate of supply of one or more metabolites. Sucrose, required to provide carbon skeletons for amino acids and many other compounds, seems to be the most likely. We reach this conclusion on the following grounds. The rates of increase in cell number and dry weight during this phase are related to total daily radiation by curves closely resembling the photosynthesis-light response curve. Removal of parts of the photosynthetic surface appreciably reduces these rates, and removal of younger competing leaf primordia increases them. On the other hand, changes in concentration of mineral elements in the external solution around roots of plants of the same size has little short-term effect. Indirect evidence of this nature suggests that the rate of growth is related to the concentration of assimilates in the sieve tubes— although this has not been explored by direct observation. Further, the relative rate of growth during this phase reaches a maximum at a time when effective vascular connections between the leaf and the stem are established. The rate of protein synthesis per unit RNA also increases at this time, suggesting that amino acids were previously in short supply. The concentration of soluble sugars and the ratio of amino acids to proteins are always very low, indicating that transported substrates are transformed as rapidly as they arrive.

By the time the leaf has reached about one-third of its final size, it becomes a net exporter of sucrose. Net import of nitrogen as nitrate or amino acids, depending on the species continues until about the time when the leaf attains its maximum area ; in later stages, appreciable export of amino acids (from breakdown of its constituent proteins) continues. The capacity of the translocation system for export may decrease slightly, but it does not seem to become impaired until the leaf has reached a very advanced stage of senescence.

There is much indirect evidence to show that the supply of mineral elements plays an important role after the leaf has unfolded. The limitations here appear to reside in aspects of supply and demand rather than in any features of the vascular system. Part may be attributed to the low conductance for ions across the root ; certainly, even with high external concentrations the plant seems increasingly unable to meet the demands of the increasing number of sinks as it progresses through its ontogeny. Removal of some of the sinks—the expanding leaves—leads immediately to increased growth of those remaining ; increased supply of nitrogen and phosphorus increases the growth of expanding leaves and delays senescence. The changes in photosynthetic activity are closely correlated with the contents of nitrogen and phosphorus

There is good indirect evidence of the involvement of growth substances in leaf growth but these seem generally to be formed *in situ* rather than imported from elsewhere. However, as pointed out earlier, lack of evidence to significant roles of translocated growth substances in this system reflects mainly the absence of adequate experimentation. At least, one system—the export from leaves to the apex of that complex involved in flower initiation—now seems to be adequately established.

Transport and the growth to the cereal ear. Very little attention has been given to the sources, nature, and amounts of material translocated to the growing inflorescence prior to anthesis, although appreciable quantities must be involved, not only to supply the growing inflorescence but also the elongating internodes. Growth following anthesis has received much attention. There are sufficient fragments to allow a reasonable picture to be synthesized, although, surprisingly, there does not seem to be a complete documentation of any one crop in the same season. Summarizing, all evidence suggests that at least 90 to 95 per cent of the dry matter of the grains comes from photosynthesis after anthesis. Loss of stored carbohydrate in the stem is greater than can be accounted for by respiration, but except in varieties with a short leaf persistence, less than 5 to 10 per cent moves into the grain. Expressed as a percentage of the final grain weigh t of barley, gross photosynthesis of the ear and the flag leaf plus peduncle each contribute about 75 per cent. Respiration of the ear is about 30 per cent and of the flag leaf about 20 per cent ; *i.e.*, about 45 per cent of the net income arises from the ear and 55 per cent is translocated from the flas leaf. In awnless wheat, ear photosynthesis is much less than in barley and some awned varieties of wheat and is sufficient to balance ear respiration only ; the rate of photosynthesis in the flag leaf region of wheat is about three times that of barley, and transport therefrom appears to account for all of the net increase in grain weight. Rice appears to behave as wheat. In maize, ear photosynthesis can account for only 1 to 2 per cent of the final weight ; about half of the carbohydrate comes from the top 6 leaves, a third from the middle 4 to 5 leaves, and a sixth from the bottom 4 to 5 leaves.

The proportions stated above should be regarded as approximate only, they vary with environment, experimental manipulations, strengths of the receiving "sinks," and photosynthetic capacities of the "sources." During the week following anthesis, for example, the expanding top internode of wheat received as much assimilate, from both flag leaf and ear, as did the grain. Removal of grains and leaves leads to compensation and read-justment of the whole flow system. Although insignificant amounts of carbohydrate move to the grain from leaves below the flag leaf in intact plants, removal of the flag leaf augments the flow from the next lowest leaf.

Reduction of floret number and shading of barley ears influenced the amount of translocation into the ears but did not affect rates of photosynthesis in leaves. Similarly, removal of two-thirds of the florets from wheat spikes did not affect the total export of ^{14}C-assimilate from the flag leaf, but more moved towards the roots and less to the ears. However, in other experiments, the net photosynthesis of the flag leaf was reduced by half within 3 to 5 hours of removing the ear; this was associated with increased concentration of assimilates in the flag leaf. Inhibition of ear photosynthesis by spraying with dichlorophenyldimethylurea caused a rapid increase in the rates of photosynthesis of the flag leaf and movement of assimilates into the ears but did not influence the concentration of sugars in the leaf.

These considerations suggest that the amount and pattern of movement is dictated by events in the sources and sinks; there do not appear to be any major impedances in the translocation pathways. There is usually ample photosynthetic capacity and only small amounts of the readily mobilizable carbohydrates stored in the stem are moved to the grain. We are not aware of any detailed quantitative investigations on the movement of nitrogenous and other substances in this system; parts of the nitrogen and phosphorus come from the leaves, but the amounts involved are uncertain.

Transport and growth of potato tubers. All the tubers on a potato plant which eventually grow to marketable size are initiated over a period of about 2 weeks and thereafter grow by both cell division and expansion. Available evidence indicates that the individual tubers on the plant grow at different rates at different times, the largest at any time not necessarily being the first to commence expansion or to maintain the fastest growth rate. Nevertheless, the tubers appear to complete between themselves for photosynthate, the mean size of tubers at maturity being an inverse function of the number of stems per unit area and hence of amount of foliage and root volume. In most plant competition systems, any competitor gaining an advantage usually continues increasingly to suppress the less successful. That this does not happen among tubers on the same stem is unexplained. The amount of ^{14}C moved to any one tuber appear to be correlated with the current rate of growth of the tuber and its ability to transform sucrose to starch ; there is little correlation with absolute size of tuber and no obvious relation between source leaf and position of the receiving tuber. "Sink strengths" here appear to be predominant and vascular geometry to play little part.

Despite, these variations between individuals, the rate of growth of the whole population of tubers on a plant is remarkably constant with time over a reasonably wide range of temperature and light conditions and of photosynthetic surface. Previously assimilated material stored in the stems and leaves may contribute as much as 25 per cent of the current increase in weight of tubers and represents about 10 percent of the material transferred to the tuber over its entire life. In addition, transfer from the haulm maintains a constant concentration of mobile ions in the tuber. The low concentration of Ca in tubers may be related to the immobility of this ion in the phloem.

The net assimilation rate of the whole plant appears to be closely associated with the rate of tuber growth. It increases soon after tuber initiation and again late in the season when the total leaf area of the plant and the photosynthetic capacities of the leaves are decreasing. It is decreased if tubers are removed from the plant or if the rate of tuber growth is slowed down by exposure to low temperature. This evidence indicates that the rate of photosynthesis is controlled to an appreciable extent by the rate of tuber growth, but the factors which control the rate of tuber growth remain unknown.

Transport and growth of storage roots of sugar beet. The distribution of assimilates between the storage root and leaves of sugar beet varies as the plant progresses through its ontogeny, the root receiving an increasingly larger proportion. Although the rate of growth is markedly influenced by light intensity, there is no effect on the developmental pattern, the proportion of assimilates going to the root always being that appropriate to the size of the plant.

The detailed distribution patterns of assimilates are similar to those found in many other species, young expanding leaves exporting little or no assimilate and that exported by older leaves being partitioned between the root and young leaves.

Much of that moving into young leaves is converted into amino acids, while much of that entering the root is stored as sucrose. Sucrose appears to move against the concentration gradient from sieve tubes to storage cells without transformation, despite the presence of all the necessary enzymes. It therefore contrasts with sugar cane.

The normal restriction of movement of sucrose to the young leaves vertically above and the root segment vertically below the exporting leaf appears to reflect simply differences in lengths of pathways rather than high resistances to lateral movement ; for example, the removal of all mature leaves from one side of the plant, followed by exposure of one of the remaining leaves to $^{14}CO_2$, resulted in movement of ^{14}C into all the immature leaves.

Events in the petiole appear to play little part in controlling translocation, but competition between developing leaves and the storage root is of greater significance. Experiments with decapitated and partially defoliated plants of which different parts were kept at different temperatures or modified by varying the nutrient supply indicate generally that photosynthetic rates and amounts translocated to particular regions are determined to a considerable extent by the relative activities of the various sinks. Removal of one part, or reducing its growth by lowering its temperature, leads to enhanced growth of the other regions. Cooling the entire storage root results in increases in sucrose concentration back along the petioles and finally in the sources leaves ; although the net assimilation rate is reduced, some evidence suggests that photosynthetic rates are not, the increased rates of respiration in the source leaves compensating for the reduced translocation. Cooling segments of the petiole leads only to transient reductions in translocation. Removal of up to half of the source leaves did not affect the rate of growth of storage roots, the net assimilation rate increasing accordingly. Reciprocal grafts between the roots and leaves of sugar beet and spinach beet showed that sugar beet roots with either sugar beet or spinach beet leaves always had a much higher net assimilation rate than spinach beet roots with either spinach beet or sugar beet leaves.